Builder's Guide to New Materials and Techniques

Paul Bianchina

McGraw-Hill
New York San Francisco Washington, D.C. Auckland Bogotá
Caracas Lisbon London Madrid Mexico City Milan
Montreal New Delhi San Juan Singapore
Sydney Tokyo Toronto

Library of Congress Cataloging-in-Publication Data

Bianchina, Paul.
Builder's guide to new materials and techniques / Paul Bianchina.
p. cm.
"A McGraw-Hill builder's guide."
Includes index.
ISBN 0-07-015763-4.—ISBN 0-07-006052-5 (pbk.)
1. Wood. 2. Fiberboard. 3. Building, Wooden. I. Title.
TA409.B53 1997
691'.1—dc21 96-49070
CIP

McGraw-Hill

A Division of The ***McGraw·Hill*** *Companies*

1 2 3 4 5 6 7 8 9 0 BKP/BKP 9 0 2 1 0 9 8 7

ISBN 0-07-015763-4 (HC) 0-07-006052-5 (PBK)

The sponsoring editor for this book was Zoe Foundotos, the editing supervisor was Ruth W. Mannino, and the production supervisor was Claire Stanley. It was set in Garamond by Joanne Morbit of McGraw-Hill's Professional Book Group Hightstown composition unit.

Printed and bound by Quebecor Book Press.

It is important to comply with all manufacturers' instructions, restrictions, and specific guides for the use of a given product. All products must be installed in accordance with local building codes.

This book is printed on recycled, acid-free paper containing a minimum of 50% recycled, de-inked fiber.

For Rose
For twenty wonderful years as my wife, my editor, my best friend, and my constant inspiration. You recharge me, and I love you for it.

Contents

Preface *xi*

Acknowledgments *xiv*

1 Today's lumber products, from forest to mill *1*

Fluctuating lumber costs *2*

The changing face of American logging *6*

Engineered lumber *8*

What it means to our forests: Some statistical information *16*

Where these trees are coming from *18*

New builders are planning for remodeling *20*

Cost comparisons *21*

A positive situation for everyone? *23*

2 An introduction to engineered lumber products *27*

Dealing with the learning curve *29*

Material Safety Data sheets *30*

Engineered lumber products *30*

Specialty engineered lumber products *43*

3 An introduction to engineered panel products *57*

Plywood *57*

Other engineered panel products *67*

4 Today's plywood and wood-based siding products *77*

An ounce of prevention . . . *77*
What this chapter covers *81*
Plywood and wood-based panel siding *82*
Board-style sidings *90*

5 A selection of other wood-based and nonwood products *99*

Aluminum shingles *100*
Fiber cement siding *101*
Fiber cement shingles *102*
Fiber cement underlayment *103*
Fiberglass doors *104*
Fiber-reinforced gypsum underlayment *104*
Fiber-reinforced gypsum wallboard and sheathing *106*
Hardboard doors and trim *106*
Insulated structural building panels *111*
Medium-density fiberboard (MDF) molding *112*
Polymer composite trim *115*
Polystyrene building elements *116*
Vinyl fencing *117*
Vinyl moldings and architectural products *119*
Vinyl siding *123*
Vinyl windows *124*
Wood-polymer lumber *127*

6 The engineered lumber manufacturing process *131*

Laminated veneer lumber (LVL) *132*
Parallel strand lumber (PSL) *139*
Laminated strand lumber (LSL) *140*
Oriented strand board (OSB) *144*
I-joists *145*

7 Hangers, connectors, and fasteners *153*

Hanger terms *156*
Hanger codes and abbreviations *157*

Hangers for composite wood products *159*
Plated truss connectors *185*
Other compatible hangers and connectors *192*
Options *196*
Installation *205*
Fasteners *206*
Epoxy-Tie adhesive system *207*

8 Computerized design services for engineered lumber *211*
TJ-Xpert software *212*
The computerized design process *214*
Custom designs *218*
Pricing and material lists *219*
Other Trus Joist MacMillan software *220*
Software from other manufacturers *221*

9 Designing and specifying structural engineered lumber *225*
What makes a strong floor? *226*
Using engineered lumber *228*
Selecting and specifying engineered lumber *229*
Live-load deflection *232*
Sizing engineered lumber I-joists for floor framing *234*
Sizing engineered lumber I-joists for roof framing *245*
Other structural engineered lumber applications *251*

10 Storing, handling, and installing structural engineered lumber products *267*
Job-site storage *268*
Handling during installation *271*
Installing I-joists for floor framing *272*
Hangers *279*
Cantilevers *280*
Web stiffeners and blocks *281*
Bracing I-joists *285*
Installing subflooring *287*
Engineered lumber wall framing *288*

Garage-door headers *289*
Installing I-joists for roof framing *290*
Holes and cutouts in engineered lumber *295*

Appendixes *301*

A Products by manufacturer's product name *303*
B Products by type of product *307*
C Engineered lumber products listed by manufacturer *311*
D Sample manufacturer's warranties *315*
E For more information *317*

Glossary *325*

Index *335*

Preface

As builders and consumers, we have to face facts. Times change. Tools change. The industry and the markets change. And now the building products we use every day on every job are changing.

Driven by the need to meet demands for lighter, cleaner, cheaper, stronger, and straighter wood in the face of dwindling forest reserves, mill closures, and environmental pressures, the lumber industry is undergoing a revolution of major proportions. In the last decade alone, hundreds of new products from beams to hardware to trim have been introduced, and with new products and technology comes a new learning curve. What's available? What are the proper applications for these new products? How do I cut and hang them? What adhesives and sealers and caulks and paints and nails and hardware do I use?

That's the purpose of this book. In one practical, easy to read volume is everything today's builder needs to know to answer all those questions, and more: how the products are produced and how they're specified and sold, what's available and where to use it, how to safely store and handle materials, what hardware to use where, and which fasteners work best—the entire gamut from mill to final trim.

The entire building industry is changing, and changing fast. Grandpa wouldn't recognize today's lumber yard, but to stay competitive, today's builder has to. This book will provide the information you need to thrive and succeed in a changing market.

Grandpa would be proud.

. . .

The art and practice of building is perhaps as useful and noble a way to make a living as any that has ever been practiced. Those of us who work in the field of construction are as proud of what we build today as the generations before us were of what they created and left behind.

But while the pride remains the same, the trade is changing, more so in the last 20 or 30 years than in the hundreds of years prior. Technology is making its way onto the job site in numerous ways, and that's a fact of life that will not be changing anytime soon. Products are appearing and evolving rapidly, altering the way we design, build, and even think about the structures we construct today.

This book is a guide. No one book is going to be able to list every new product, and the stuff that's brand new to one builder might be old hat to another. Instead, the purpose of the book is to offer an introduction to some of the technology that's shaping the way we build, and hopefully to open your eyes to some of the many new, useful, and truly innovative products that are now available. Some are products you may have seen and wondered about, some you may have used but never knew if you were working with correctly, and others you may not be aware of at all.

For the most part, I have avoided using brand names and product names in writing this book; there are so many of them that they tend to get a little confusing. Instead, most of the products are listed under their generic names, with descriptions of the materials used to manufacture the product, and some basic information on sizes, colors, availability, and other such information. Where the product appears to be truly unique, at least as of the time of this writing, the product and/or the manufacturer's name has been included as well.

To help clear up a little more of the confusion, in Appendixes A through C you will find a number of products listed by both their trade name and their manufacturer. This will, hopefully, help you in a couple of different ways: you can find specific products you're looking for if you happen to know the trade name; you can find out what generic category specific products fall into; you can find out what material a specific product is manufactured from; or you can track down the manufacturer of specific products in order to get some additional information.

Also, in Appendix E, "For More Information," you will find a number of manufacturers of engineered lumber and other new products, along with their phone numbers and addresses. Manufacturers have a product to sell, of course, but they also have a stake in the continuing growth of the industry. As such, they are excellent sources of information, from technical support and product specifications to video tapes, literature, and product demonstrations. They are a source not to be overlooked!

There is one other thing worth mentioning. The engineered and composite wood and nonwood product markets are growing and changing rapidly, and, as mentioned above, this book is by no means a complete listing of every product or every manufacturer. The successful builder of today—and tomorrow—needs to keep abreast of the field, and needs to spend a little time reading in addition to the time spent with a hammer. You need check out the trade shows and trade magazines periodically, and perhaps cruise the information superhighway via your computer and the incredible resources of the Internet. You'll also want to work closely with your material suppliers and product representatives to find the products that will work best for your particular application.

Life is about learning.

Paul Bianchina

Acknowledgments

There are a number of people who were extremely cooperative in taking the time and expending the energy to get information for me, and without all of them this book would never have been possible.

I especially want to thank Leon Pantenburg, Marketing Communications Manager for Trus Joist MacMillan, for the wealth of product information he sent, for all of his time on the phone, and for arranging a tour of one of Trus Joist's incredible manufacturing plants. I also want to thank Bob McFarland and Jeff Gottfried of Trus Joist MacMillan's Eugene, OR, plant for the excellent tour and the wealth of technical information they provided; Beth Hunt at Trus Joist's main office for smoothing out a couple of glitches in the otherwise smooth flow of information; and the folks at Oliver, Russell & Associates for the great CD and other product information.

Another big debt of gratitude is owed to Tawn Simons, Composite Wood Industry Manager for Simpson Strong-Tie Company, who provided me with everything I could think of and more— about the company's products, and then proofread and corrected part of the manuscript. Many thanks for all your time.

My deep appreciation also to Marilyn H. LeMoine, Public Relations Department Manager for the American Plywood Association, who has helped out on some of my previous book projects and was there to lend a hand once again, and to Kathleen M. Arndt, the Advertising and Product Promotion Manager for Louisiana-Pacific Corporation, and her co-workers, all of whom were quite generous in keeping me well stocked with literature.

Some others who were there to help out an information-starved writer and who definitely deserve some recognition for their efforts are Kim Drew of Georgia-Pacific; Peggy L. Hunt at Masonite; Timm Locke of the Western Wood Products Association; Jeff Olson of Boise Cascade; Linda Kukura of Thermal Industries, Inc.; Joe Knife from Classis Products, Inc.; and Michael Lupo of ABT Building Products Corporation. Especially warm thanks go to Denny, Janet, Doug, Dane, and all the rest of the great staff at Parr Lumber's Bend, OR, yard.

Finally, special thanks to my agent, Deborah Schneider of Gelfman Schneider Literary Agents in New York, and to my editor at McGraw-Hill, April D. Nolan. And, as always, my deepest debt of gratitude is to my wife Rose, who edits, proofreads, types, and, with extreme patience, graciously forgets that she even has a husband during the long months of getting a book like this from idea to reality.

1

Today's lumber products, from forest to mill

For decades upon decades, builders have built homes and offices and stores and virtually everything else, from storage sheds to apartment buildings, with solid lumber. They not only built with solid lumber, but they also remodeled with solid lumber. They thought and planned and measured and created in solid lumber. It was an indisputable fact that builders, like all the generations of builders before them, went to the lumber yard and purchased pieces of solid wood that were milled from solid logs.

At the job site, workers used hand saws to cut the lumber to length. They used hammers to nail the lumber together. The odd piece of wood that wasn't dry, straight, and virtually free of defects was tossed onto the scrap pile. It had always been this way, and would be this way forever.

Enter the electric circular saw onto the building scene. Builders quickly learned that there was a faster, easier, and more profitable way to get the job done, and the switch was on. Circular saws were perceived by just about everyone to be a better way of working, and they became ever more powerful and affordable. In a remarkably short period of time, hand saws, which had once been the mainstay of every carpenter's toolbox, began to disappear from the job site, left to survive as something of a specialty tool for unique applications.

At least the hammer was sacrosanct. After all, nothing could ever replace the hammer as the only means of driving a nail into a piece of wood. Nothing, at least, until a great new generation of air compressors, more powerful and portable and affordable than ever, began to roar on every job site. Again, within just a small span of time, everyone from framers to finish carpenters began to hang up their hammers, and the switch to air-driven nailers and staplers of every size and variety was on.

Time continues to march on and, as it does, the building industry continues to change. Markets shrink, competition for clients and lender dollars becomes fiercer by the day, and every builder has to look for ways to cut cost and improve productivity while still maintaining a top-quality product. Where else to look but at the lumber products they employ every day on every job?

To generations past, the term *old growth* (Fig. 1-1) was one they had never heard of. Trees were trees, and the forests were full of them. They were big, solid, and plentiful, and if lumber was needed, it was simple enough to cut and mill additional logs from a seemingly endless supply in order to meet the demand. It has actually been in only the past couple of decades that our awareness of environmental issues has increased. We as a society have seen the effects of clear-cutting (Fig. 1-2) and poor forest management in some areas (Fig. 1-3) and we've come to realize that, without care, the supply is not as infinite as we had always believed. In fact, the U.S. Forest Service estimates that only about 10 percent of the original ancient forests in the United States are still standing, and they further estimate that as much as half of what's remaining could be cut within the next 50 years.

And environmental issues are only part of this new recognition of forests. As cities and suburbs grow and land falls to development, we have also become increasingly aware of a need and a desire to enjoy the outdoor environment by way of pursuits such as camping, fishing, and hunting. We've become more and more aware of the role that plants, animals, insects, and all the other components of the world around us play in everyday life.

Fluctuating lumber costs

New regulations at the state and federal levels, aimed at protecting valuable wilderness areas and wildlife habitats, have restricted logging in many areas. It's estimated, for example, that of the roughly 7½ million acres of old-growth trees in Oregon and Washington, more than half is protected in one or more ways from being logged. The supplies of harvested logs available for conventional milling into solid lumber have dwindled considerably, making wood more expensive and of generally lower quality than what was previously available.

1-1 *Old-growth trees of this size are seeing an increasing amount of protection and a decreasing amount of logging.*

Grading rules have changed in the face of availability, and most any builder will tell you that today's Number 1 and Number 2 lumber would have been somebody's firewood just 50 years ago. One change in grading rules that occurred about two to three years ago is

1-2 *Example of unsightly and increasingly unpopular clear-cut logging practices.*

1-3 *Poor forest management practices have resulted in damage such as this in some forest areas.*

a great example of how much the quality of lumber has declined. That one change alone resulted in a downgrading of span capabilities by an average of about two feet! In other words, if a board of a certain size and grade was rated as having the strength to support a given load over a span of 16 feet under the old grading rules, the new rules decrease the rating of that same board to a span of around 14 feet. To you, the builder, that means you have to increase your lumber sizes or decrease either your spans or on-center spacing, and any of those three alternatives equates directly to higher costs for you and your home-buying clients.

Good-quality solid lumber has become less the norm and more the exception. Builders have become increasingly tired of wet, heavy, "pond-dried" lumber, and of the increase in knots, splits, checks, wanes, and rivers of oozing pitch. In some cases, the changes have been gradual. Builders who have been around the trades for a decade or two might be shocked to go back in time and look at the lumber they used to be able to take off the stack at the lumber yard.

Solid sawn lumber from old-growth trees used to have as many as 18 to 25 rings per inch, compared to today's lumber, which has as little as four to six rings per inch. A smaller number of rings means that the trees are young and small when harvested, not having reached a high degree of structural strength. It also means that knots and other defects, which are found primarily at the outer edges of the tree, are more prevalent and more of a problem.

This is especially exasperating to builders of quality homes. It equates to a constant battle against heavy, hard-to-cut lumber, lumber that warps and twists and makes quality drywalling and finish carpentry a real struggle. Each warp, twist, squeak, and nail pop is a problem that has to be corrected, either during the building stage or at some point later as a call-back, and every correction is money out of the builder's pocket—and eventually the homeowner's as well.

Added to these factors was the frustration for builders in not being able to accurately bid a job and then maintain that cost estimate through the project. During one three-month period in 1995, for example, the cost of a green Douglas fir 2 × 4 jumped by almost 35 percent! Homeowners and homebuyers tend to ignore these lumber cost fluctuations as not being something having a direct effect on them, but consider the following excerpt from an article in the June 3, 1996 edition of *Nation's Building News*, a trade newspaper published by the National Association of Home Builders (NAHB):

Accelerated increases in framing lumber prices over the past several weeks are starting to have an adverse impact on the affordability of new homes.

As of May 24, the composite price of framing lumber was $418, the highest price for lumber in more than a year," said NAHB President Randy Smith. The recent increases could add as much as $2,000 to the cost of a 2,000-square-foot house, and knock as many as 850,000 families out of the pool of qualified home buyers.

For every $1,000 added to the cost of a median-priced $130,000 home, NAHB estimates that 427,762 home buyers—5% of whom will actually buy a home at any given time—no longer qualify to buy a home. A $2,000 increase would thus knock out 42,700 home buyers.

The changing face of American logging

Builders aren't the only ones to suffer as the result of these changes. The logging and milling of lumber to feed America's voracious appetite for building was once considered an honorable profession, but today it's viewed with growing derision as something akin to piracy. Loggers are vilified as the rapists of America's public lands, yet they will look you in the eye and tell you that they're just earning a living, like generations of their family and friends before them.

The loggers can't help but see the writing on the wall. Most of America's sawmills, especially in the heavily timber-reliant Pacific Northwest, are being faced with a steadily declining supply (Fig. 1-4). Mills are being closing at an alarming rate, and going with them are good-wage jobs that families have depended on for longer than they care to remember. Some small mill towns have seen as much as half of their workforce suddenly without employment as large mills dramatically scale back production—or close their doors altogether.

The buggy whip is gone, the steam locomotive is a now just a crowd-pleaser at Sunday afternoon museum visits, and the old-growth logging practices of the past century are going the same route (Fig. 1-5). Loggers bemoan the facts, but facts they are. The world is no longer what it was (Fig. 1-6). We all, builders, loggers, and American consumers, need to make changes.

But the demand for lumber—quality lumber—is still there as the construction industry, despite its own ups and downs, continues to churn out new buildings. Loggers, mill workers, and mills need to

1-4 *Declining timber supplies are forcing the closure of many lumber mills.*

1-5 *Many states have regulated or banned a number of previously acceptable types of logging, lowering the availability of raw timber.*

1-6 *Mill closures and the changing face of the timber industry have brought about new types of lumber.*

work. The people are there, the equipment is there, and the demand is there; it just took some innovative thinking to bring it all together. Something had to change in order to meet the lumber demand, save jobs, and keep the mills running.

Times change, tools change, jobs change, and building techniques change. And now the very lumber that we've used and relied on for years is changing, which is as it should be. Enter the concept of engineered lumber.

Engineered lumber

It's interesting to note that only an estimated 50 percent of the builders in America are familiar with the term *engineered lumber*, while in Europe, where waste is not an option, alternative building products have been in use for decades. Even more interesting is the fact that the idea of engineered lumber is actually far from new. The concept of taking two pieces of wood and joining them together in a way that overlaps the weaknesses of each board and creates a whole that is stronger than the sum of the parts has been with us since ancient times.

For example, evidence of rudimentary forms of plywood has been discovered in the Egypt of 1500 B.C. Plywood in its modern form really became known here in the early 1930s. Its use grew during World War II, and the demand for quick and affordable housing in the years immediately following the war caused this technology to take off. More and more forms and grades of plywood were developed and marketed, in everything from concrete forms to subfloors to siding.

Going back even further than our modern plywood is the glue-laminated beam, or *glulam* as it's more commonly known in the industry, which has been familiar to builders since the turn of the century. Many of the concepts employed in the creation and manufacturer of today's engineered lumber products got their start with the manufacture of glue-laminated lumber.

The basic guiding principle behind glulams is an engineered concept called *repetitive usage*. This simply means that, in an assembly of several pieces of wood, as is the case with the glulam, the structural defects in one piece of wood are covered and reinforced by the solid portions of the wood above and below it. This concept explains how smaller, thinner pieces of wood can be combined and assembled to make up a single, larger beam having greater strength than a piece of solid timber of the equivalent size. Repetitive usage creates a whole that is stronger than the sum of its parts.

The glulam, however, was essentially a response to the need for framing members capable of spanning large, open interior spaces, and they became more common and more readily available with the open-space architecture that began coming into vogue in the 1950s. Little else—with the exception of venerable old plywood—was done in the field of engineered wood.

It was not until the 1960s, however, as concern for the nation's forests began to increase, that we turned our eyes toward the lumber that was going into the framework of the houses and small commercial buildings we were constructing. Old-growth timber was becoming harder to get to and harder to log. More and more tracts were being set aside for recreational uses or wildlife habitats. Second-growth trees (Fig. 1-7) began to be harvested in increasing quantity, and often with less than satisfactory results.

More and more research was directed at the problems of quality and supply. It was soon learned that, despite the decreasing supply of old-growth timber available for cutting, timber supply was not really at the heart of the problem. In fact, since the 1940s and the advent of improved forest management, better logging practices, and a dramatic increase in forest land reclamation, trees are actually being planted at a faster rate then they're being harvested.

1-7 *Second-growth timber seen in previously logged areas. These trees are smaller and there is more underbrush.*

The problems were with the quality of the timber more than with the quantity. Second-growth and now third-growth trees are smaller, yielding less usable lumber per tree. The younger trees are wetter and weaker, having less tensile strength than old-growth lumber. These younger trees, due to their smaller size, also have more defects that result in generally poorer-quality lumber.

To complicate the problem, it is the younger, newer wood—called juvenile wood—near the center of the tree that is the weakest wood (Fig. 1-8). Mature wood toward the outside of the tree, found in greater abundance in large-diameter, old-growth trees, provides the strongest lumber. Smaller trees have less mature wood at the outside, and what they do have is often wasted in the sawmilling process, which cuts off much of the stronger, outer, curved portion of the log in order to square off the lumber. And since smaller-diameter trees have less distance from the outside of the tree to the center of tree—which in a large piece of solid sawn timber is the same as the distance from one edge of the board to the other edge—the differences between the mature wood at the edge and the juvenile wood at the center are more pronounced, greatly increasing the dimensional instability of the sawn lumber. In other words, it's a lot more likely to warp.

The challenge, then, became finding out how to take these smaller, weaker trees, trees that had the very definite advantage of being both

1-8 *Juvenile wood near the center of a second-growth tree results in wood that is weaker and less dimensionally stable. Note the large size of rings.*

plentiful and renewable, and improving on their natural deficiencies, in much the same way that plywood makes a strong panel out of much weaker veneers. Put another way, how can you take a tree apart and then put it back together again in a way that makes the best use of its strengths while minimizing or eliminating its weaknesses (Fig. 1-9).

The solutions that were first examined with serious commercial concerns were then, as they still are today, multifaceted:

- Using smaller logs grown from smaller trees (Fig. 1-10).
- Using species of trees that are fast-growing, with the goal of farming the trees as a cash crop (Fig. 1-11).
- Engineering out the defects and weaknesses inherent in every tree, but even more prevalent in smaller, less mature trees.
- Producing lumber that was larger and more structurally sound then what could be produced from a log using conventional sawmilling practices. The finished product had to be superior to the original log.
- Using more of each log than what was currently used with conventional sawmilling practices (Fig. 1-12).
- Creating a product that could be produced in quantities and sizes not related to the original size and quality of the logs being processed.

1-9 *The family of engineered lumber products (left to right): I-joist, TimberStrand LSL, Microllam LVL, Parallam PSL, and Parallam PSL posts.* Courtesy Trus Joist MacMillan

1-10 *Smaller, fast-growing trees such as yellow poplar are perfect for engineered lumber.* Courtesy Trus Joist MacMillan

1-11 *Aspen, shown here, is another fast-growing species for the manufacturer of engineered lumber that might prove valuable in the future as a crop tree.* Courtesy Trus Joist MacMillan

- Creating lumber that was dry, solid, and dimensionally stable, consistent in both size and performance.

The first of these attempts at engineered lumber came about with the introduction of Microlam, the first generation of laminated veneer lumber (LVL) that was introduced by Trus Joist MacMillan in 1969. Working with trees that were too small for use in traditional sawmilling operations, engineered lumber technology began to fully use the strength of the tree in whatever part of the tree that strength was found. Using a production technology similar to that used in the manufacture of plywood, veneers are peeled off the logs, allowing as much as 30 percent more usage of the log when compared to processing the log through a standard sawing operation. The veneers are sliced and glued in parallel sheets, which is the main difference between LVL and plywood, which alternates the veneer layers perpendicular to each other. The veneers are then pressed into solid members that can be as much as 4 feet wide, 3½ inches thick, and an incredible 80 feet in length.

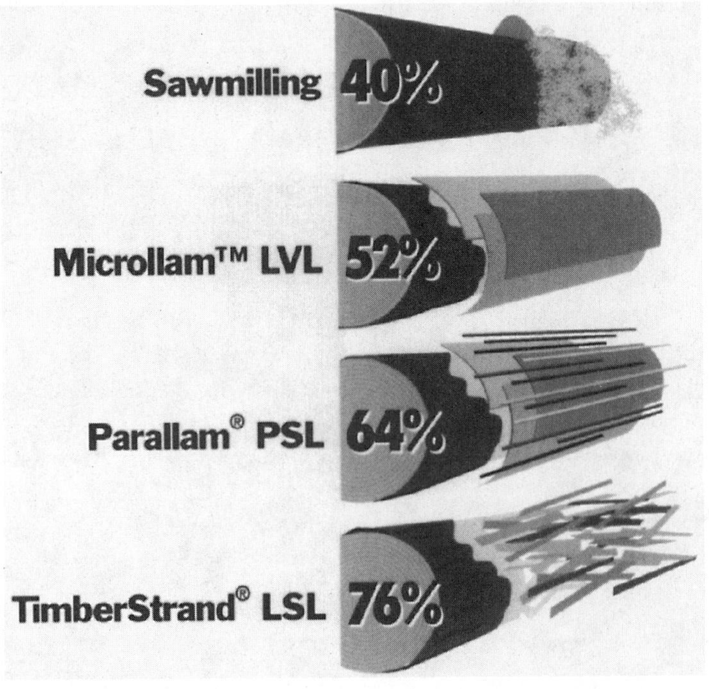

1-12 *Engineered lumber is a big improvement over conventional sawmilling practices when it comes to using more of the tree.* Courtesy Trus Joist MacMillan

LVL members are quite strong, and are a perfect example of how this technology of taking a tree apart and putting it back together again can greatly improve over mother nature. A piece of LVL has a higher design rating than a solid sawn piece of Number 2 Douglas-fir of the same dimensions. Looked at another way, LVL has 50 percent more structural value per cubic foot of raw material than conventional solid sawn members.

Following on the technology of LVL was the next generation of 2-foot to 8-foot-long strands of veneer, all of which can be taken from young, small-diameter second- and third-growth trees that not so long ago were burned as fuel wood or processed into low-value chips. The strips are dried, pressed, and formed into huge members that are capable of handling heavy structural loads. LVL members of up to 11 inches thick, 20 inches deep, and 66 feet long are ideal for beams, posts, headers, and industrial applications such as girders and bridges—all from small pieces of "waste" material once considered to be virtually without value.

To better understand what this means to the forest, consider a PSL billet (a pressed-up member of PSL or other engineered wood) that is 66 feet long, 19 inches wide, and 12 inches thick. The billet is made from strands of wood processed from trees that are between 8 and 11 inches in diameter, with an average age of 40 to 60 years. To mill that same member from a solid log, the tree would have to be a massive five feet in diameter at the base and be at least 200 years old. And even at that, the solid piece of lumber would not have the same strength as the PSL member.

The 1980s and 1990s have brought a new set of priorities and problems, and another new generation of engineered wood products to meet them. Environmental concerns about the dwindling supply of trees versus the steady demand for lumber have raised increasing questions about our forest policies, as have the effects of continued logging on the health of the watershed and wildlife of all types. The northern spotted owl, an endangered species whose habitat was reported to be threatened by the logging of old-growth timber, became the beleaguered rallying point on both sides of the environment-versus-timber issue.

In response to these concerns, yet another new engineered lumber product entered the marketplace. The product of research by Trus Joist MacMillan, laminated strand lumber, or LSL, was introduced in 1990 under the trade name TimberStrand. LSL uses even smaller logs than PSL and LVL products, logs that would otherwise be considered too small, too weak, or too crooked to have any other manufacturing use.

LSL members are made up of strands of lumber approximately 12 inches long. The strands are coated with resin, aligned parallel with each other, and pressed into billets. Current manufacturing methods allow the pressing of members up to 8 feet wide, 5½ inches thick, and 48 feet long. Uses include beams, headers, and decking, and LSL is also quite popular for use as a rim board.

As this technology continues to evolve under the encouragement of millions of dollars of research and development investment, the use of extremely fast-regenerating wood species such as aspen and yellow poplar—which previously had no timber use whatsoever—is becoming a reality. Already on the horizon—and, in fact, already in use—is the promise of tree farms upon which fast-growing trees are raised simply as crops. As fields on one side of the tree farm are being harvested, fields on the other side are being planted, a continuous process of growth and use that is no different from raising a crop of corn or wheat.

Another interesting offshoot of the engineered lumber revolution is combining these products to make yet another engineered lumber product, as is the case with the I-joist. I-joist technology uses a web of oriented strand board (OSB), a sheet material formed from pressed layers of resin-coated wood chips (see Chapter 2), combined with top and bottom flanges of LVL. The result is a structurally solid joist with excellent dimensional stability and considerably less weight then conventional 2 × 10 or 2 × 12 joist material.

I-joists are available from several different manufacturers, and can be found in depths from 9½ inches to 36 inches, and in lengths up to 80 feet. Try finding a nice straight 80-foot-long 2 × 12 at your local lumber yard. The joists are also prepunched with knockouts to simplify wiring and plumbing. Here again, both the LVL flanges and the OSB web are manufactured from small-diameter logs and other low-value, renewable materials.

What it means to our forests: Some statistical information

Numbers are interesting things and can be bandied about in a number of ways, but they sometimes help us to understand how one thing can affect another, and so from that standpoint are certainly worthwhile. In terms of how new generations of engineered lumber technology are helping our forests, the following facts and figures that have come about as part of the ongoing forest technology research being done by Trus Joist MacMillan are very enlightening:

- Using standard, solid, dimensional lumber, it takes approximately 10 trees to frame the average 2100-square-foot, single-family home. Each tree is an average of 20 inches in diameter, 100 feet tall, and about 80 years old.
- Roughly two to three of those trees go into the floor framing alone.
- Using engineered lumber in the floor of that home, as opposed to solid 2 × 10 or 2 × 12 lumber, would require about one tree, or a net savings of approximately one to two trees in the floor system alone.
- Since their introduction, engineered lumber I-joists have been installed in approximately one million homes. This alone represents a savings of between 2 and 3 million trees.
- The National Academy of Sciences estimates that it takes 5.75 million British thermal units (BTUs) of energy per ton to extract, transport, process, manufacture, and build with laminated veneer lumber. In contrast, steel requires 50.32 million BTUs per ton for the same manufacturing and construction process, or the equivalent of almost 9 times as much energy.

Another statistic tossed out is that wood is 15 times more efficient of an insulator than concrete, 400 times more efficient than steel, and 1770 times more efficient than aluminum—not that there are many homes being built of aluminum. All in all, though, wood is certainly one of the best insulators being used in construction today, which helps conserve our fossil fuels.

Given the poor quality of lumber being delivered to the job site these days, another statistic of particular interest is the waste factor. According to a study done by Trus Joist, as much as 11 percent of the lumber that's purchased and delivered to a construction site for use in building an average home is tossed away, either for defects or simply as drops in the process of cutting the lumber to length. (As you know, solid lumber is supplied only in even 2-foot increments, so cutting an 11-foot joist from a 12-foot board is going to a leave a foot of waste that will probably end up in the trash pile.)

Most engineered lumber products, on the other hand, are delivered rough-cut to length, so the waste factor is minimized. There is also no waste due to defects, since all the lumber is manufactured to the same standards. The result is a waste factor of only around

1 percent. Take the 10 trees that go into the average house, factor in the 10 percent difference in waste between solid lumber and engineered lumber, then multiply by the approximately 1.25 million housing starts in 1995 alone. The result is about one tree per house going into the trash, or the equivalent of one and a quarter million wasted trees.

While old-growth forests have dwindled to a point where only about 10 percent of this country's original stands of ancient trees are left, there really is no need to continue the cutting. We are currently planting about 37 percent more timber than we are harvesting, and if we continue to advance the technology that allows for cutting and processing small, rapid-growth trees, the old-growth stands can be left for the health of the ecosystem and the enjoyment of future generations.

Where these trees are coming from

The timber used for the manufacture of engineered lumber is derived from a wide variety of sources, and is dependent on the policies of the manufacturing company, the location of the manufacturing plants, the size and species of trees being used, and the general availability of the trees. There are actually a variety of species being used for engineered lumber products, including yellow poplar, southern pine, cedar, Douglas-fir, aspen, western hemlock, and gum. The three most common sources for these trees include:

Company-owned lands

Some of the manufacturers, especially those that have been in the timber industry for a long time, own tracts of timberland. Some of this is old growth, although, as previously discussed, those supplies are dwindling because of overcutting and strict regulation, and old-growth timber is almost exclusively reserved for milling solid sawn lumber. Much of what is being cut and processed on company lands is natural second- and third-growth lumber, along with some harvesting of trees that were planted as part of reforestation projects in the 1950s and 1960s and are now reaching usable size.

As an example, Louisiana-Pacific Corporation reports in their 1995 Corporation Annual Report that it owns about 1.6 million acres of timberland. The bulk of their land holdings, about 900,000 acres, are reported to be in Texas and Louisiana, and are comprised mostly of relatively fast-growing southern pine. The company also reportedly owns about 485,000 acres in northern California.

Public lands

Each year, the Forest Service places selected tracts of state-owned forest land up for bid. The state conducts the bidding process and sets the rules for how, where, and when the logs can be cut and transported. Manufacturers, through contract buyers or through their own buyers, bid on these lands in a sealed bidding process, with the highest bidder being awarded the timber rights. Similar bidding is done for timber growing on federal lands.

Here again, some of this timber is old growth, which is destined for milling into solid lumber. Very few old-growth tracts are ever sold in their entirety; instead, a certain amount of old-growth timber is included as part of a sale of second- and third-growth materials. After the buyer of the timber completes logging operations, the state conducts reforestation of the harvested areas, using either state employees or, more often, contract tree-planting labor.

Private lands

Growing timber on private land for eventual sale is a relatively new and rapidly increasing arena. Many landowners are turning over sections of their acreage for growing timber, finding it to be a crop that is relatively easy to grow and profitable to sell. Several studies have indicated that this segment of the timber market will continue to play an increasingly important role in future engineered lumber production.

Trus Joist MacMillan, for example, does not own any of its own timberlands, but instead purchases huge quantities of logs from private sources. In Kentucky and West Virginia, as much as 95 percent of the logs that supply TJ's engineered lumber manufacturing plants in those areas come from private timber purchases. The company currently has about 100,000 privately owned acres in eastern Kentucky under contract, and hopes to have upward of a quarter of a million acres under contract by the end of 1996.

The timber that is grown for harvest on private land is not governed by the strict management policies that regulate logging and reforestation on state and federal lands. There are, however, voluntary programs called *best management practices* (BMPs). BMPs are strongly encouraged by state forestry associations and engineered lumber manufacturers, since they help promote healthy trees and sustained, profitable yields for landowners. Some universities, recognizing the potential for creating strong and lucrative new markets for the farmers in their areas, are getting involved as well. The University of Kentucky, for example, has a landowner assistance program that helps small landowners learn to manage their timberlands.

Timber is obtained from private owners in one of two ways. First, the buyer can work directly with a private timber owner and simply negotiate a private sale of all or part of the available timber. In the second situation, owners of private timber tracts can put the trees up for a sealed-bid sales process, in which the sale is typically made to the highest bidder. Purchase prices are based on careful evaluation of the type, age, size, and quality of the timber, as well as the costs associated with cutting, transporting, and milling the logs.

New builders are planning for remodeling

In the past, it was not all that common to find the builder of a brand-new house already thinking about remodeling. With the increased usage and awareness of engineered lumber products, however, that way of thinking is changing as well. No longer are new-house builders thinking only of which products will perform the best in the new house; they're also thinking about which products will perform best when the house needs to be remodeled.

There are an uncounted number of homeowners who have had a contractor examine their home and give an estimate on a room addition or perhaps an attic conversion. While the figures for the addition are acceptable, the homeowner is often shocked to discover the additional, unanticipated costs of remodeling that must be done before the addition can get underway. There's the sag in the old roof that has to be repaired, along with resupporting the rafters. Or there's that squeaky, uneven floor that needs to have new supports placed under the joists. Or the fact that the windows and doors don't operate correctly because of sagging and uneven wall and floor framing.

The framework of a house is literally the skeleton over or on top of which all the other elements of the home—from drywall and paint to cabinets and siding—are placed. Quality framing makes a new home square and solid, while poor framing practices and materials cause headaches from the first day of occupancy all the way to the remodeling 15 years later.

Standard sawn lumber varies greatly in quality, dimensional sizes, moisture content, and dimensional stability. With that in mind, a growing number of builders are using engineered lumber not only as a means of simplifying and strengthening the framework of their new homes, but also as a selling point for future remodeling. Many remodeling problems can be addressed by some farsighted thinking from day one, and an increasing number of homeowners—especially ones who have already been through a remodeling in the past—see that, and select their contractors and specify their materials accordingly.

Cost comparisons

On April 5, 1996, lumber prices as reported in *Nation's Building News* stood at $345 per 1000 board feet. As of May 10, the figure had risen to $419, and the paper commented that "concerns are rising along with lumber prices." Lumber prices are driven by demand, of course, but also by factors such as dwindling supply, cost of cutting and processing, and governmental regulations. For several years now, prices for solid sawn lumber have been on a roller-coaster ride, and that presents real problems for builders who estimate the cost of a building project to a client, and then need to stand behind that price in the face of unstable lumber costs.

Manufacturers of engineered lumber products are also touting the relative price stability of their products as another selling feature. What follows are excerpts from a report published by Trus Joist MacMillan, a company that has probably put more research effort into the field of engineered lumber than any other I'm aware of, as to the cost differences between solid lumber and engineered lumber. (The methodology followed was to conduct an installed cost survey to determine the overall cost of framing a typical house. Trus Joist MacMillan devised two sets of building plans for a 2090-square-foot, two-story home with a basement. One set specified the company's FrameWorks building system of engineered lumber products, and the other used 2 × 10s, 2 × 12s, and glulam beams in all comparable structural applications. Builders were asked to estimate a variety of labor costs to build both plans, a process akin to bidding an actual project. Material costs were based on a 12-month, national average for the products specified on the plans.)

> *Engineered lumber may cost more—but only at the lumberyard. In the field, the balance shifts. Builders responding to Trus Joist MacMillan's recent Installed Cost Survey report saving an average of about $370 per house with the FrameWorks building system compared to using traditional lumber, citing labor efficiencies, less material waste, and far fewer customer call-backs.* [Fig. 1-13]
>
> *From Seattle to Montreal, ranging from small shops to companies that sell hundreds of homes a year, the 37 builders responding to the survey built more than 2,500 homes combined*

	Engineered lumber	Solid sawn lumber	
Materials	$3061.65	$2548.04	
Waste	30.62	280.06	
Labor	705.87	1118.97	
Call-backs	0	221.04	
Total	$3798.14	$4168.11	(+9.75%)

1-13 *Results of a study showing the difference in installed cost between engineered lumber and solid sawn lumber.* Courtesy Trus Joist MacMillan

	Engineered lumber system	Standard lumber system
Materials	$3061.65	$2548.04
Waste*	30.62	280.06
Labor	705.87	1118.97
Call-backs+	0.00	221.04
Total‡	$3798.14	$4168.11

* Calculated at 1 percent for engineered lumber, 11 percent for standard lumber.
+ Calculated as cost per house, related to floor performance only.
‡ The standard lumber costs $369.97 or 9.75 percent more.

1-14 *Cost comparison chart.* After Trus Joist MacMillan, Cost Comparison Chart, Installed Cost Survey.

in 1994 and generated nearly $830 million annually in entry-level, move-up, luxury, and custom-home markets.

Trus Joist MacMillan asked these builders to estimate labor and other association costs to frame two homes, one with engineered lumber, the other with traditional lumber products. [The survey methodology is described below.] The result: Builders surveyed estimated a $369.97 savings with the FrameWorks system, nearly 10 percent less than the cost to *build the same plans with traditional lumber. The difference offsets a gap in a materials-only cost calculation.* [Fig. 1-14]

Labor savings tipped the balance in favor of an engineered lumber system. Builders were asked to estimate costs for 10 different tasks (or phases of work), including materials handling and installation of floor joists and framing connectors. On average, the ability to frame a house more quickly with engineered lumber saved them more that $400 per house. Builders say it takes about two-thirds the time to build with engineered lumber than with comparable solid sawn materials. Installing the Silent Floor system was the most efficient phase of work, requiring only half as much time in some regions than buildings with comparable 2× 10 and 2× 12 floor joists.

. . . TJI joists weigh less than solid-sawn materials—often allowing a single framer to carry and install them—but they are engineered to a uniform quality that simplifies product selection and installation at the job site. In addition, TJI joists are cut to specifications, thus eliminating the time spent sawing traditional lumber to the proper lengths.

Less material waste and fewer customer service call-backs also saved money for builders who used the FrameWorks system—on average, almost $500. Service call-backs alone were estimated at more than $200 for a floor system framed with solid sawn lumber; in the East and Midwest regions, builders estimated the cost to repair the nail pops, floor squeaks, and ceiling drywall cracks that can occur with traditional lumber at between $466 and $562 per house.

A positive situation for everyone?

Engineered wood represents probably the closest thing that we in the building trades are likely to see in the way of a win-win situation. For the builder, these engineered wood products represent structural members of dependable size and quality, as well as dependable price and availability as their popularity grows (Fig. 1-15). Consumers should benefit from a structure of more consistent quality (Fig. 1-16), and, hopefully, housing prices and profit margins alike will begin to reflect a decrease in labor and material costs.

For the environment, and in turn for all of us as a society, the benefits could well be incalculable. Second- and third-growth trees are perfectly suited for these applications, sparing a dwindling and

1-15 *Stacks of I-joists in a retail lumberyard are clear evidence of the growing popularity of engineered lumber.*

increasingly expensive supply of old-growth timber. Fast-growing species such as aspen, cottonwood, yellow poplar, and sweet gum are being raised and harvested like wheat, corn, or any other renewable crop, then chipped or peeled for use in engineered wood products. In the west, even weed trees such as the juniper, which have been choking range lands in some areas, are being considered for chipping.

"We're trying to use every piece of the tree," said one Georgia-Pacific representative. "That's our goal, and we feel it's the most positive approach to both forest management and quality building products."

Environmental issues, as well as manufacturer hype, product literature, and promises aside, builders are always going to focus first and foremost on what works in the houses they build. With that as a criteria, it appears that a growing number of builders are accepting the value of engineered lumber. Louisiana-Pacific, Trus Joist MacMillan, Georgia-Pacific, and Weyerhaeuser have all been reporting very encouraging sales figures, with some claiming a doubling and redoubling of engineered wood sales in the last three to five years.

One study, conducted by Leonard Guss Associates of Bellevue, Washington, forecasts that sales of laminated veneer lumber will triple, up to a billion dollars by the year 2002. And a report from George Carter Affiliates in Oradell, New Jersey, showed a 47 percent increase in the sales of wood I-joists in two years—during a period

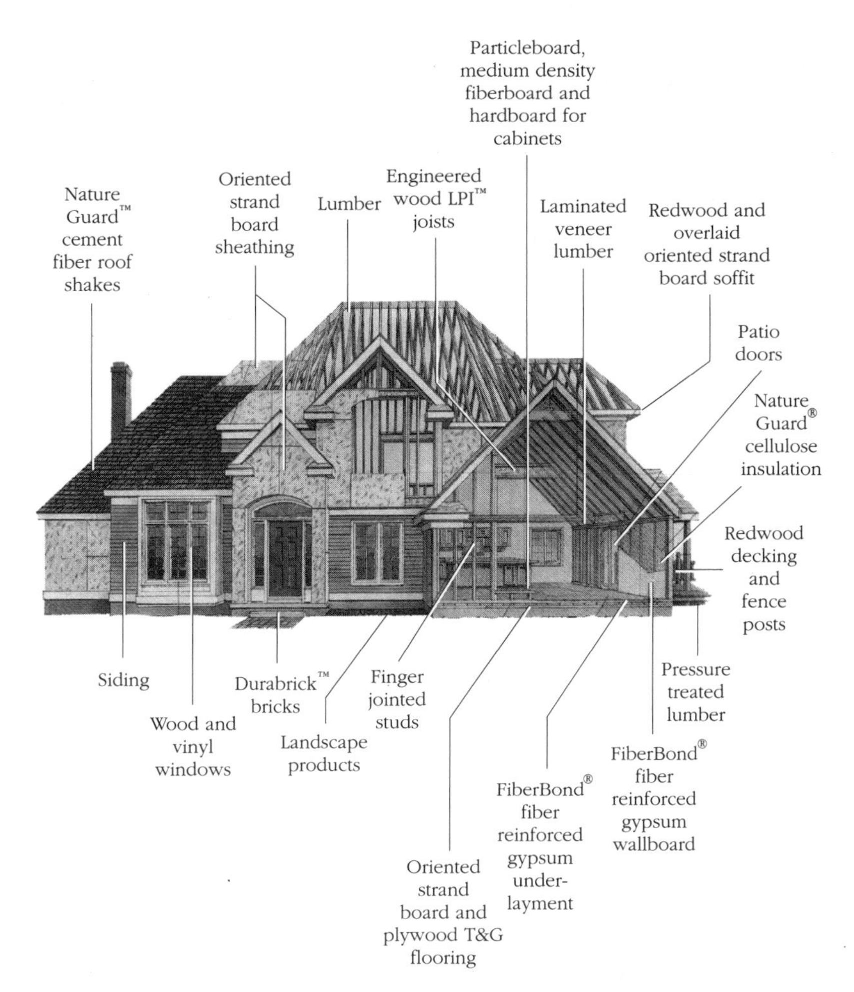

1-16 *Engineered lumber products are finding a use in virtually every part of today's home.* Courtesy Louisiana-Pacific

when housing starts rose only 1 percent. Their report also projected that the I-joist demand will grow 140 percent by the year 2000.

Interviews with custom home builders also show an increasing awareness and acceptance, but point out some problems as well. The fact that many of these products are not stocked by the lumber yards and need to be ordered is a fairly common complaint, and back orders and occasional long lead times for ordering are the biggest irritants.

But most accept that the industry is still in its infancy, and recognize that this truly is the wave of the future. "I remember when no one had ever heard of that stuff, let alone used it," one builder remarked, pointing to a huge stack of Louisiana-Pacific's phenomenally popular InnerSeal lap siding sitting in a lumber yard, "and now it's everywhere. They stock TJIs and glulams here, too. It probably won't be long before engineered wood is all you see in the yards, and Doug fir lumber is the special-order product."

It's a changing industry in the United States—Europeans recognized the need for alternative materials several decades ago—and the days of the "natural" all-wood house are over. Dwindling forest reserves, environmental concerns, and constant pressure on builders to provide affordable housing in order to keep the American Dream alive has moved the building industry into new arenas, and the changes are permanent.

2

An introduction to engineered lumber products

As you read in the last chapter, the concept of engineered lumber is not new, and one ancient engineered wood product—plywood—has probably been a staple on your job sites for many, many years. What *is* new is the wealth of research and technology that has been poured into the industry in the last couple of decades, resulting a huge number of new products to arrive in the lumber yards, seemingly overnight. New products present new opportunities for cost savings and improved quality control on your projects, but they also present new problems: ordering, specifying, training, and all the rest of the learning curve that we as contractors would just as soon do without.

The products, however, are here to stay for as long as anyone cares to predict, so the sooner you start that learning curve for both you and your employees, the sooner you can start reaping the benefits—because these products really do represent a better way to build.

One of the problems is that the sheer number of engineered lumber and panel products and their offshoots can get extremely confusing (Fig. 2-1). Add to that the fact that you'll often see the same products listed or specified in different forms—plywood, plywood sheathing, and plywood siding, for example—and you might wonder what the differences really are. Then there are all the trade names for the various products, many of which have the same underlying product composite as all the others. The generic wood I-joist, for example, is called a TJI joist by Trus Joist MacMillan, an LPI joist by Louisiana-Pacific, and, in a descriptive if slightly less imaginative labeling process, a Wood I Beam joist by Georgia-Pacific.

2-1 *Part of the growing volume of different engineered lumber products is seen at a Trus Joist MacMillan plant in Eugene, Oregon.*

Finally, there are the grade and product identification stamps that are present on virtually every product that local lumber yards supply. But do you really understand the stamp, and the importance of what the information it contains is telling you? How do you know the panel or board you've selected is the right one for the specific application you're using it for?

All this confusion and onslaught of information is enough to cause many builders to shy away from trying anything new, preferring instead to remain with the handful of products they recognize and are familiar with, "the stuff we've always used." There are advantages and disadvantages to this way of thinking, for while there is certainly quite a bit of comfort in familiarity, there is also the danger of ignoring a variety of great, innovative new products that can make your jobs stronger, safer, more attractive, and more profitable.

In this chapter, I'll examine many of the new engineered lumber products now on the market, including the abbreviations, terminology, and some common uses. This includes the basic engineered lumber products, as well as some of the more innovative uses of these products. All individual product descriptions are listed alphabetically by their most generic name (e.g., I-joist instead of TJI). Some of the more common product names are also listed in some areas, and are cross-referenced to the generic heading. Engineered panel products such as plywood and OSB are described in Chapter 3, and siding, roofing, and decking products, as well as interior and exterior trim, are all covered in Chapter 8.

Dealing with the learning curve

Before looking at all of the specific products, I'm going to throw in a few important words about employee training. Most contractors have no organized training program for their crews, and often with good reason. It's hard enough just getting everyone organized, loaded up, and out the door in the morning, without worrying about planning a lecture or a demonstration. Building a house or remodeling an office is a demanding, time-consuming process, and profit margins are often slim enough as it is. Who has the time or the resources to pay a crew to sit around the shop or the job site and drink coffee while they look at the latest I-joist to hit the market?

Actually, *you* do. And you should. Whether you have crew members who have been swinging a hammer for centuries or ones who aren't even sure what a hammer is, they need to learn how to use the products you're providing them with. Time is money, definitely, but look at that time as an investment in making money, not wasting it. A well-trained crew is much more efficient, getting your jobs done faster and with less mistakes. They make better use of the materials you provide them with, creating less waste and making fewer recuts. They're safer, the job site is safer, OSHA is happier, and your job gets done faster, cleaner, better, and with a healthier bottom line.

Training doesn't need to be a formal process, or one that eats up a lot of time. Consider having a crew meeting in the mornings once or twice a month, lasting for a half an hour or 45 minutes. You can pass out some product brochures and take a few minutes to go over them. Or you can pass around a few samples, perhaps give a brief demonstration. Many manufacturers will give or lend you video tapes that show what their products are and how they should be used correctly, and these professionally produced tapes are both interesting and informative. You might also give some serious thought to combining these crew-training sessions with those safety meetings you're supposed to be having but probably never get around to.

Besides video tapes, many companies offer training aids such as posters, brochures, product samples, and booklets. If you have a number of employees, some manufacturers will even send a product representative out to talk with you; it doesn't hurt to ask. You might even want to team up with one or two other builders you know and combine all your workers into a larger group for the morning, and invite a manufacturer to speak.

However you choose to organize and present your training, don't ignore it. Continuous training is crucial if you want to use these new products correctly, safely, and cost-effectively.

Material Safety Data Sheets

If you're like most builders, you've probably never really thought about Material Safety Data Sheets, or MSDSs (Fig. 2-2), or perhaps even been aware of their existence. MSDSs are an OSHA requirement, and they are intended to give the user of any particular material some important information about that material, including chemical makeup, toxicity, flammability, safety equipment and precautions, how to deal with leaks, how to safely dispose of the material, special handling requirements, antidotes, emergency phone numbers, and a variety of useful stuff.

Not surprisingly, given the state of our government today, the MSDS program, which started as a useful safety tool, has gotten completely out of hand. An MSDS is now required on virtually everything you use on the job site. This is valuable safety information if you're working with certain chemicals or other hazardous materials, but runs to the absolutely ridiculous when you realize that there are MSDSs for things like sand, bricks, lumber, wood glue, even—are you ready for this?—water.

Without editorializing further, this is merely to warn you that, as a builder, you need to provide MSDSs to yourself and all your employees. Most if not all of the new engineered lumber and nonlumber products have MSDSs available for them, and it's up to you to get ahold of them and keep them available and up to date. Failure to keep this material current and readily available can result in some serious monetary fines from OSHA.

For more information, contact your local OSHA representative, or ask the agent who provides your worker's compensation insurance; they usually have programs in place to help you understand how to accumulate, update, file, organize, and distribute MSDSs and other safety information.

Engineered lumber products

It is the engineered and composite lumber products that are at the real heart of the engineered wood revolution, and actually at the heart of this book. It is also here where the most confusion and misunderstanding lies, due for the most part to the radical newness of the products and the confusing plethora of names and slogans associated with these products.

There is also some distrust of these new products among builders, especially the ones who have been around the trade for awhile. And perhaps rightfully so, for these are the men and women on the front lines who stake their reputations, the employment of their crews, and the financial stability and security of their companies on the products they select and install.

"I've used solid lumber in my houses for 18 years," one builder commented when asked why he didn't use more engineered lumber products in his custom homes. "I know what it is, where it comes from, and how to cut it and nail it. In other words, I know exactly what it will and won't do, and I can depend on that knowledge and that consistency."

He has a point, and the engineered lumber manufacturers are well aware of it. They're not trying to replace sawn lumber, I've been told, simply to augment, improve, and accompany it with new products that perform constantly and are a little easier on our forest resources. They're content to let the market regulate itself and to allow these products to prove themselves over time and use, and all indications are that they'll do just that.

So if this book is unable to do anything about your trepidation concerning these products, it should at least clear up some of the confusion. What follows is a listing of the common engineered lumber products currently on the market, listed alphabetically by their accepted generic industry descriptions, and not by product or manufacturer name. You'll probably be shocked to discover that there are actually only five you should be aware of. In the following section there is a listing of four products that are interesting composites of engineered wood technology, from the increasingly familiar I-joist to an innovative, new, fully engineered, bolt-together truss system for gaining new living space over a garage.

To further clarify the differences in products and perhaps clear up a little more of the confusion, refer to Appendix A at the end of the book for a listing of these products by trade name, and a second reverse listing in Appendix B, which shows you which trade names go with which products.

Glulams

Glue-laminated beams (Fig. 2-3), or *glulams* as they have come to be known in the industry, are certainly nothing new. In fact, one of the first known uses of glue-laminated timber was in 1893, in the

MATERIAL SAFETY DATA SHEET

L-P ORIENTED STRAND BOARD (OSB) PRODUCTS

SECTION I - General Information

Manufacturer's
Name: Louisiana-Pacific Corporation
Address: 111 SW Fifth Avenue
Portland, Oregon 97204
Emergency
Telephone Number: (503) 221-0800
Preparation/
Revision Date: April 1996

SECTION II - Hazardous Ingredients

Chemical Name	Common Name	CAS No.	ACGIH TLV TWA	ACGIH TLV STEL	OSHA PEL TWA	OSHA PEL STEL
Wood Dust	Wood Dust	N/A	5 mg/m³**	10 mg/m³	5 mg/m³	10 mg/m³
Resin Solids as PNOC	Resin Solids	N/A	10 mg/m³ (inhalable) 3 mg/m³ (respirable)	-- --	15 mg/m³ (inhalable) 5 mg/m³ (respirable)	-- --
Zinc Borate as PNOC*	Zinc Borate	138265-88-0	10 mg/m³ (inhalable) 3 mg/m³ (respirable)	-- --	15 mg/m³ (total) 5 mg/m³ (respirable)	-- --
Formaldehyde	Formaldehyde	50-00-0	0.3 ppm (ceiling)	--	0.75 ppm	2.0 ppm

* ONLY L-P SIDING OSB MATERIALS (e.g. panel siding, lap siding, and soffits) CONTAIN ZINC BORATE; NO OTHER OSB PRODUCTS CONTAIN THIS MATERIAL.
** For OSB products composed of hard woods (e.g. beech, aspen, and oak), the TWA is 1 mg/m³.

SECTION III - Physical and Chemical Characteristics

Composition: OSB products are a cured composite of wood strands bound together using resins. These products may contain more than 1% methylene diphenyl diisocyanate (MDI) as resins. However, chamber tests conducted on commercial OSB boards did not detect airborne levels of MDI, indicating that no MDI off gassing occurs above the method detection limit of 20 parts per trillion (ppt).

Appearance: Light brown wood panels.

Odor: Slight to none.

pH: N/A

Bulk Density: 34-44 PGF

Melting Point: N/A

Vapor Pressure: N/A

Vapor Density: N/A

Evaporation Rate: N/A

Solubility: N/A

SECTION IV - Fire and Explosion Hazard Data

Flammability Characteristics: OSB products may ignite if exposed to temperatures exceeding 400°F. These products are combustible and may burn if exposed to open flames, high temperature objects, or oxidizing chemicals.

Explosion Hazards: Airborne concentrations of finely divided wood and resin dust, when combined with an ignition source, can create a fire or explosion hazard if the concentration of wood dust exceeds 40 g/m³.

Flash Point: N/A

Extinguishing Media: Normal firefighting methods appropriate for surrounding fire, such as water or CO_2 extinguishment, may be used in case of fire.

Special Fire Fighting Procedures: Firefighters should wear self-contained breathing equipment and full protective clothing.

SECTION V - Reactivity Data

Stability: These products are stable under normal conditions.

Incompatibilities: Oxidizers.

Hazardous Decomposition or Byproducts: Toxic constituents which may be found in wood smoke include carbon monoxide, aldehydes, and polycyclic aromatic hydrocarbons.

Hazardous Polymerization: Will not occur.

SECTION VI - Health Hazards

Route(s) of Entry: Inhalation.

Health Hazards: Chronic exposure to wood dust in OSB products may cause dermatitis, respiratory irritation and increased risk of upper respiratory tract disease. Susceptible individuals may experience watery eyes and rhinorrhea (runny nose). Zinc borate dusts may cause respiratory irritation at concentrations greater than 10 mg/m³. Polymerized diisocyanate containing resins in OSB panel dust may irritate the eyes, skin, throat and upper respiratory system. Low concentrations (less than 0.05 ppm) of formaldehyde off gassing from OSB panels may also irritate the eyes and respiratory system.

2-2 *Sample of a Material Safety Data Sheet (MSDS) for engineered lumber, in this case, oriented strand board.* Courtesy Louisiana-Pacific.

Carcinogenicity: Wood dust has been classified by the International Agency for Research on Cancer (IARC) as carcinogenic to humans in furniture and cabinet-making, and possibly carcinogenic to humans in carpentry and joinery operations. Formaldehyde has been classified by the IARC, the National Toxicology Program (NTP), and the ACGIH as either carcinogenic to humans or as a potential carcinogen. Formaldehyde is also regulated by OSHA as a human carcinogen.

Signs and Symptions of Exposure: Potential symptoms of exposure to OSB dusts are similar to those for wood dust. These may include: Irritation (redness) of the eyes; dermatitis (abnormality of the skin); cough, wheeze and inflammation of the sinus passages; possible epistaxis (nosebleed); and abnormal sensitivity of the respiratory system.

First Aid: Avoid prolonged exposure to dust levels in excess of the published standards. If symptoms of exposure are experienced or accidental over-exposure occurs, move the affected individiual to fresh air. If eye or skin irritation is experienced, flush both eyes or affected area with copious amounts of fresh water for at least 15 minutes. If irritation persists, or if foreign matter remains in the eye, seek medical attention.

Section VII - Precautions for Safe Handling and Use

Steps to be Taken in Case Material is Spilled: Shovel or sweep up spills and place in container for disposal.

Waste Disposal: This product is not a hazardous waste under federal law.

Note: This product is not regulated by DOT per 49 CFR 172.01.

Section VIII - Control Measures

Respiratory Protection: Use a NIOSH-approved dust respirator if ventilation is inadequate.

Ventilation: Provide ventilation adequate to keep airborne concentrations below the exposure limits.

Eye Protection: Wear vented safety goggles.

Protective Clothing: Wear gloves for routine handling.

Work Practices: Good housekeeping practices should be used to minimize dust levels in the air and to reduce the possibility of slipping on dust collected on floor surfaces.

Section IX - Additional Information

Disclaimer: This MSDS is intended solely for safety education and not for use in relation to specifications or warranties. The information presented in this MSDS was obtained from usually reliable sources and is provided without any representations or warranties regarding the accuracy or correctness. Since the handling, use, and storage of this product is beyond our control, Louisiana-Pacific assumes no responsibility and disclaims liability for any loss, damage, or expense arising therefrom.

Comments: Louisiana-Pacific has attempted to provide a readable and informative MSDS for use with L-P products. Should you have any comments or suggestions regarding this MSDS, please send them to Louisiana-Pacific Corporation, 111 SW Fifth Avenue, Portland, Oregon 97204-3601, Attn: MSDS Information Coordinator (telephone number (503) 221-0800).

2-2 *(Continued)*

construction of an auditorium in Switzerland. In 1934, glue-laminated beams were used, appropriately enough, in the new U.S. Department of Agriculture's Forest Products research laboratory in Madison, Wisconsin, which is still in use today.

In 1942, phenol-resorcinol adhesives were developed, which are fully water-resistant. This advancement in adhesives was quickly applied to the manufacture of glulams, allowing them to be used in exposed outdoor applications. The manufacture of glulams was first standardized in the United States in 1963, and those standards continue to be refined and upgraded as technology and research dictate. The latest manufacturing standards have been in place since 1993.

2-3 *Glulam beam supporting the upper floor above a large garage building, leaving a clear span below.*

A glulam beam is simply a large beam made up of several smaller layers of wood, called *laminations*. Each lamination is selected for its structural properties and also, depending on the grade of the beam, for its appearance. The laminations are arranged with the grain of all pieces running parallel with the long dimension of the beam, and with the strongest pieces on the top and the bottom of the beam to better resist tension and compression stresses in these areas. The laminations are coated with adhesive, then pressed together and cured into a single structural member.

Glulams are available in both balanced and unbalanced configurations (Fig. 2-4), which has to do with the arrangement of the various laminations in regard to the location of the bending stress placed on the beam. Balanced beams are symmetrical in quality, outward from the mid-height of the beam. They are used in situations where both the top and the bottom of the beam could end up in tension, depending on where the beam is installed and the type and location of the loads applied to it.

Unbalanced beams have laminations of higher quality on the tension side of the beam than on the compression side. This carefully engineered and calculated assembly of laminations changes the bending-stress rating of the beam, and can be adjusted according to the beam's intended use. Unbalanced beams are most commonly used in long, single-span applications, where the beam is supported only at each

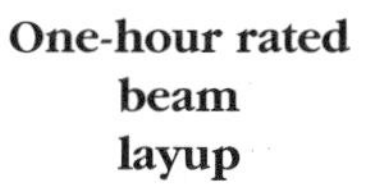

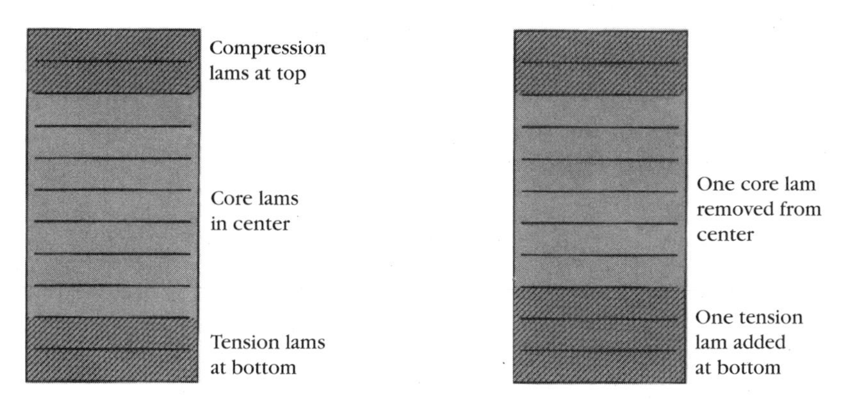

2-4 *Sample of how different laminations are used in a glulam beam to change its structural characteristics.* Courtesy American Plywood Association.

end. The top side of an unbalanced beam is clearly stamped with TOP to ensure that the beam is properly oriented as it's installed.

Glulams are also constructed with a very slight upward curve in them, as viewed from the side. This curve, called a *camber*, helps prevent the beam from sagging downward as loads are applied to it. There is no actual standard camber for a beam, since it varies with manufacturer and intended use. On the average, however, the camber is approximately equal to the curvature of a circle having a 1,600- to 2,000-foot radius.

A glulam is also available in three appearance grades: industrial, architectural, and premium:

- *Industrial.* Industrial-grade beams are lightly surfaced on the two wide sides, and might have visible glue and other defects that affect the appearance of the beam without altering its strength properties. They are used primarily in concealed locations, or in industrial and other applications where appearance is not a consideration.
- *Architectural.* This is probably the most common of the appearance grades, combining a cleaner, more finished appearance than the industrial grade with a lower cost than the premium grade. Architectural-grade beams are suitable for exposed applications in most circumstances, and this is the stock beam for most lumberyards and suppliers.

- *Premium.* Premium-grade beams have laminations that are selected for appearance as well as strength, and are the highest of the appearance grades. They are typically specified only where the finished look of the beam is extremely important. Premium beams are available by special order only, and are the most expensive of the three grades.

Glulams are available in a virtually unlimited range of sizes and configurations, including custom curves and extremely long, unsupported spans. Stock beams, which are suitable for most construction uses, have a fairly standard amount of camber and are available in standard widths of 3⅛, 3½, 5⅛, 5½, and 6¾ inches. Glulams need to be sized for the span and the loads that are specific to where they'll be installed (Fig. 2-5). Most lumberyards can assist you with sizing beams for specific use, or you can order *Glued Laminated Beam Design Tables* (EWS S475) from the American Plywood Association (see Appendix E at the end of the book).

(1) Indicates structural use:
B-Simple span bending member
C-Compression member
T-Tension member

(2) Mill number.

(3) Identification of ANSI Standard A190.1, Structural Glued Laminated Timber.

(4) Applicable laminating specification.

(5) Applicable combination number.

(6) Species of lumber used.

(7) Designates appearance grade. INDUSTRIAL, ARCHITECTURAL, or PREMIUM.

2-5 *Sample grade stamp for a glulam beam.* Courtesy American Plywood Association.

Laminated strand lumber (LSL)

Laminated strand lumber, or LSL, is actually the newest and latest entry into the world of engineered lumber (Fig. 2-6). The process uses about 76 percent of the log, as opposed to the 40 percent that's used in conventional milling procedures. It also makes extensive use of smaller logs and species that were previously thought to be too weak for structural use in construction.

In the LSL process, logs are steamed and cleaned, then cut into strands up to 12 inches long. The strands are dried and treated with resin, then aligned by machine to be parallel with each other. The glued strands are then cut into huge solid blocks called *billets*, which are 8 feet wide, 35 or 48 feet long, and up to 5½ inches wide. These billets are then cut and sanded to final dimensions.

LSL has a number of job-site applications, including rim boards, beams and headers (Fig. 2-7), and as a core material for a number of trim and millwork applications. Currently, the only brand-name LSL product available is TimberStrand LSL from Trus Joist MacMillan.

Laminated veneer lumber (LVL)

LVL products are akin to plywood, but with one notable exception: All the veneer strands in LVL run parallel to one another, rather than perpendicular, as is the case with plywood.

Veneers are peeled from logs of two species of wood, Douglas-fir and southern pine (fast-growing yellow poplar will soon be used as well). The veneers are dried, sorted, glued, and hot-rolled to their final thickness, then trimmed to final width and length. According to the manufacturers, the process uses approximately 52 percent of the log, compared to the 40 percent used in regular sawmilling operations, and results in a structural member that is 40 percent stronger than a solid sawn piece of Number 2 Douglas-fir lumber of the same dimensions.

The LVL process results in members that are uniform in size and thickness, and perform consistently with little if any dimensional movement. It also allows manufacturers to produce structural members in sizes that are currently difficult or completely impossible to get with conventional trees and sawmilling practices: lengths of up to 80 feet, and cross-sectional dimensions of up to 3½ × 48 inches.

You can use LVL wherever you would use solid sawn lumber. Its primary applications on the construction site are for beams and headers, and also for concrete forms and scaffold planks in certain high-load situations. LVL is also commonly used for the top and bottom

2-6 *Cross-section of a piece of TimberStrand laminated strand lumber (LSL), showing tightly compressed wood strands.* Courtesy Trus Joist MacMillan.

2-7 *LSL being used as a short-span header.* Courtesy Trus Joist MacMillan.

flanges in I-joist manufacture, and is growing in use for ridge and perimeter beams in the manufactured housing industry.

Machine Stress Rated lumber

In these days of inconsistent lumber quality, one way of ensuring that the lumber you purchase will perform to certain standards is to order lumber that is designated Machine Stress Rated, or MSR. While not technically an engineered lumber product, in that it is solid sawn lumber and not a composition of strands or veneers, I am nonetheless including it here as another example of how technology is being used to improve the performance of the wood products we use in everyday construction.

Under the rules and guidelines of the Western Wood Products Association (WWPA), participating lumber mills subject the lumber they cut to a mechanical stress-rating procedure. A system of rollers exerts bending stress on the lumber, which is then measured and evaluated electronically (Fig. 2-8). Each piece is automatically identified and marked with the appropriate grade stamp (Fig. 2-9).

Parallel strand lumber (PSL)

PSL is similar to LSL, in that the basic process involves laminating parallel strands of lumber (Fig. 2-10). PSL, however, starts with wood veneer that is then clipped into strands four to eight feet in length, rather than the short wood strands used in LSL. The strands are coated with adhesive, aligned parallel with each other, then pressed and trimmed into solid structural members that are much stronger than the original tree that the veneers came from.

As with all the other engineered lumber products, PSL makes use of small, second- and third-growth trees, typically Douglas-fir, hem-fir, and southern pine, and at least one manufacturer will soon be using yellow poplar as well. PSL members are available in finished sizes of up to 11 × 17 inches, and in lengths of up to 66 feet. PSL is exceptionally strong, and is used for beams (Fig. 2-11), columns, headers, posts, and in other similar structural applications. It is also large and stable enough for use in bridges and other industrial and civil construction projects. This type of lumber has a rather dramatic appearance and lacks the knots, splits, and pitch pockets common to large-dimension sawn lumber, making it ideal for many exposed uses. PSL is also available in a pressure-treated form, which is good

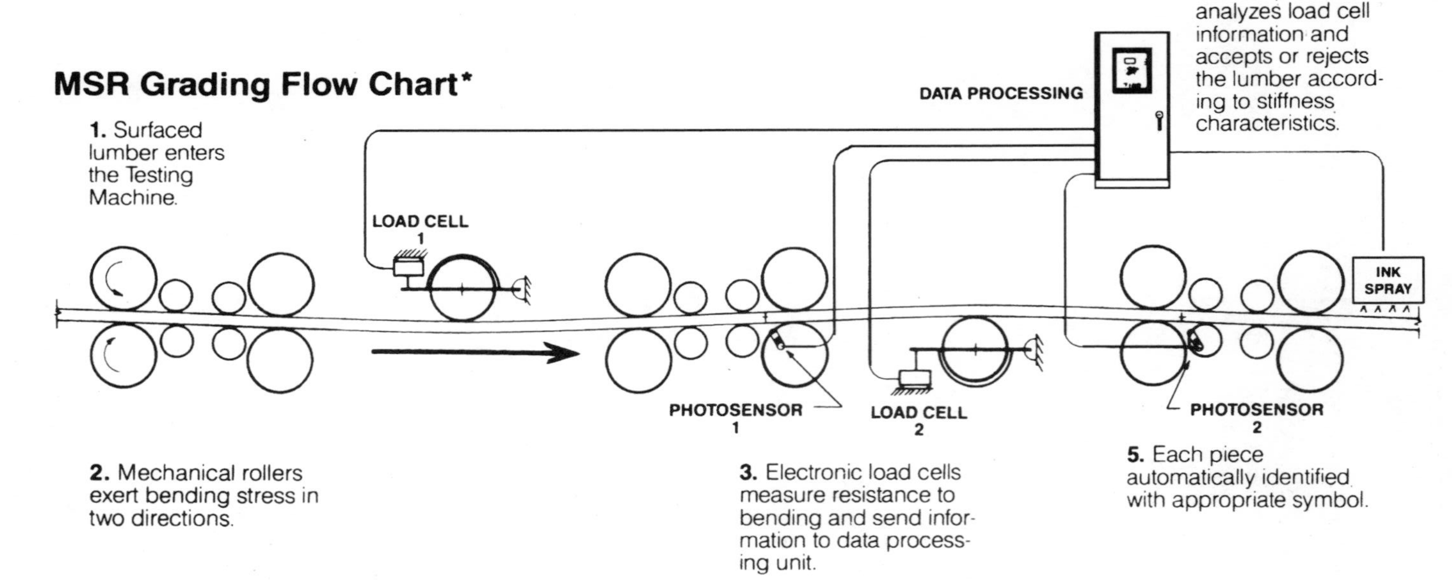

2-8 *Process for testing and evaluating lumber for Machine Stress Rating.* Courtesy Western Wood Products Association

Grade Stamp Facsimiles

All MSR lumber must be gradestamped by an American Lumber Standards-approved agency. The grade stamp must include a MSR designation such as "MACHINE RATED", the registered trademark of the grading agency, the mill number, moisture content designation, species identification, and the F_b and E rating. The F_t value will appear when the material has been quality control-tested for tension.

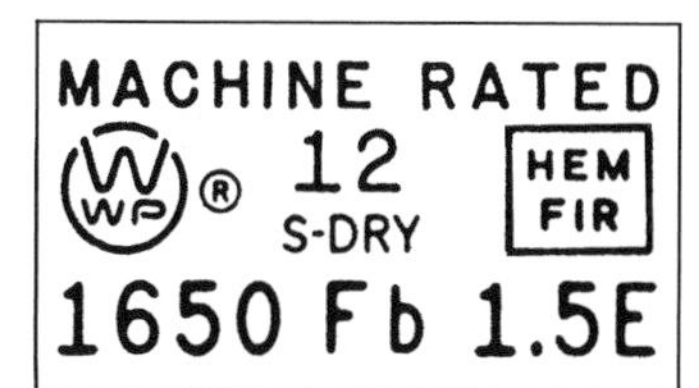

Typical MSR Stamp

2-9
Grade stamp for Machine Stress Rated lumber. Courtesy Western Wood Products Association

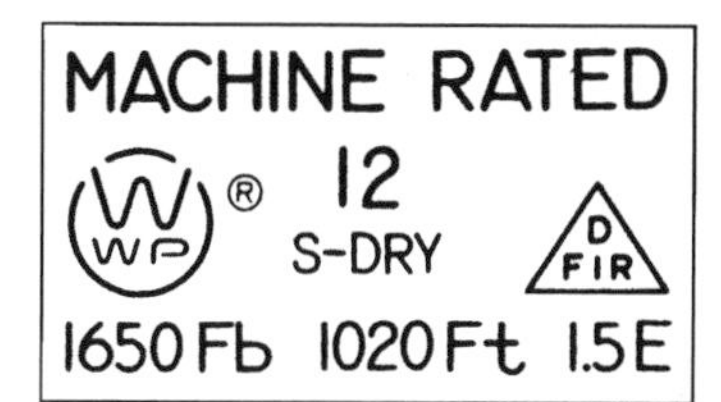

MSR Stamp with F_t
(Grade stamp when tension quality control is provided.)

for deck supports, carports, balconies, and other uses where weather exposure and contact with concrete is a consideration.

Trus Joist MacMillan was the originator of this product, and as of this writing their Parallam PSL is the only brand-name, structural-grade, parallel strand lumber on the market.

Structural-glued lumber

Structural-glued lumber takes the engineered lumber concept in a slightly different direction. Instead of taking wood strands or veneers and pressing them into new lumber shapes, structural-glued lumber

2-10 *Parallam parallel strand lumber (PSL), showing thick strands oriented to the long dimension of the beam, resulting in an intricate grain pattern that looks good exposed.*

2-11 *I-joists and an exposed Parallam PSL beam in a vaulted ceiling.* Courtesy Trus Joist MacMillan

uses smaller sizes of standard sawn lumber and assembles them into larger pieces for a variety of structural and nonstructural uses (Fig. 2-12).

Structural-glued lumber is manufactured from dry lumber of various species, including Douglas-fir–larch, hem-fir, and other western softwoods, and can be found in five common grades: Stud, Select Structural, No. 1, No. 2, and Standard and Better. The lumber is glued up in three different forms: end-jointed or finger-jointed, which is used to create lumber in longer lengths; edge-glued, which creates lumber in greater widths; and face-glued, which is done to increase the lumber's thickness. Standard sizes are 2 × 3 through 2 × 12 inches, and 2 × 6 through 5 × 8 inches for structural laminate sizes.

Structural-glued lumber is manufactured by several different mills, and is graded and certified under the rules of the Western Wood Products Association. Grade stamps on the individual lumber pieces give specific details about the design and certification of that piece (Fig. 2-13).

Specialty engineered lumber products

Given the unique size and load-bearing characteristics of engineered lumber, it seems like the next logical step in their evolution is to combine

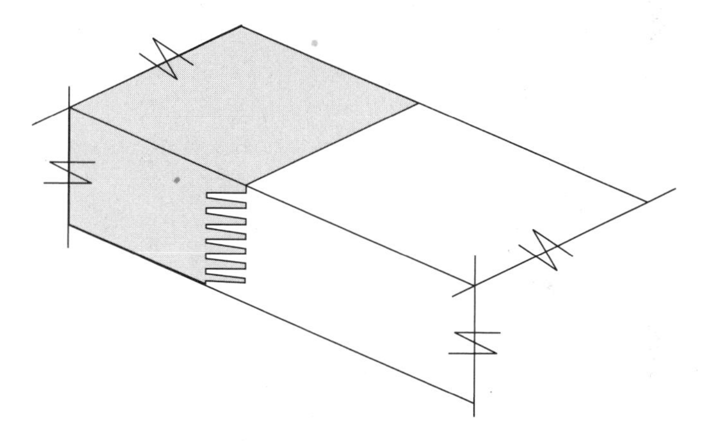

WWPA's quality control procedures
for structural-glued end-jointed (i.e. finger-jointed)
lumber ensure that it will meet or exceed the performance capabilities
of solid-sawn lumber. It is approved under all model building codes to be used
interchangeably with solid-sawn lumber of the same size, grade and species.

2-12 *An example of the finger-jointing used to produce structural glued lumber.* Courtesy Western Wood Products Association

(1) Structural-glued lumber assembled with an exterior-type adhesive (meeting the requirements of ASTM Standard D 2559) is gradestamped CERT EXT JNTS.

Grade stamp facsimile

Exterior-type adhesives are suitable for the bonding of wood, including treated wood, structural end-jointed and laminated wood products for general construction, for marine use, or for other uses where a high-strength, waterproof adhesive bond is required. These structural-glued lumber products can be used interchangeably with solid-sawn lumber for joists, rafters, plates, studs, etc.

2-13
Grade stamp for structural-glued lumber. Courtesy Western Wood Products Association

them into special products to meet specific needs. This section lists some of the most interesting things currently on the market.

I-joists

The I-joist (Fig. 2-14) is probably the most visible symbol of the engineered lumber revolution. I-joists are actually a manufactured composite of two engineered lumber products: LVL for the top and bottom flanges, and OSB for the web. First the flanges are cut to width, then a groove is routed in one face. The web material is cut to size and trimmed to fit the groove in the flange, then the two components are glued and pressed into a finished joist.

Combining different widths and thicknesses of flanges (Fig. 2-15) with different heights of web material allows joists of exceptional strength and long lengths to be produced. In fact, I-beams are produced in lengths of up to 80 feet, which is obviously impossible

2-14 *A clear example of the improved dimensional stability of an I-joist (left) compared to standard solid sawn lumber.* Courtesy Trus Joist MacMillan

with standard sawn lumber. And here again, the environment benefits from I-joists, which use smaller second- and third-growth trees to manufacture the two components that make up the joist. One

TJI®/15 DF, TJI®/25 DF, TJI®/35 DF and TJI®/55 DF JOISTS

TJI®/15 DF, TJI®/25 DF, TJI®/35 DF and TJI®/55 DF joists are available in 9½" through 16" depths from your local Trus Joist MacMillan dealer. Other joist depths and series are also available to support almost any load condition you require. Contact your lumber yard or your local Trus Joist MacMillan representative for assistance or product information.

> Southern Pine (SP) TJI® joists may be directly substituted for the corresponding depth and series of Douglas Fir (DF) TJI® joists listed within this guide.

TJI®/15 DF JOIST
9½" and 11⅞" Depths

TJI®/25 DF JOIST
9½", 11⅞", 14" and 16" Depths

TJI®/35 DF JOIST
11⅞", 14" and 16" Depths

TJI®/55 DF JOIST
11⅞", 14" and 16" Depths

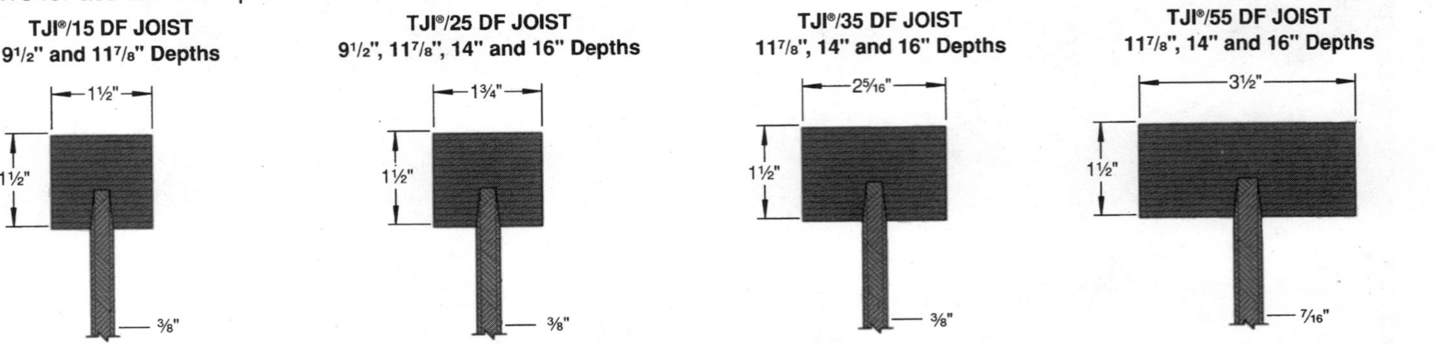

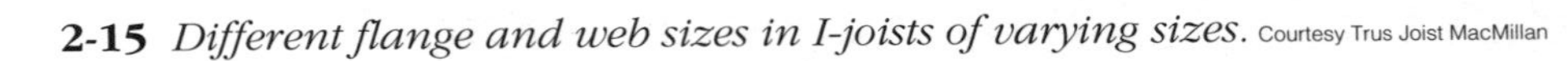

2-15 *Different flange and web sizes in I-joists of varying sizes.* Courtesy Trus Joist MacMillan

manufacturer estimates that one tree goes into the I-joists needed to frame the floor for a 1,000-square-foot house, as opposed to two or three trees needed to make up the equivalent amount of solid sawn lumber.

The I-joist is probably the most readily accepted of the new generation of engineered lumber products. It's been on the market for several years now, and has been used by enough builders in enough situations to prove its worth. I-joists are consistent in size and in dimensional stability. They are easy to cut and install (Fig. 2-16), and much lighter in weight than conventional sawn lumber of equal size. The webs are prestamped with knockout holes for wiring and plumbing, which also saves labor costs for those two trades. And with the growing number of specialized hangers and connectors designed specifically for use with these products (see Chapter 5), installation is also easy and consistent.

I-joists are used in floor joist, ceiling joist, and roof rafter applications. They can be used in multiple assemblies to meet varying span, spacing, and load demands, or as headers in certain framing situations. Most manufacturers offer full computerized load calculation and framing design services that greatly simplify layout, span, and connection details.

At the moment, at least five manufacturers are building one or more versions of the I-joist: Boise Cascade, Georgia-Pacific, Louisiana-Pacific, Tecton Laminates, and Trus Joist MacMillan.

Engineered lumber systems

Several manufacturers have taken the engineered lumber concept a step further and have designed complete engineered lumber framing systems. A framing system is, of course, an additional chance for manufacturers to present and promote—and hopefully convince you to buy—their products. But these systems are really a lot more than just that.

A well-designed engineered lumber framing system provides the builder with a number of useful tools. It helps you to better understand how the different products interact, and how you can use them together to effectively form the structural shell of a structure. It shows you what goes where, and what product is approved and appropriate for what usage. There is fully researched and documented computer software available for designing and sizing the components of the systems, and predrawn photo libraries available free for adding structural details to your construction plans and drawings (see Chapter 8).

2-16 *Fitting a lightweight I-joist into a hanger, which is in turn attached to another engineered lumber product, Microllam LVL.*
Courtesy Trus Joist MacMillan

Another very valuable aspect of an engineered system is the installation instructions and illustrations. Each manufacturer provides booklets and instruction sheets that clearly detail each step in the construction process, from blocking and bracing to nails and hanger hardware. This is a great help for novice and old pro alike, and assures you that you're using the products safely and properly. They also make great training aids for you and your crew members.

Here are some examples of engineered lumber framing systems that are currently available. Contact the manufacturers for complete details:

- FrameWorks and Silent Floor, from Trus Joist MacMillan (Figs. 2-17 and 2-18)
- Simple Framing System, from Boise Cascade
- Solid Start, from Louisiana-Pacific

Open-web trusses

Open-web trusses are a structural framing product from Trus Joist MacMillan that combines the benefits of wood and steel, primarily for commercial and industrial applications. They use a wood top and bottom flange that are made from either solid sawn lumber or engineered lumber. The flange is combined with open, tubular-steel web chords to form a very strong, solid, and relatively lightweight truss (Fig. 2-19).

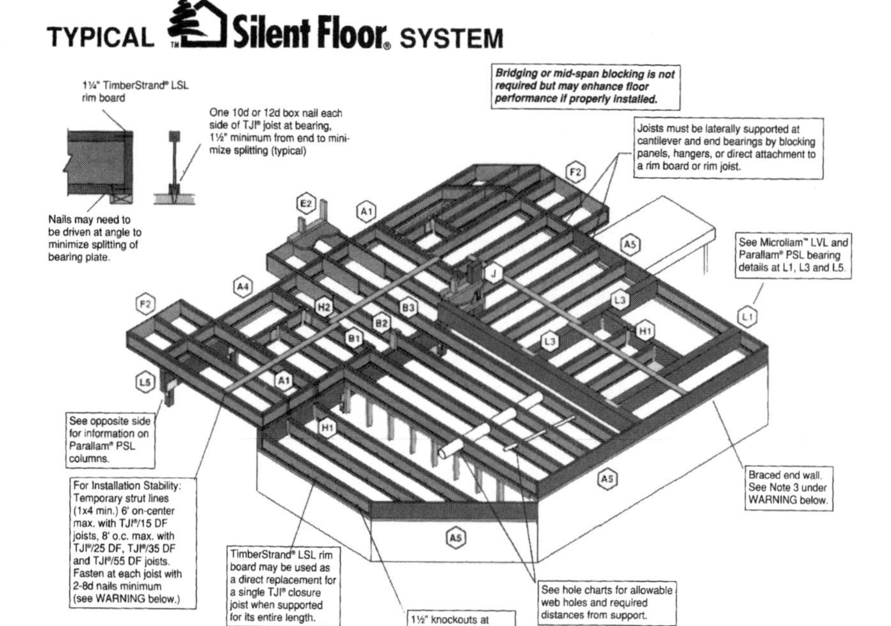

2-17 *Illustration showing a typical engineered lumber floor framing system. Letters and numbers are keyed to detail drawings provided by the manufacturer.* Courtesy Trus Joist MacMillan

Open-web trusses provide an alternative to all-steel trusses in long-span or high-load applications, and use less lumber than conventional all-wood trusses. For example, open-web trusses were used recently to construct a roof cover over the Lower Norman Bypass Dam near Los Angeles. According to the manufacturer's calculations, the open-web trusses used approximately 6 board feet of lumber per linear foot of truss, compared to an estimated 23 board feet of lumber per linear foot if a glulam system had been used instead.

You can contact the manufacturer for specific span and construction details, or ask your material supplier.

Parallam PSL wood bridges

This is an interesting use of the brand-new parallel strand lumber (PSL) technology, and is a great example of how these engineered and manufactured lumber products can be used in demanding exterior applications. The product of a limited partnership between Trus Joist MacMillan's industrial division in Minnesota and Hughes Brothers, Inc., of Nebraska,

2-18 *A whole-house engineered lumber system, including joists, studs, rafters, and headers.* Courtesy Trus Joist MacMillan

these are complete bridge packages constructed from pressure-treated PSL for use in a variety of short-span situations (Fig. 2-20).

You are first provided with a full set of shop drawings for a bridge that is designed for your specific length, width, and load requirements. The bridge components are then carefully shop-fabricated and pressure-treated, again to your specifications. The entire package, including all girders, guardrails, and all necessary hardware, is delivered to your job site, and the factory provides you with technical support to erect the components and monitor the bridge's performance after the installation is complete.

SpaceMaker trusses

The SpaceMaker Truss System is designed and built by Trus Joist MacMillan specifically to create additional living space over a standard garage. The system consists of a series of unique, bolt-together trusses (Fig. 2-21) that are manufactured from the company's TimberStrand LSL material. Each truss is assembled on site (Fig. 2-22) using

2-19 *Open-web trusses on a large commercial building.* Courtesy Trus Joist MacMillan

instructions and a hardware package provided with the trusses, and forms a solid, open truss system consisting of a bottom chord/floor joist, upper chord/rafters, a horizontal collar tie that forms the flat ceiling, and two vertical kneewall members that form the walls.

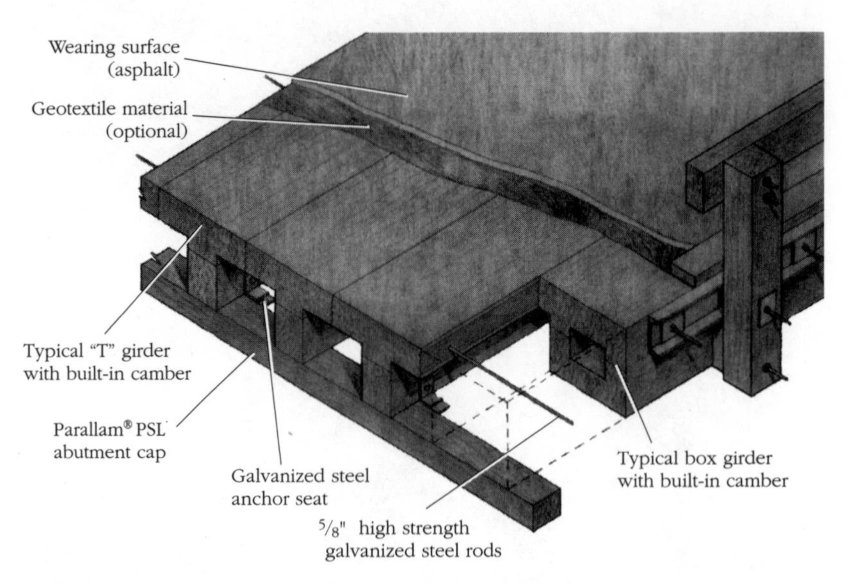

2-20 *Cross-sectional drawing of a Parallam PSL engineered lumber bridge.* Courtesy Trus Joist MacMillan

The truss system is designed to fit standard garage widths of 22, 24, 26, and 28 feet, in either 10:12 or 12:12 pitch, and can be cantilevered as much as 2½ feet on each end to accommodate garages of odd dimensions. The truss system provides a minimum of 200 square feet of additional usable floor space.

Usable room widths range from 11 up to 14 feet, and ceiling heights range from 6 feet, 6 inches to 9 feet, 9 inches. Kneewall heights also vary with the size and pitch of the truss, ranging from 3 feet, 5 inches up to 6 feet, 1 inch. The trusses can be laid out in different on-center configurations to allow for dormers, and to create the necessary floor openings to accommodate a stairwell.

TimberStrand 1.3E Premium Studs

If you're completely fed up with trying to frame a house using wet, heavy studs that warp, twist, and are cut with so much wane that the edges often end up only an inch or less wide, you might be ready to try this new Trus Joist MacMillan product manufactured from laminated strand lumber (LSL).

As with the company's other LSL products, Premium Studs (Fig. 2-23) begin as rapid-growth aspen and yellow poplar trees, which are then stranded, formed into billets through a steam injection process

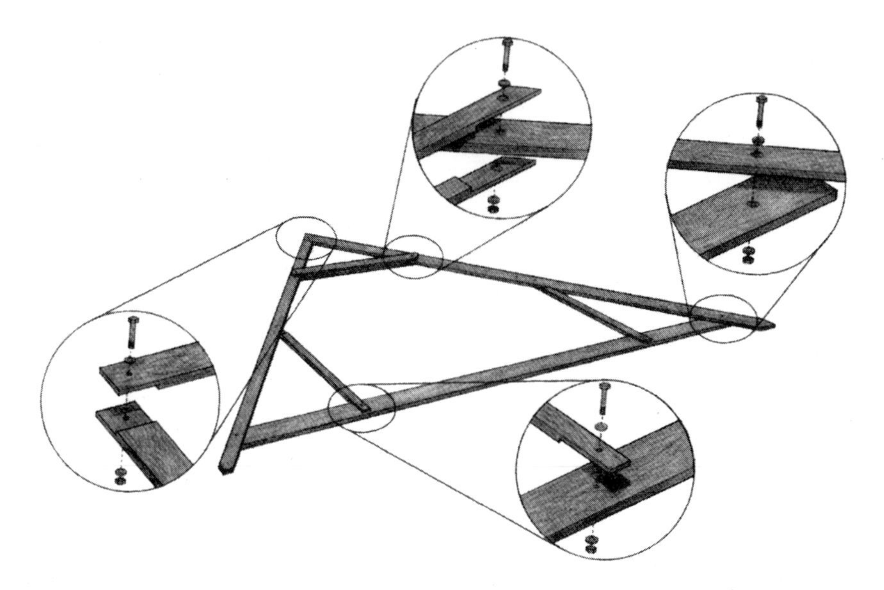

2-21 *An example of how the SpaceMaker trusses are bolted together.* Courtesy Trus Joist MacMillan

(see the description of laminated strand lumber, earlier in this chapter), and then gang-sawn to final size. Premium Studs are available in 2 × 4 and 2 × 6 nominal sizes (1½ × 3½ inches and 1½ × 5½ inches, respectively) to match up with standard-dimension solid sawn lumber. The manufacturer offers a warranty against defects in material and workmanship for the normal and expected life of the building, something that's a little tough to find with normal studs.

2-22 *Bolt-together SpaceMaker trusses being erected over a residential garage.* Courtesy Trus Joist MacMillan

2-23 *Long, straight studs are easy when you use LSL engineered wood.* Courtesy Trus Joist MacMillan

3

An introduction to engineered panel products

In Chapter 2, you looked at some of the new engineered lumber products on the market. In this chapter, it's time to examine engineered panel products. Some, such as plywood, you'll be familiar with from long use on the job site; others, such as oriented strand board, you will probably know less about. Wherever possible, I will provide grade-stamp information so you can begin to see what the stamps mean, and learn how to get the information you need off them.

The products are arranged into two general sections: plywood and other engineered panel products.

Plywood

Plywood (Fig. 3-1) is thought to be the oldest form of engineered lumber, with examples of an earlier form of it dating back to Egypt, 3500 years ago. The modern form was developed in the early 1930s, and in 1933 the Douglas Fir Plywood Association was formed to represent manufacturers and establish consistent grading rules and manufacturing processes. That original association became the American Plywood Association (APA) in 1963, which in 1994 became known as The Engineered Wood Association—clear recognition of the advances and growth in the engineered wood industry.

Plywood is an engineered panel comprised of an odd number of layers (Fig. 3-2), each laid perpendicular to the ones above and below it, with the face (front and back) panels always oriented so the grain is parallel with the long dimension of the panel. When you

3-1 *Plywood is an all-around excellent material for wall sheathing.*

take into consideration the various grades, wood species, face textures, edge treatments, structural ratings, panel sizes and thicknesses, and all the other variations possible in the manufacturer of plywood, there are literally thousands of possible types of plywood available. For the purposes of the construction industry and this book, however, I will describe only the most commonly used types. For further information or for specific usage information on any type of plywood or other engineered panel, contact the APA (see Appendix E).

Grading plywood

Plywood panels are rated according to four different factors, each of which affects the panel's use, appearance, durability, and how suitable it is for specific applications. These ratings include:

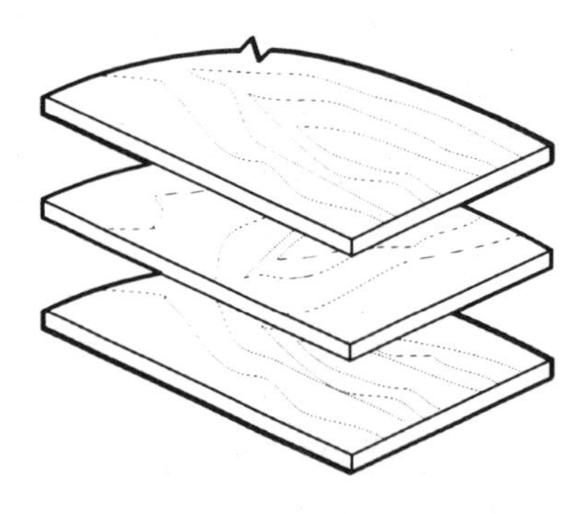

3-2 *Perpendicular laminations in a sheet of plywood give the material its strength.* Courtesy American Plywood Association

Grades

Plywood grades refer to either the grade of the veneer or to the grade of the panel itself, which is also a reflection of the veneer grades. (Fig. 3-3) Veneer is graded by its appearance, and by the number and size of allowable repairs in that veneer. There are five veneer designations—A, B, C, C-Plugged, and D, with the highest-quality veneer designated as an A grade and the lowest as a D. A combination of these veneer grades for the face and back veneers of the panel, respectively, give the panel its grade rating. For example, a panel with an A veneer on the face and a C veneer on the back would be designated as an A-C grade. For exterior use, veneers must be C grade or better; D-grade veneer is allowed only on plywood rated for interior or protected use.

Exposure

The next rated criteria are the exposure classifications, of which there are four:

- *Exterior.* Exterior-rated plywood has a fully waterproof glue, and is designed for applications where the plywood will be continually exposed to the weather or to moisture conditions.
- *Exposure 1.* Exposure 1 panels have an exterior glue, but might lack the other characteristics, such as bond quality or stability, necessary to quality the panel for an exterior rating. Exposure 1 panels are designed for protected applications, such as roof sheathing that will be covered with roofing materials.
- *Exposure 2.* An Exposure 2 panel has an intermediate glue (not fully waterproof), and is primarily intended for interior applications where occasional exposure to moisture or leakage might occur.
- *Interior.* Interior plywood uses an adhesive that lacks waterproofing qualities, and is intended for dry interior applications only.

Species group number

There are over 70 species of wood (Fig. 3-4) used in manufacturing plywood, some of which have very strong structural properties. Others are used strictly for their appearance characteristics. For this reason, plywood is rated according to five species groups, designated Group 1 through Group 5, of which Group 1 is the most structurally strong. Examples of wood species in the Group 1 category include Douglas fir, larch, and maple, while Group 5 consists solely of basswood, poplar, and balsam.

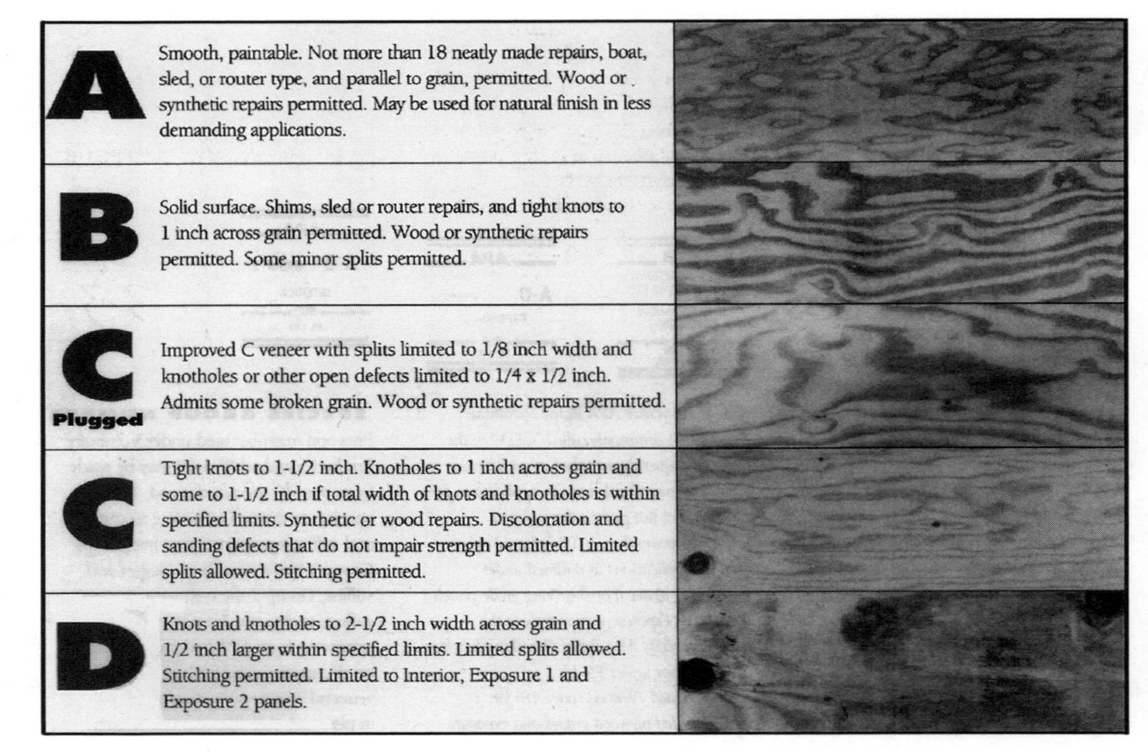

3-3 *Examples of the grading requirements for each of the five plywood veneer grade ratings.* Courtesy American Plywood Association

Group 1	Group 2		Group 3	Group 4	Group 5
Apitong[a][b]	Cedar, Port Orford	Maple, Black	Alder, Red	Aspen	Basswood
Beech, American	Cypress	Mengkulang[a]	Birch, Paper	Bigtooth	Poplar, Balsam
Birch	Douglas-fir 2[c]	Meranti, Red[a][d]	Cedar, Alaska	Quaking	
Sweet	Fir	Mersawa[a]	Fir, Subalpine	Cativo	
Yellow	Balsam	Pine	Hemlock, Eastern	Cedar	
Douglas-fir 1[c]	California Red	Pond	Maple, Bigleaf	Incense	
Kapur[a]	Grand	Red	Pine	Western Red	
Keruing[a][b]	Noble	Virginia	Jack	Cottonwood	
Larch, Western	Pacific Silver	Western White	Lodgepole	Eastern	
Maple, Sugar	White	Spruce	Ponderosa	Black (Western Poplar)	
Pine	Hemlock, Western	Black	Spruce	Pine	
Caribbean	Lauan	Red	Redwood	Eastern White	
Ocote	Almon	Sitka	Spruce	Sugar	
Pine, Southern	Bagtikan	Sweetgum	Engelmann		
Loblolly	Mayapis	Tamarack	White		
Longleaf	Red Lauan	Yellow Poplar			
Shortleaf	Tangile				
Slash	White Lauan				
Tanoak					

(a) Each of these names represents a trade group of woods consisting of a number of closely related species.
(b) Species from the genus Dipterocarpus marketed collectively: Apitong if orginating in the Philippines, Keruing if originating in Malaysia or Indonesia.
(c) Douglas-fir from trees grown in the states of Washington, Oregon, California, Idaho, Montana, Wyoming, and the Canadian Provinces of Alberta and British Columbia shall be classed as Douglas-fir No. 1. Douglas-fir from trees grown in the states of Nevada, Utah, Colorado, Arizona and New Mexico shall be classed as Douglas-fir No. 2.
(d) Red Meranti shall be limited to species having a specific gravity of 0.41 or more based on green volume and oven dry weight.

3-4 *A list of the different types of wood used in the manufacture of plywood.* Courtesy American Plywood Association

Span rating

All APA panels are rated and marked with a span number, which indicates the maximum recommended center-to-center distance (in inches) for the structural members that support the panel. All span rating numbers are based on installing the panel with the grain and the long dimension running across (perpendicular to) the supports, and supporting the panel with at least three members.

Span ratings appear on a panel's identification stamp (Fig. 3-5) as two numbers separated by a slash mark, such as 24⁄16. The first number indicates the panel's span rating when used as a roof sheathing,

HOW TO READ THE BASIC TRADEMARKS OF APA – *THE ENGINEERED WOOD ASSOCIATION*

Product Standard PS 1-95 is intended to provide for clear understanding between buyer and seller. To identify plywood manufactured by association member mills under the requirements of Product Standard PS 1-95, four types of trademarks and one typical edge mark are illustrated. They include the plywood's exposure durability classification, grade and group, and class or Span Rating. Here's how they look, together with notations on what each element means.

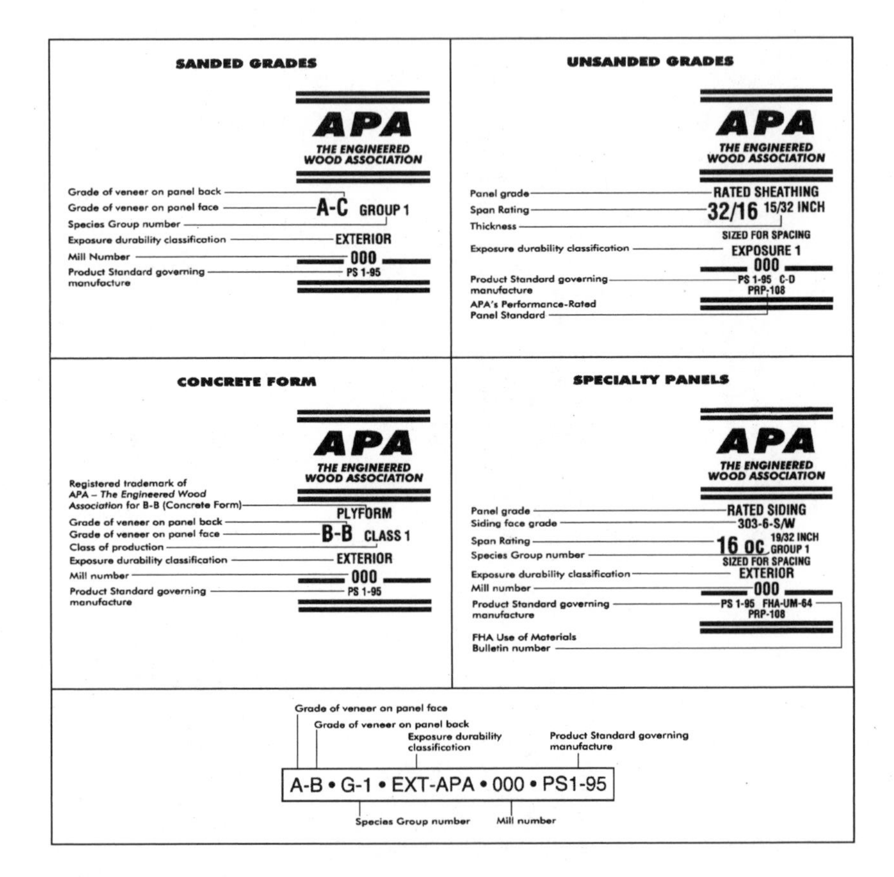

3-5 *American Plywood Association standard grade stamps for four different types of plywood.* Courtesy American Plywood Association

and the second number indicates the span rating if the panel is used for subflooring. A plywood panel with the grade stamp of 24⁄16, for example, could be used as roof sheathing over supports that are a maximum of 24 inches on center, or as a subflooring over supports that are a maximum of 16 inches on center. Once again, that rating assumes that the long dimension of the panel runs perpendicular to the supports, and that it spans a minimum of three supporting members.

Performance-rated panels

The American Plywood Association has further designated four grades of plywood and other types of engineered wood panels that meet specific criteria for construction use. These ratings simplify the selection, specification, and identification of panels for specific uses in various areas of the structure, and help ensure consistency and quality in the construction process. These four grades are:

APA Rated Sheathing

Rated Sheathing panels are designed primarily for use as subflooring, roof sheathing, and wall sheathing, but they are also used in a wide range of general construction uses. Sheathing panels are not limited to just plywood, but also include oriented strand board (OSB), composite boards, and combinations of OSB with conventional plywood veneer faces. APA Rated Sheathing is available as follows:

Common thicknesses: 5⁄16, 3⁄8, 7⁄16, 15⁄32, 1⁄2, 19⁄32, 5⁄8, 23⁄32, and 3⁄4 inch.

Exposure ratings: Exterior, Exposure 1, and Exposure 2.

Span ratings (roof sheathing/subfloor): 12/0, 16/0, 20/0, 24/0, 24/16, 32/16, 40/20, 48/24, 60/32, Wall-16 OC, and Wall-24 OC.

APA Structural I Rated Sheathing

This is an unsanded panel for use in applications requiring greater strength, stiffness, or shear properties, including panelized roof construction and the construction of shear walls. Structural I sheathing can be plywood, composites, or OSB. For conventional plywood panels in this category, all veneer plies are graded for special improved species, and panels marked PS 1 are limited to veneers from Group 1 wood species. APA Structural I Rated Sheathing is available as follows:

Common thicknesses: 5⁄16, 3⁄8, 7⁄16, 15⁄32, 1⁄2, 19⁄32, 5⁄8, 23⁄32, and 3⁄4 inch.

Exposure ratings: Exterior and Exposure 1.

Span ratings (roof sheathing/subfloor): 20/0, 24/0, 24/16, 32/16, 40/20, and 48/24.

APA Rated Sturd-I-Floor

Panels in this category are specially designed to act as both a subfloor and a smooth underlayment beneath carpet and pad installations. They are available as conventional plywood, OSB, or composite boards, and with square or tongue-and-groove edges. APA Rated Sturd-I-Floor panels are available as follows:

Common thicknesses: $^{19}/_{32}$, $^{5}/_{8}$, $^{23}/_{32}$, $^{3}/_{4}$, $^{7}/_{8}$, 1, and 1$^{1}/_{8}$ inches.
Exposure ratings: Exterior, Exposure 1, and Exposure 2.
Span ratings: 16, 20, 24, 32, and 48.

APA Rated Siding

These are panels manufactured specifically for siding, fencing, and similar applications. They are available in a variety of surface textures and configurations, and as conventional plywood, composites, or as an overlaid oriented strand board. APA Rated Siding is available as follows (see Chapter 7 for additional details and product descriptions):

Common thicknesses: $^{11}/_{32}$, $^{3}/_{8}$, $^{15}/_{32}$, $^{1}/_{2}$, $^{19}/_{32}$, and $^{5}/_{8}$ inch.
Exposure ratings: Exterior only.
Span ratings: Typically 16 OC and 24 OC for wall-stud spacing (exact span rating is listed in the trademark stamp on the panel).

Sanded and touch-sanded plywood

Plywood having a B grade or better veneer face will typically be used in the manufacture of cabinets, furniture, and other similar applications. As such, these panels are sanded smooth during the manufacturing process, and are known as *sanded panels*. Panels with a C-grade face are most commonly used only in construction applications, and are generally either hidden from view or covered by another material. These panels are lightly sanded only as necessary to achieve the final desired thickness, and are described as *touch-sanded*.

There are several grades of plywood that fall into the sanded or touch-sanded categories and are used in both general construction and finish work applications. Only the most commonly used grades are included here:

A-A grade

This type of panel has an A-grade veneer on both the front and back faces, and is used only where the appearance of both sides is important, as with cabinet and furniture building. A-A grade plywood is available as follows:

Common thicknesses: $^{1}/_{4}$, $^{11}/_{32}$, $^{3}/_{8}$, $^{15}/_{32}$, $^{1}/_{2}$, $^{19}/_{32}$, $^{5}/_{8}$, $^{23}/_{32}$, and $^{3}/_{4}$ inch.
Exposure ratings: Interior, Exposure 1, and Exterior.

A-C grade

This plywood has an A-grade face and a C-grade back, resulting in a less expensive panel for use where the appearance of only one side is important, as with soffits, some types of cabinets, linings, fences, etc. A-C grade plywood is available as follows:

Common thicknesses: ¼, 11/32, ⅜, 15/32, ½, 19/32, ⅝, 23/32, and ¾ inch.
Exposure rating: Exterior.

C-C Plugged grade

C-C Plugged plywood has face of C-grade veneer that's been plugged, repaired, and touch-sanded, and a back veneer that's a standard C grade. This is a very commonly used grade of plywood for construction applications where one side of the panel will be exposed for painting, as with eaves and soffits, or where a smooth face is needed below another surface material, such as underneath ceramic tile or carpet and pad. C-C Plugged panels are available as follows:

Common thicknesses: 11/32, ⅜, ½, 19/32, ⅝, 23/32, and ¾ inch.
Exposure rating: Exterior.

C-D grade

One of the most commonly used grades of plywood for general construction work is C-D, which has a C-grade face and a D-grade back. You will often hear contractors speak of CDX plywood, which is a very common roof and wall sheathing material. CDX is actually an Exposure 1 plywood, however, not a true exterior grade, and as such it should not be used in exposed exterior applications. It's fine, however, in the previously mentioned sheathing applications where it will be covered with shingles, siding, or other weather-resistant finish materials. C-D grade panels are available as follows:

Common thicknesses: ⅜, ½, 19/32, ⅝, 23/32, and ¾ inch.
Exposure ratings: Interior and Exposure 1.

Sanded shop grade

When plywood panels are damaged during manufacturing, shipping, or on site at the lumber yard, they might lose some of their structural properties, as well as being diminished in appearance and in their ability to resist moisture. Many of these panels will be touch-sanded to clean up the surface (which reduces the original manufactured thickness of the panel), and sold as shop-grade plywood for general utility purposes. Shop plywood is clearly stamped with "Shop Cutting Panel, All Other Marks Void," indicating that the manufacturer and the APA no longer consider the panel as meeting the standards of the original grade stamp.

Shop plywood is a little less expensive than some of the other grades, and is of generally good quality, although you might have to work around whatever defects demoted the panel to shop grade in the first place. Shop-grade panels are great for shelving, temporary bracing, and general shop uses in indoor or limited-exposure situations. All grades and thickness can be found as shop plywood, although ½-, ⅝-, and ¾-inch panels seem to be the most common.

APA specialty panels

APA Marine

These are specially designed plywood panels with solid-jointed cores, very limited face repairs, and core voids or gaps, and they're limited to only two species: Douglas fir and western larch. This type of plywood is used in the manufacture of boat hulls and other marine and underwater applications. APA marine plywood is available as follows:

Common thicknesses: ¼, ⅜, ½, ⅝, and ¾ inch.

Exposure rating: Exterior only.

APA B-B Plyform, class I

Plywood panels in this classification are designed for use as highly reusable, concrete form panels. Both faces are sanded smooth and oiled at the mill for durability and easy release. Where extremely smooth finishes are required, the panels are also available with a high-density overlay (HDO) face. Plyform panels are available as follows:

Common thicknesses: 19⁄32, ⅝, 23⁄32, and ¾ inch.

Exposure rating: Exterior only.

APA Decorative

This is plywood that's manufactured with any one of a number of decorative face-veneer options, including resawn, grooved, and brushed. In Interior and Exposure 1 grades it is used as interior paneling, display cases, and in some types of cabinet and furniture making; in Exterior grade it's commonly used for siding, fences, gable ends, and other exterior appearance applications. Decorative panels are available as follows:

Common thicknesses: 5⁄16, ⅜, ½, and ⅝ inch.

Exposure ratings: Interior, Exposure 1, and Exterior.

APA Plyron

Plyron is a specialized panel with a plywood core and tempered or untempered hardboard faces. It is used for counters, shelving, cabinet doors, and other similar applications. Plyron is available as follows:

Common thicknesses: ½, ⅝, and ¾ inch.

Exposure rating: Interior, Exposure 1, and Exterior.

Other engineered panel products

In addition to plywood, there are a number of other panel products on the market, including plywood-based products such as MDO and a variety of composite wood products, including MDF and hardboard. This section lists, in alphabetical order, some of the other panel products not commonly associated with the plywood group.

Fiberglass-reinforced-plastic plywood (FRP)

FRP is an APA-trademarked product, consisting of one or multiple sheets of plywood with a fiberglass-reinforced resin coating bonded to both surfaces. There are a variety of options in the thickness, glass fiber content, surface texture, and even the color of the overlaid layer, depending on the desired application. The overlay is usually from 25 to 60 mm thick (1 mm = ⅟1000 of an inch), making it quite tough and durable.

The core of the FRP panel is usually a C-C grade, exterior-glue plywood, in thicknesses of ¼ to 1⅛ inches or more. The fiberglass and resin coating is applied over this core under heat and pressure, or by spraying and curing, usually with heat. FRP plywood can be made up in single 4-foot × 8-foot sheets, or several sheets of plywood can be seamed together prior to application of the coating layer, resulting in panel sizes of 8 to 10 feet wide and 45 feet or more in length, depending on the application.

FRP plywood has a number of commercial and industrial uses, including truck, van, and trailer bodies; shipping containers; various types of coolers and tanks; and reusable concrete forms.

Hardboard

Hardboard, most commonly known by the trade name Masonite, is the generic name for a type of panel material made from wood fibers. It is yet another product that makes good use of waste wood materials.

To make hardboard, wood chips and shavings from sawmills and other sources are first screened for size. Any chips over ¾ inch are rechipped to a smaller size prior to processing. In the next stage, in a process called *defibration,* high-pressure steam is used to break the chips and shavings apart into soft fibers. Water is then added to the fibers to form a slurry, which also helps remove impurities.

The water/fiber slurry flows over a screen mat, where excess water is drained away. A series of roller presses then squeeze out the rest of the water as the mat moves along the line. The mat is cut into 16- or 18-foot lengths, and is then fed into steam-heated hydraulic presses. The wood fibers, along with lignin, the wood's natural bonding agent, and other resins, are pressed under as much as 2,000 pounds per square inch of pressure, at temperatures ranging up to 550°, interlocking the fibers together in a process known as *felting* or *interfelting*.

If the pressed fibers are left untreated, the resulting hardboard product is called standard or untempered hardboard. In one of the most common variations of the manufacturing process, drying oils are added to the fibers and the felted mat is baked, producing tempered hardboard. Tempered hardboard is harder, heavier, more moisture-resistant, and a darker brown color than the standard version. Manufacturers can create other versions of hardboard by adding materials that improve hardness, improve finishing properties, increase resistance to moisture and abrasion, and increase strength and durability.

Hardboard in any of its various forms has a number of uses in the construction, commercial, industrial, and manufacturing worlds. As a contractor or homeowner, you'll be most familiar with it as door skins, cabinet skins and facings, prefinished paneling, and underlayment. Another popular and common use is perforated hardboard, known by the common trade name of Pegboard. Other hardboard uses include laminate substrate, tool and die blanks, electrical circuit boards and transformers, industrial and laboratory bench tops, flooring for commercial and industrial applications, and a paint-grade trim material.

Hardboard is approved by several flooring manufacturers for use as an underlayment, providing the material meets class 4, service-grade standards. This is not a common lumberyard material, however, and, given this lack of availability and the problems associated with swelling and puckering of the wood fibers around the heads of nails and other fasteners, most builders have traditionally shied away from using hardboard for this purpose. If you intend to use it for underlayment—and it certainly has some advantages in certain remodeling situations—be sure to check with the flooring manufacturer for specific installation and preparation details prior to putting it down.

Hardboard is produced in several standard thicknesses ranging from 1⁄10 of an inch up to 6 inches, depending on the intended use, but the ⅛- and ¼-inch versions are the only two you'll routinely find stocked in the lumber yard. Standard sheet size is 4 × 8 feet, but other dimensional material is also routinely produced. Most hardboard is smooth on one side only, designated S1S, with the back side showing a lightly waffled surface that is left by the metal screens that support the mat during the manufacturing process. Another version that is smooth on both sides (S2S) is also available.

High-density overlaid plywood (HDO)

HDO is an APA-trademarked product with a number of heavy-duty uses on construction sites as well as many other commercial and industrial applications. HDO consists of a core sheet of Exterior-grade plywood, covered on both sides with a resin-impregnated fiber surface that is bonded to the plywood with a combination of heat and pressure. The resulting overlaid sheet withstands severe weather exposure without the application of additional finishing materials, and also resists abrasion, moisture penetration, and the deteriorating effects of many common solvents and chemicals. HDO is commonly used for concrete forming in high-stress areas, or where the forms will be reused repeatedly. It's also ideal for industrial tanks, marine applications, and many agricultural uses (Fig. 3-6).

According to APA manufacturing guidelines, the core plywood used in the manufacture of HDO must have a minimum of a C-grade or C-Plugged-grade core veneers, with a minimum of B-grade or better face veneers. All glues used in the manufacturing process are 100 percent waterproof. HDO panels are most commonly available in a light-tan, semi-opaque color. Black, brown, olive drab, and other colors are also available. Common panel sizes are 4 × 8 feet, but both 9- and 10-foot-long panels are also commonly available. Standard panel thicknesses are ¼, 5⁄16, ⅜, ½, ⅝, and ¾ inch.

HDO (and MDO; see later in the chapter) can be worked with most common hand, portable, and stationary woodworking tools. Table and circular saws are best equipped with blades having little or no set to the teeth. Feed the panel or the saw so the teeth enter the panel from the face side. When drilling HDO, use a high-speed bit and high drilling speeds, and provide a backing board on the rear of the sheet to prevent the opposite face from chipping out. Holes should be drilled a minimum of ¼ inch from the edges of the panel.

HDO (and MDO) can be fastened in the same manner as standard grades of plywood. You can achieve the best holding power by using spiral nails or ring-shank nails, or flathead screws having a thread design specifically intended for use with plywood (countersink the fiber surface of the panel before installing the screws). Bolts, staples, and air-driven fasteners are also fine. For gluing, roughen the surface first and then use a resorcinol or phenolic-type glue.

Chemical Resistance of Overlays

REAGENT	EFFECT ON HDO	EFFECT ON MDO
Amyl Acetate	N	N
Acetic Acid 10% and 99.5%	N	N
Acetone	N	N
Amyl Alcohol	N	N
Benzene	N	N
Calcium Hypochlorite 30%	N	D-red-brown
Carbon Tetrachloride	N	N
Chloroform	N	N
Cresol	N	N
Formalin 37%	N	N
Formic Acid 88-90%	S, D-grey	R, S, D-yellow-red
Hydrochloric Acid 10%	N	S, D-yellow-brown
Hydrochloric Acid 37%	S, R, D-pink	S, R, D-red-brown
Hydrogen Peroxide 30%	N	D, yellow
Methyl Alcohol	N	N
Monochlorobenzene	N	N
Nitric Acid 1%	D-brown	D-yellow
Nitric Acid 5%	S, R, D-brown	S, D-yellow-brown
Nitric Acid 30%	S, R, D-brown	R, S, D-yellow-brown
Nitric Acid 70%	S-to plywood, R, D-brown (surface gone)	R, S, D-orange-yellow
Phosphoric Acid 85%	S, R	R, S, D-yellow-red-brown
Soapless Detergent (Dreft)	N	N
Sodium Carbonate 25%	D-brown	D-red-brown
Sodium Chloride 10%	N	D-yellow-brown
Sodium Chloride 25%	N	N
Sodium Hydroxide 1%	D-red-brown	R, S, D-red-brown
Sodium Hydroxide 30%	S, R, D-brown	R, S, D-red-orange
Sulfuric Acid 10%	N	R, S, D-yellow-purple
Sulfuric Acid 35%	N	R, S, D-yellow-purple
Sulfuric Acid 50%	D-Pink-orange	R, S, D-yellow-purple
Sulfuric Acid 70%	S, R, D-brown	R, S, D-yellow-purple
Sulfuric Acid 97%	S, R, D-black	R, S, D-yellow-purple
Zinc Chloride 50%	N	D-brown

3-6 *The effects of various chemicals on MDO and HDO plywood products.* Courtesy American Plywood Association

Medium-density fiberboard (MDF)

MDF is another one of those composite panel products that has grown tremendously in popularity in recent years, primarily because of its quality as a paintable substrate, and the high cost of quality natural-trim boards for use in paint-grade trim applications.

MDF is a cousin to particleboard (see *Particleboard* later in the chapter), in that it is composed of a mix of very finely ground sawdust and glue that is compressed under strict conditions of heat and pressure into a solid panel. The advantage to MDF is that the sawdust is much finer, resulting in a very hard, dense sheet that machines well, holds a shaped or routed edge much better than regular particleboard, and paints out exceptionally well.

A wide variety of wood species are used to make the sawdust used in the manufacture of MDF, including recently introduced, fast-growing Malaysian species such as acacia and bamboo, and lesser-known Asian species such as batai and yamane. Resins used in the manufacturing process include urea, melamine, and phenol formaldehyde polymers, either individually or in combination. As of this writing, a number of different wood and adhesive combinations are being tried and developed, and MDF shows every indication of continuing to grow in usage and popularity.

MDF panels are extremely heavy when compared to other types of panel materials: around 110 pounds for a sheet of ¾-inch thickness, as opposed to about 65 pounds for a similar size sheet of plywood. While hard and dense, it is not particularly strong, and must be well supported in whatever use it's applied to (shelving, for example). It is somewhat susceptible to moisture, and should be sealed on all faces and edges.

It is recommended that you use carbide-tipped tools when working with MDF since the resins and hardness of the panels will quickly dull high-speed steel blades and bits. MDF glues up quite well with standard woodworking glues, but use waterproof glue if the finished assembly will be exposed to moisture or high-humidity conditions, such as kitchen or bathroom cabinets. Air-driven nails and staples work for assembly, but you will achieve the best results by predrilling the material and assembling with long, coarse-threaded wood screws.

Common uses for MDF include cabinet and furniture frames and carcasses, underlayment, mobile-home decking, and the cores, skins, and molded panels for doors and other decorative uses. Due to its stability and density, it is also an excellent material for audio speaker cases. In the construction industry, MDF is used extensively to manufacture paint-grade moldings and trim, and you will find an extensive and growing selection of standard molding patterns milled from MDF, usually preprimed for convenience.

Since MDF is used quite extensively in the manufacture of cabinets, the common sheet size is one inch larger in each direction than the standard 4 × 8-foot panel you usually see (49 × 97 inches, as opposed to 48 × 96 inches) to allow for waste in the cutting operations for the end-use manufacturer. Standard thicknesses are ¼, ½, and ¾ inch.

Medium-density overlaid plywood (MDO)

MDO, also an APA-trademarked panel, is similar in makeup to HDO (see the previous section). It is plywood with an inner core of C-grade veneers and B-grade or better face veneers, covered with a resin-treated fiber overlay. The overlaid layer on the MDO panel has a very lightly roughened surface, which makes it ideal for the application of paint and other coatings. Because of its high weather resistance and excellent paint adhesion, MDO plywood is commonly used for exterior siding (both sheet and lap), signs, accent panels, soffits, and other similar applications.

The standard color from most manufacturers is a light tan, woodtone shade, and it is commonly available in both a smooth and a wood-textured finish, with or without a factory-applied primer. Unlike HDO, MDO panels can be ordered with the fiber overlay on one or both sides, depending on the intended application. Sheet sizes and thicknesses are the same as for HDO, as are the methods of cutting and assembly.

Melamine

Melamine is the brand name for a composite panel product, and while I intended to stay away from brand names in favor of generic descriptions (e.g., hardboard instead of Masonite), Melamine is how most everyone refers to this product; there really isn't a good generic name that everyone seems to agree on. Sorry for the inconsistency.

Melamine is simply particleboard or MDF (see previous section) that has a surface coating of foil applied to one or both sides to somewhat imitate plastic laminate. The foil coating is available in a wide variety of solid colors that match many of the standard plastic laminate colors currently available from Wilsonart, Formica, Nevamar, and other leading laminate manufacturers, although what you'll find commonly stocked in the lumber yards and homes centers is white and almond. The foil covering is decorative and a little easier to clean than the surface of standard particleboard or MDF, but it does not add to the strength or stability of the panel.

Melamine has a clean, finished look to it, and has gained a tremendous amount of popularity is recent years as a cabinet facing and interior material, following on the heels of the popular European style of cabinetry. It is also widely used for shelving, office furniture, and a variety of other uses where an economical alternative to actual plastic laminate is desired. Melamine is quite heavy compared to plywood, and the foil surface is somewhat prone to nicks and chips. The material can be cut and worked with standard woodworking tools, although, again, carbide-tipped blades and cutters are highly recommended. You must remove the foil coating by cutting, dadoing, routing, or sanding in areas where the panel is to be glued, as the coating prevents a good bond from occurring. Use any good woodworking glue, and secure the panels using long, coarse-threaded wood screws in predrilled holes.

Standard thicknesses are ¼, ½, and ¾ inch, and 48 × 96-inch or 49 × 97-inch panels are standard. Other common sizes are 12, 18, and 24-inch widths in various lengths for use as shelving. The panels might be supplied with bare edges or with the edges banded with a heat-sensitive, color-coordinated, edge-banding tape. Additional iron-on or glue-on tape is readily available for site or shop finishing of cut edges.

Oriented strand board (OSB)

OSB is easily the most popular of the new generation of engineered wood panel products, and its use equals or even exceeds that of plywood on most construction sites (Fig. 3-7). There are several reasons for this, including strength, economical cost, availability, ease of use, and a resource-friendly composition.

3-7 *Oriented strand board (OSB) is an extremely popular structural sheathing material for walls and roofing.*

OSB, as the name implies, is made up of thousands of strands of wood, oriented in specific directions and pressed up into panels. OSB uses small, fast-growing, easily renewable species of trees such as poplar and gum, which are processed through a chipper to create the individual chips. Approximately 85 to 90 percent of the tree is used in the manufacturing process.

The individual chips, or strands, are arranged by machine into three perpendicular layers (Fig. 3-8) that overlay one another at right

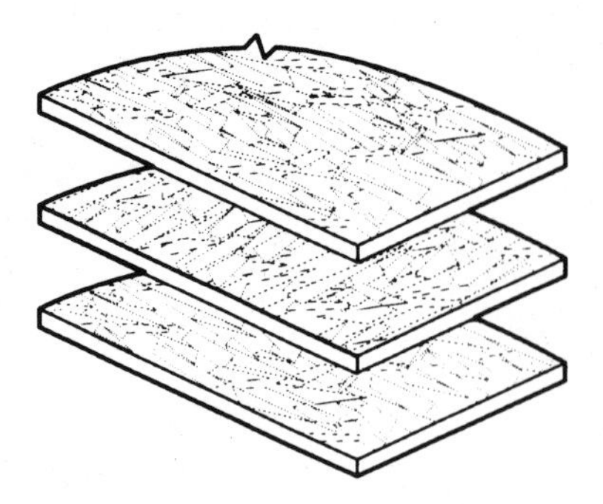

3-8
OSB is comprised of three layers of glued wood strands. Each layer is oriented with the wood fibers in the same direction, and each layer is perpendicular to the next one. Courtesy American Plywood Association

angles, similar to the construction of conventional plywood. The strands are coated with a waterproof phenolic resin adhesive, then pressed into sheets of uniform thickness and cut to a uniform 4 × 8-foot sheet. The panels might be smooth on both sides or, as is often the case with sheathing panels, one side will be smooth and the other will have a slightly textured surface to improve traction when the panel is used as a roof sheathing. Edges are either square or tongue and groove, and are typically coated to help prevent moisture penetration.

Standard thicknesses are ¼ inch, considered a nonstructural "hobby" panel, ⅜, 7⁄16, ½, and ⅝ inch, which are sheathing-rated panels (see the previous section *APA Rated Sheathing*), and 19⁄32, 23⁄32, and 1⅛ inches, which are all rated as Sturd-I-Floor panels (see *Sturd-I-Wood*, earlier in the chapter).

Particleboard

Particleboard is a commonly used, nonstructural panel material that has been around manufacturing plants and construction sites for many years. It is made of relatively coarse sawdust, wood particles, and wood fibers from various species of wood, mixed with any of a number of adhesive resins and pressed up under heat and pressure into solid sheets.

Particleboard is available in a variety of grades, with the industrial-rated grades being the best suited for construction and quality manufacturing uses. Another factor in the selection and use of particleboard is the type of resin used in its manufacture. For many years, resins containing urea formaldehyde (called UF particleboard) were the most common.

In recent years, however, it has been discovered that emissions of urea formaldehyde gas, called *outgassing*, in the first few weeks after installation have posed an indoor air-quality problem, and the gas has created some mild to serious health risks for people who are sensitive to it, particularly the elderly. This has proven to be a particularly troublesome issue for the mobile-home industry, where the extensive use of particleboard in the manufacturing process and the relatively tight construction of the factory-built homes have combined to create some very high concentrations of urea formaldehyde gas. This problem has mostly been solved, however, through the use of phenol formaldehyde resins, which do not have the same potentially dangerous outgassing characteristics.

If you are using particleboard with urea formaldehyde resin, remember that the concentration levels are primarily dependent on increases in temperature and humidity, and lack of ventilation. Keeping the structure as well-ventilated as possible after installing the particleboard is adequate to prevent health problems for most people. How long you need to ventilate it is based on factors such as temperature, humidity, quantity installed, tightness of the structure, etc. You might want to specify low-emitting UF bonded particleboard or phenol formaldehyde particleboard, however, especially in bathrooms, basements, and other high-humidity areas. If in doubt about anything having to do with the use or installation of particleboard, or for details and instructions on how to monitor air quality, contact the manufacturer of the panels.

Particleboard has a number of uses, and for many years perhaps the most common of those on the construction site has been as an underlayment. Particleboard has fallen into some disfavor in recent years, however, due to problems with some of the new generations of thin, highly resilient vinyl flooring that is being laid over it. Particleboard, though quite dense and manufactured from waterproof adhesives, still absorbs moisture quite readily. Absorption begins from the edges inward, causing the seams to swell slightly and form ridges, which can "telegraph" through some types of thin vinyl flooring.

The National Particleboard Association (NPA) claims that a number of underlayment-related failures for particleboard come from poor installation practices, and this is probably true—although, in the defense of most installers, it's difficult to sort through all the contradictory installation instructions to decide exactly how to install this stuff correctly. The NPA, for example, recommends gluing the particleboard down with white carpenter's glue, but some flooring manufacturers state that gluing the underlayment might void the warranty for the flooring.

A recent report from the Building Materials and Wood Technology division of the University of Massachusetts sums the situation up quite well by noting that "specifying and installing underlayment for resilient floor covering seems like a game of dodge-the-bullet. Everyone wants to pass the responsibility for a failed floor onto someone else." The best advice before selecting and installing particleboard as an underlayment is to contact the manufacturer of the flooring product you'll be installing over it and request *specific underlayment installation recommendations in writing.* It might very well save you an expensive call-back.

On a more positive note, particleboard has a very smooth surface, is relatively easy to machine (carbine-tipped cutters and blades work best), and it bonds well with laminates, all of which combine to make it a very popular substrate for use in cabinet, furniture, and countertop construction. It is available in a number of thicknesses up to 2 inches, and also in a number of panel sizes, most commonly 4 × 8 feet. You can also commonly purchase 25-inch-wide material to manufacture countertops.

One new version of particleboard recently on the market is ECP, from Boise Cascade. ECP—Electrically Conductive Particleboard—is used to discharge static electricity, making it an ideal product for use in computer and other business and office furniture, clean room construction, assembly areas, and other places where the accumulation of static electricity presents a problem.

There is another offshoot of standard particleboard, which is denser and smoother, known as medium-density fiberboard (MDF). See the separate section earlier in the chapter for additional information.

4

Today's plywood and wood-based siding products

Virtually any builder who has either constructed a new home or remodeled an older one in the last ten years or so is familiar with the revolution in the residential and light commercial siding market. Engineered lumber and panel products have really made their marks here, and contributed tremendously to the understanding and popularity of the entire engineered wood concept.

Lower material and installation costs, reduced call-back trips and expenses, improved maintenance, and a general improvement in overall performance when compared to natural wood siding are all standard features of the many composite sidings now on the market. Another major bonus of immeasurable value and importance is that these products help save the natural wood resources of our forests.

An ounce of prevention . . .

As with anything, there are some drawbacks to engineered and composite wood sidings. One product in particular, Louisiana-Pacific's tremendously popular Inner-Seal lap siding, has taken more than its share of hits lately, due to the alleged moisture-related failure of the product on a number of homes. As of this writing, the intricacies of this huge class-action suit—which, as usual, has more to do with legalistic game playing and profit motives than it does with how the product performed or how many homeowners were adversely affected—were still being worked out.

It's interesting to note that if you were to walk down the streets of any residential area in any town in America, you would see natural sawn-wood siding that has failed. Check the number of splits and cracks and warps on a house along the Maine shore or in the fierce Oklahoma sun. Watch some beautiful natural cedar bevel siding being installed on a home in Oregon or Washington, then check the cupping and staining in the same material just a few years later. No one has sued the logger who cut down those trees, or the mill who cut the lumber, or the contractor who installed the boards—although that's probably on the horizon as well.

We are, unfortunately, far too litigious of a society, and if we can point a finger at someone with money, we'll do it. Lawsuits have little to do with right and wrong. Products of all sorts fail every day for everything from poor design to normal wear and tear, and the sad part is that we've lost our ability to judge what is and what is not a genuinely good or bad product. As such, this lawsuit will certainly cause a setback in the engineered wood siding industry, without, in all likelihood, our ever being able to judge accurately the qualities of the product.

Inner-Seal siding did indeed fail in some situations, as have other engineered wood and composite wood products. Some are pointing to poor installation, some to poor design of the product, some to bad manufacturing practices, and still others to unreasonable storage and handling requirements.

The upshot of all this editorializing is that while the problems are, in my opinion, certainly not large enough to discourage the use of engineered wood siding products, you do need to be aware of some differences between these materials and the sawn-board siding you're used to installing. An ounce of prevention early in the installation will pay back big benefits in terms of improved performance and longer life for any of these siding materials.

Engineered and composite wood sidings, whether in panel or board form, are easier to damage than natural wood, and with any kind of surface or edge damage comes an increased risk of problems from moisture. You need to be a little more careful with how you stack, store, and handle these materials, and even with how you nail them. Most manufacturers issue strict instructions for how and where to place the nails in the boards, both to avoid damage to the board and to lessen the chances of moisture getting in.

I do have to point out that sometimes these installation requirements can be too stringent, to the point of being ridiculous, but deviation from them can affect your warranty. One manufacturer noted in their instructions that the nail was to be installed so that the head was left 1⁄16 inch from the face of the board, which we all know is impossible to do consistently in actual job-site installation conditions. Yet failure to install the boards according to their instructions voids the warranty!

Weather protection is another big issue with these siding materials, which are definitely moisture-sensitive if not properly covered (Fig. 4-1). There are usually requirements for storage on site to protect the

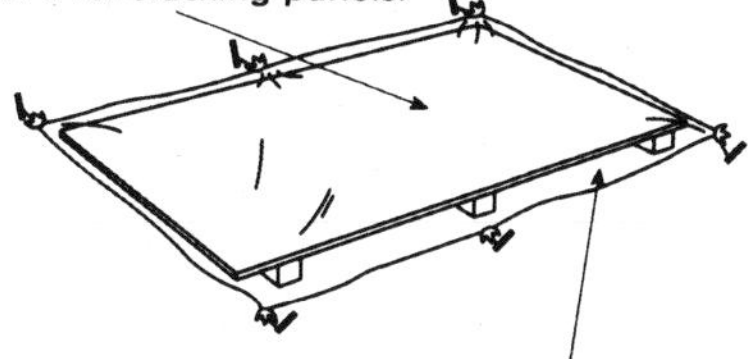

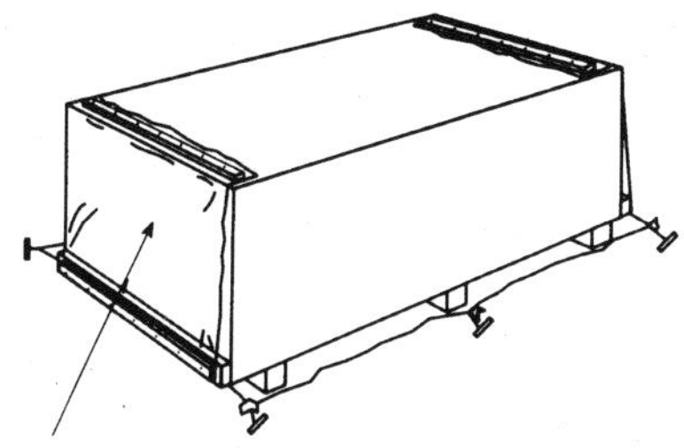

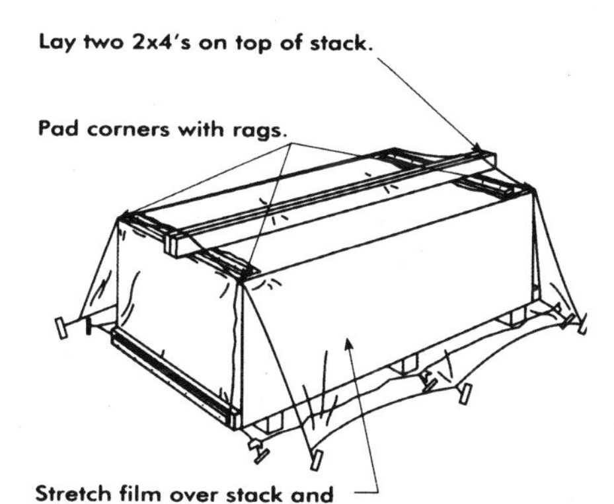

4-1
It is crucial to properly protect all engineered lumber products from the weather, since they can be damaged by excess moisture. Courtesy American Plywood Association

boards and panels from both ground and atmospheric moisture—and this applies to solid plywood as well. You also need to prime over any surface dings, chips, or other imperfections as soon as possible, again to prevent moisture from getting into the inner fibers of the material.

Finally, there are some important instructions issued by every manufacturer concerning caulking and painting (Fig. 4-2). All gaps, penetrations, and other possible moisture entry sources need to be caulked or otherwise sealed with an appropriate material right after installation of the boards, and the material needs to be painted as soon as possible, usually within 30 days of installation. In order to protect both the material and your warranty, be sure and follow these instructions carefully.

FINISHING REQUIREMENTS FOR WARRANTY COVERAGE

14 MINIMUM 2.5 DRY MILS PAINT

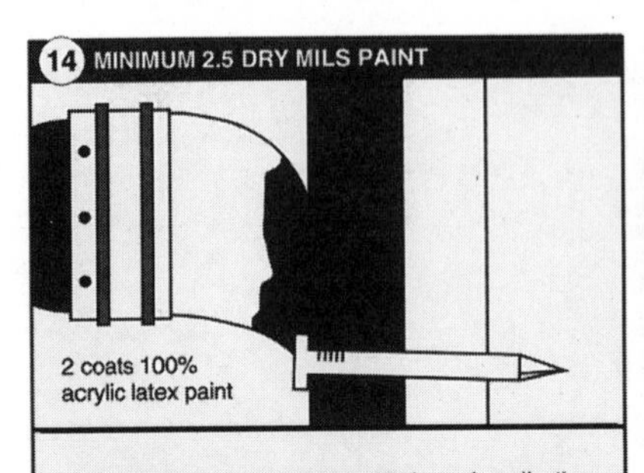

- Siding must be painted within 90 days of application to protect the substrate. If the siding cannot be finished within 90 days, use prefinished siding or completely reprime with a quality 100% acrylic latex wood primer.
- Surface preparation is extremely important. Prior to painting, insure that the surface is free of dirt, mildew or other contaminants.
- Use a high quality, exterior 100% acrylic latex paint, that is specifically recommended for primed wood sidings.
- Note: Alkyd oil primers and top coat paints must not be used.
- Field applied paint must be a minimum of 2.5 dry mils in addition to the factory primer. This is best achieved by brush applying TWO COATS at the manufacturer's required spread rate. (Research has indicated that the optimum thickness for the primer and top coats, is 4-5 dry mils).
- The paint can be applied by sprayer, roller or brush as long as all exposed edges are completely covered, and the resultant film is a minimum of 2.5 dry mils.
- L-P recommends back brushing when spraying or rolling the top coat.

15 PAINT ALL BOTTOM EDGES

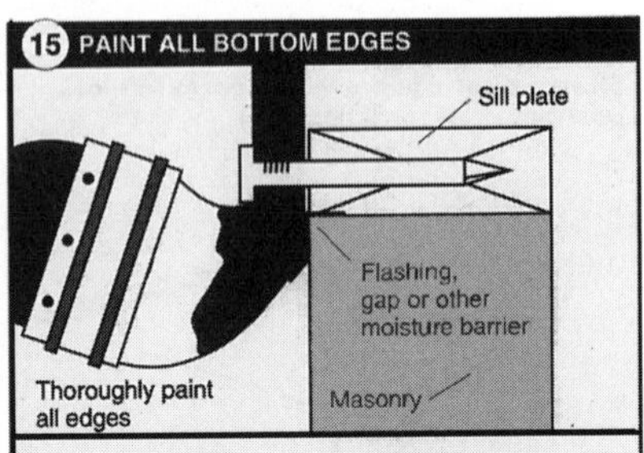

- PREFINISHED SIDING meets painting requirements if the 100% acrylic finish (top coat) is applied according to the paint manufacturer's specifications and is a minimum of 2.5 dry mils and carries a minimum 5-year written warranty on the 100% acrylic finish.
- ALL FINISHES need to be uniform and properly-cured for designed performance. IT IS IMPORTANT TO GET COMPLETE COVERAGE ON ALL EDGES AND GROOVES. Paint panel ends while still in unit.
- ALL FINISHES must be formulated to be mildew resistant and applied according to the paint manufacturer's requirements for proper surface preparation, spread rate, application methods, TEMPERATURE AND MOISTURE CONDITIONS.
- DO NOT USE: Semitransparent stains; shake and shingle paints; vinyl based resin combinations (vinyl acetate, PVA, vinyl acrylic, vinyl acetate/acrylic copolymer paints), or alkyd (oil) based products.

LP IS NOT RESPONSIBLE FOR THE PERFORMANCE OF FIELD OR MACHINE APPLIED FINISHES.

FAILURE TO FOLLOW L-P'S FINISHING REQUIREMENTS VOIDS THE WARRANTY AND MAY LEAD TO EXPENSIVE DAMAGE TO THE SIDING.

4-2 *To protect the siding and preserve the material's warranty, OSB siding must be properly painted within 90 days of application, as shown in these product instructions.* Courtesy Louisiana-Pacific

Speaking of painting, there is one other drawback that, while fairly obvious, still bears mentioning. All the engineered wood and wood composite sidings now on the market—with the exception of plywood panel and board products and composite panels with a natural wood veneer face—must be painted. They are sold factory-primed and in some cases even factory-painted, which certainly saves a step or two and further reduces the installed and finished costs, but they're still not natural wood. They don't have a stainable surface grain, and are not intended to be finished with a clear sealer.

A final word of advice. It's a very good idea to obtain the installation instructions, finishing instructions, and warranty information before you select a siding product, and make sure that you're familiar and comfortable with all three before selecting that product. If you aren't, either contact the manufacturer's representative for clarification or simply choose another product. If you do select another product, you should still contact the manufacturer of the first product and let them know the reasons you choose to go with another manufacturer. It's only through feedback from contractors about the realities of job-site conditions and real-world installation practices that these products will continue to improve.

What this chapter covers

Walk through a lumber yard or home center or flip through the pages of a trade publication or do-it-yourself magazine and you'll find an incredible array of siding products manufactured from plywood, wood composites, wood by-products, and engineered wood. In this chapter, you'll get an introduction to some of the many new products that are out there for siding any type or style of home or commercial building.

As was mentioned earlier in the book, this is by no means a complete listing of every siding product or every manufacturer. The field of engineered and composite wood products is changing daily, so you need to work closely with your material suppliers and product representatives to find the products that will work best for your particular application.

For clarity and easier reference, the products covered in this chapter are broken down into two main groups:

Plywood and wood-based panel sidings. For many years, contractors and homeowners alike have built homes and garages and additions and just about every other type of structure using plywood siding—an engineered lumber product we often fail to think of as an engineered lumber. Plywood also remains a very popular siding product in today's market, so I have included product descriptions for some of the more popular plywood sidings in this section. Also included here are composite and engineered panel siding products such as hardboard.

Board-style sidings. This section takes a look at some of the many new composite wood and engineered wood siding materials that are manufactured in board form as opposed to panel form.

Plywood and wood-based panel siding

The general reference to any house that is sided with panels or sheets of siding has always been "Oh, that house has plywood siding," and for many years that would have been accurate. Siding manufactured from solid plywood—solid meaning that it has a series of wood veneer laminations—has been used successfully for decades. Plywood siding is solid, attractive, long wearing, easy to install, and relatively inexpensive.

Good-quality plywood siding also takes good-quality veneers, the supply of which is constantly shrinking. "Clear" grades of plywood siding—those without knots or repairs—are often difficult to find and correspondingly expensive. Lesser grades, those having defects and repairs in the face veneer, are fine for painting, but have somewhat lost their appeal for staining because of the visibility of the repairs.

For these and other reasons, engineered and composite panel siding is becoming more and more common, and what someone might call "plywood siding" is today more than likely either a composite panel—one having plywood veneer faces and a core of other materials—or a totally engineered wood product.

Whether manufactured from plywood or other materials, panel or sheet siding (the two terms have become pretty much interchangeable in the construction industry) is generically grouped together. The panels are typically 4 feet wide and 8, 9, or 10 feet in length. They have square or, more commonly, shiplap edges and any of a number of face patterns. Many if not most are manufactured under the stringent grading and quality control standards of the American Plywood Association (APA), a nonprofit trade association whose member mills produce about 80 percent of the structural panel products made in the United States.

American Plywood Association (APA) 303 siding

The most common and popular of all the plywood sidings are those manufactured under the specifications for APA 303 siding (Figs. 4-3 and 4-4). Within the 303 siding group are 13 different face grades having smooth, overlaid, rough-sawn, and other textures; a variety of different groove patterns; and siding manufactured from a variety of different wood species (Fig. 4-5). The different classifications within the 303 series have different face grades and different rules and standards for how many patches are permitted. The table in Fig. 4-6 shows the different class numbers, along with the details of the patching. APA 303 series face textures, species, and sizes are:

Rough-sawn

Has a slightly rough surface, and is available with or without grooves.

Thicknesses: 11⁄32, 3⁄8, 15⁄32, 1⁄2, 19⁄32, and 5⁄8 inch.

Species: Douglas-fir, redwood, cedar, southern pine, and some others.

APA Texture 1-11

T-1-11 as it's commonly known is one of the most popular of the plywood sidings. It has a slightly rough face, shiplapped edges, and

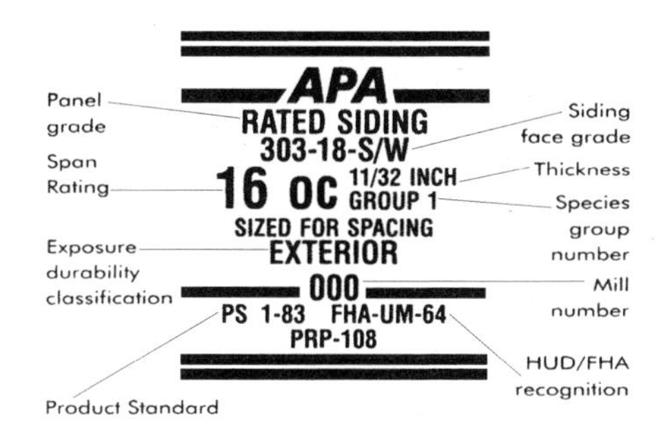

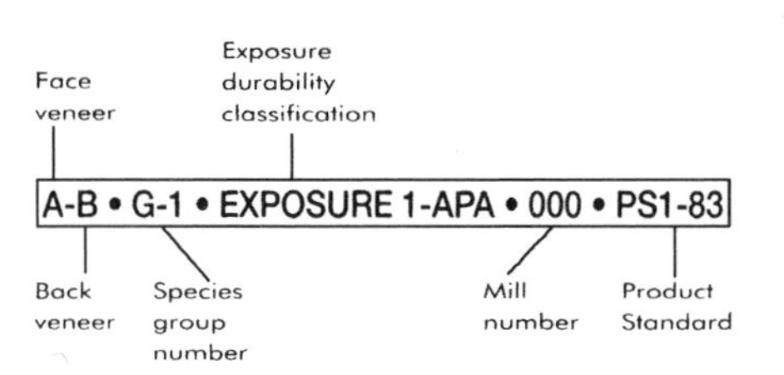

4-3 *Typical grade stamp for siding that's manufactured in compliance with the grading rules of the American Plywood Association.* Courtesy American Plywood Association

Recommended Finishes			
303 Series Plywood Siding Grades	Stains		House Paints Minimum 1 Primer Plus 1 Top Coat (acrylic latex)
	Semi-Transparent (oil)	Solid Color (oil or latex) 1	
303 -0C	2	2	2
303 -0L	Not Recommended	4	2
303 -NR	2	2	2
303 -SR	3	2	2
303 -6-W	2	2	2
303 -6-S	3	2	2
303 -6-S/W	3	2	2
303 -18-W	3	2	2
303 -18-S	3	2	2
303 -18-S/W	3	2	2
303 -30-W	3	2	2
303 -30-S	3	2	2
303 -30-S/W	3	2	2

1. Except for over laid panels, use stain-resistant primer with light-colors latex stains, since the wood extractives may cause a discoloration of the finish.
2. Recommended with provisions given in text.
3. Finish may be semi-transparent oil stain if color contrast between repairs and surrounding wood is acceptable. (See text.)
4. Some panel manufacturers recommend only solid-color acrylic latex stain. Consult the manufacturer's recommendations.

4-4 *Recommended paint and stain finishes for plywood siding.* Courtesy American Plywood Association

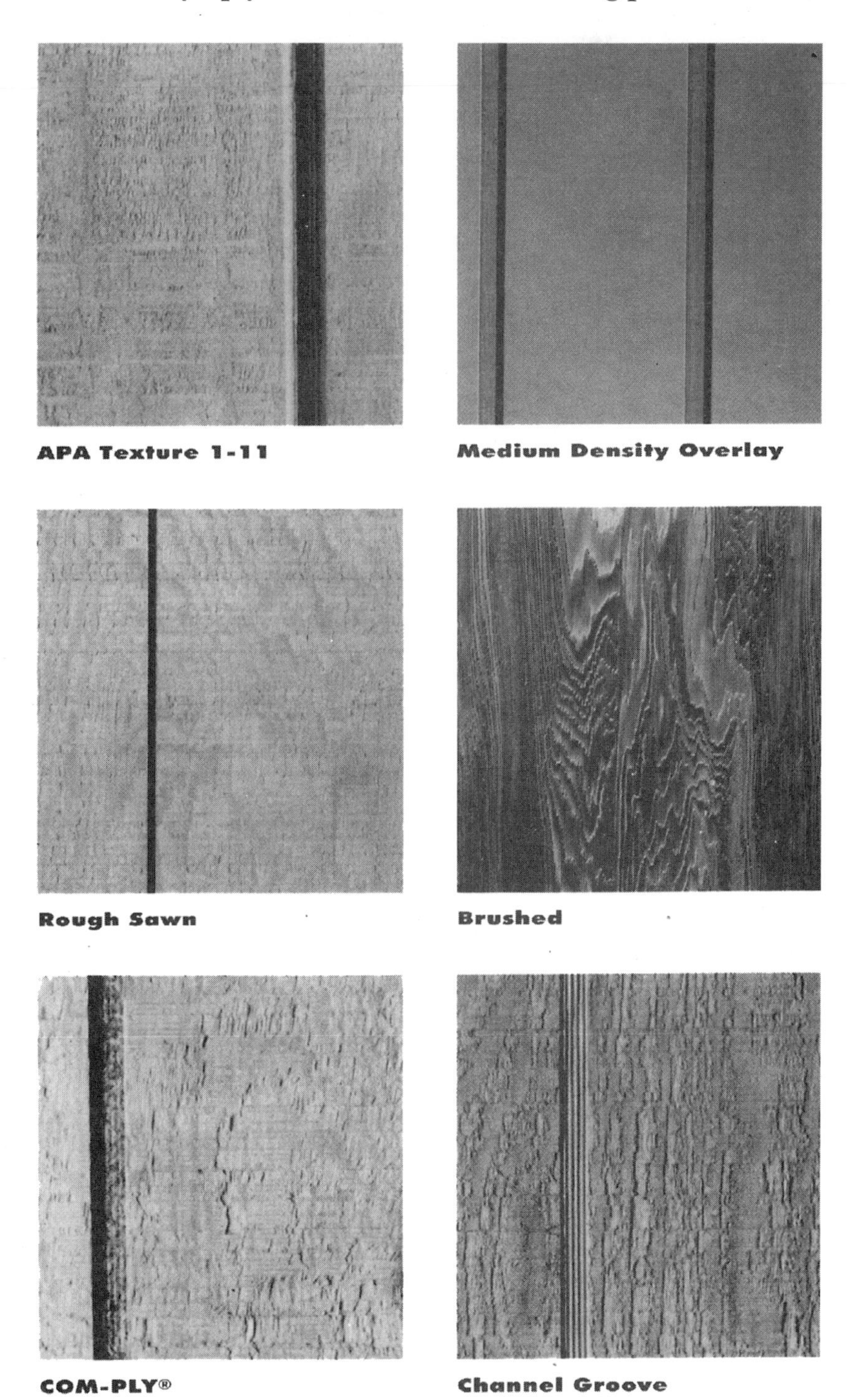

4-5 *An example of six common types of plywood siding.* Courtesy American Plywood Association

303 Siding Face Grades

Class	Grade*	Patches	
		Wood	Synthetic
Special Series 303	303-0C (Clear)	Not Permitted	Not Permitted
	303-0L (Overlaid, e.g. MDO Siding)	Not Applicable for Overlays	
	303-NR (Natural Rustic)	Not Permitted	Not Permitted
	303-SR (Synthetic Rustic)	Not Permitted	Permitted as Natural-Defect Shape Only
303-6	303-6-W	Limit 6	Not Permitted
	303-6-S	Not Permitted	Limit 6
	303-6-S/W	Limit 6—Any Combination	
303-18	303-18-W	Limit 18	Not Permitted
	303-18-S	Not Permitted	Limit 18
	303-18-S/W	Limit 18—Any Combination	
303-30	303-30-W	Limit 30	Not Permitted
	303-30-S	Not Permitted	Limit 30
	303-30-S/W	Limit 30—Any Combination	

4-6 *Face-grade veneers approved for use on American Plywood Association Series 303 siding panels.* Courtesy American Plywood Association

¼-inch-deep × ⅜-inch-wide grooves, typically 4 or 8 inches on center, although other spacings are sometimes available.

Thicknesses: 19⁄32 and ⅝ inch.

Species: Douglas-fir, redwood, cedar, southern pine, and some others.

Kerfed rough-sawn

This panel has a rough-sawn face with narrow grooves resembling saw kerfs, typically 4 inches on center.

Thicknesses: 11⁄32, ⅜, 15⁄32, ½, 19⁄32, and ⅝ inch.

Species: Douglas-fir, redwood, cedar, southern pine, and some others.

Reverse board-and-batten

These panels have wider grooves, commonly 1 to 1½ inches wide × ¼ inch deep spaced 12 inches on center. A variety of different face textures are available.

Thicknesses: 19⁄32 and ⅝ inch.

Species: Douglas-fir, redwood, cedar, southern pine, and some others.

Brushed

This is a type of surface texture in which the softer wood is removed from the panel, accentuating the harder natural grain of the wood by highlighting it in relief. Brushed texture panels are available with or without grooves.

Thicknesses: 11⁄32, ⅜, 15⁄32, ½, 19⁄32, and ⅝ inch.

Species: Douglas-fir, cedar, and some others.

Channel groove

Channel groove siding is very similar to T-1-11 in the types of available face textures and surface patterns. The only difference is in the groove patterns, which on these panels are much shallower, only 1⁄64 to 1⁄16 inch deep, ⅜ inch wide, and 4 or 8 inches on center.

Thicknesses: 11⁄32, ⅜, 15⁄32, and ½ inch.

Species: Douglas-fir, redwood, cedar, southern pine, and some others.

Medium-density overlaid (MDO)

MDO plywood, described in Chapter 3, is an excellent surface for paint adhesion. In the 303 series of sidings, several variations of MDO are available, including no grooves, V grooves spaced 6 or 8 inches apart, T-1-11 patterns, or reverse board-and-batten pattern.

Thicknesses: 11⁄32, ⅜, 15⁄32, ½, 19⁄32, and ⅝ inch.

Com-Ply

Com-Ply is an APA trade name for a plywood and wood fiber composition panel that is often made up into siding. Com-Ply is made up in three or, more commonly, five layers (Fig. 4-7). The face and back of the panel are standard plywood veneers, oriented with the grain running parallel to the long dimension of the panel, just like regular plywood. A layer of wood fiber and resin is sandwiched in between

4-7 *Com-Ply tongue-and-groove subflooring. Note corner damage to one sheet (center) caused by rough handling during unloading.*

the veneers, and the sheet is pressed up under heat and pressure in a one-step pressing operation.

In a five-layer sheet (Fig. 4-8), there is also a cross-band of veneer in the middle of the sheet. This veneer is oriented with the grain perpendicular to the two face veneers (parallel with the short dimension of the panel). There are then two layers of wood fiber pressed in between the three veneer layers. Com-Ply has a couple of advantages over standard plywood. For one, the manufacturing process is a little more resource-friendly, in that it uses waste wood fibers for the cores layers instead of solid wood veneers. Also, the pressure created during the pressing process forces the wood fibers into any voids in the face and center veneers, filling them with solid material.

Most of the standard 303 series plywood siding patterns of grooves and surface textures are also available in the Com-Ply material. Standard Com-Ply sheet sizes and thicknesses are the same as for standard plywood siding in the various categories listed in the previous section.

Hardboard

Hardboard is another engineered wood material that has been around job sites for quite awhile in a couple of basic forms, but it is now being "rediscovered." Several manufacturers are producing both panel

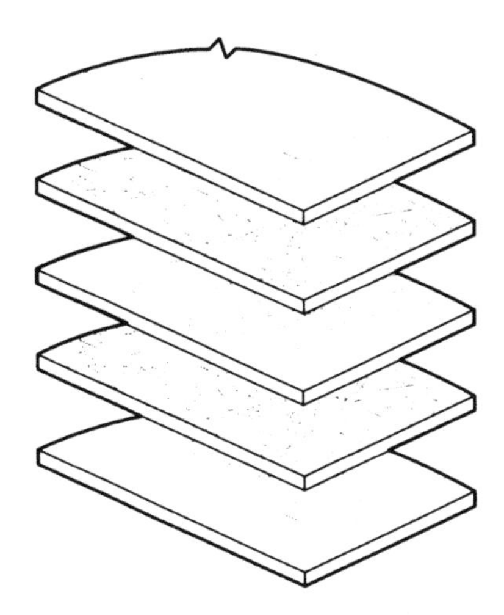

4-8 *The makeup of Com-Ply. Note the five layers, made up of a top, bottom, and center layer of wood veneer and two inner layers of wood fiber.* Courtesy American Plywood Association

and board siding products from hardboard that are beautiful, durable, and, when painted, virtually indistinguishable from natural wood.

Hardboard siding, like standard hardboard panel materials, is made from dry, clean, carefully selected wood fibers from a variety of species, pressed up under tremendous heat and pressure to form a solid mat. There are no core voids, and none of the checks, patches, or surface variations common in plywood and natural wood.

One of the other advantages to hardboard siding is the small size and fine texture of the wood fibers used in the manufacturing process. Actual pieces of natural wood or plywood are commonly used as molds to create the surface texture on the plates that the fibers are laid up and pressed on. The tiny fibers press tightly into these intricate patterns, creating surface textures that look and feel very similar to wood, mimicking even the finest of knots and grain patterns.

In panel siding, hardboard panels are most often manufactured to closely resemble standard plywood 303 patterns (see the previous section on APA 303 siding). Hardboards are available to match the T-1-11 pattern, as well as the channel and reverse batten styles. Surface textures include rough-sawn cedar and fir, cross-sawn cedar and fir, pecky cypress, knotty barn-board style, smooth plain face, and smooth grooved face.

Some manufacturers also offer hardboard panels that have a surface texture very much like traditional hand-troweled stucco. Here again, actual stucco samples are used to make the manufacturing plates, and the fine fibers of the hardboard faithfully recreate all the intricate sweeps, swirls, and shadow lines. Once painted, the panels are virtually indistinguishable from a real hand-textured stucco finish.

Standard panel sizes: 4 × 7 feet, 4 × 8 feet, 4 × 9 feet, and 4 × 10 feet.
Standard panel thicknesses: 7⁄16 and ½.
Standard finishes: Factory-applied white or gray primer, ready for field painting, or any of several factory-applied and baked-on colors.

Oriented strand board

Oriented strand board (OSB) is another relatively common engineered wood product that is now seeing use as a siding material (Fig. 4-9). The OSB used as siding is essentially the same as the OSB sheets used for sheathing and other construction applications (see Chapter 3). Strands of wood fiber are processed from small-dimension logs, then coated with a resin adhesive, oriented in three perpendicular layers, and pressed under heat and pressure into a solid mass (Fig. 4-10).

For siding use, the OSB is actually only the substrate (the underlying structural surface) of the siding. In a separate set of processing steps, a hard resin overlay is applied over the OSB substrate. This overlay is primed with a weather-resistant primer and then embossed into a cedar-grain surface texture. The panels are left as solid sheets, and can be cut with vertical grooves on 4- or 8-inch centers. Both square edge and shiplapped edge panels are available. After completing all cutting and forming operations, the final step in the manufacturing process is to prime all the grooves and laps for moisture resistance.

Sizes

All OSB siding sheets are 4 feet wide, like standard plywood, to match up to common 16- or 24-inch on-center framing practices. Sheet lengths of 8 and 9 feet are the most commonly available, although most lumberyards stock only the 8-foot material. Depending on availability and current manufacturing runs, you might also be able to special-order 7-, 10-, 12-, and even 16-foot-long sheets. Common thicknesses are 7⁄16 and 19⁄32 inch.

Board-style sidings

Plywood siding really came into vogue and popular usage in the 1950s, when the demand for housing in the still-booming postwar years made this a time- and cost-effective siding material. Its popularity grew in the 1960s, where its sleeker styling complemented many of the contemporary architectural looks of that period, and continued on into the 1970s, with the cost-saving benefits of faster in-

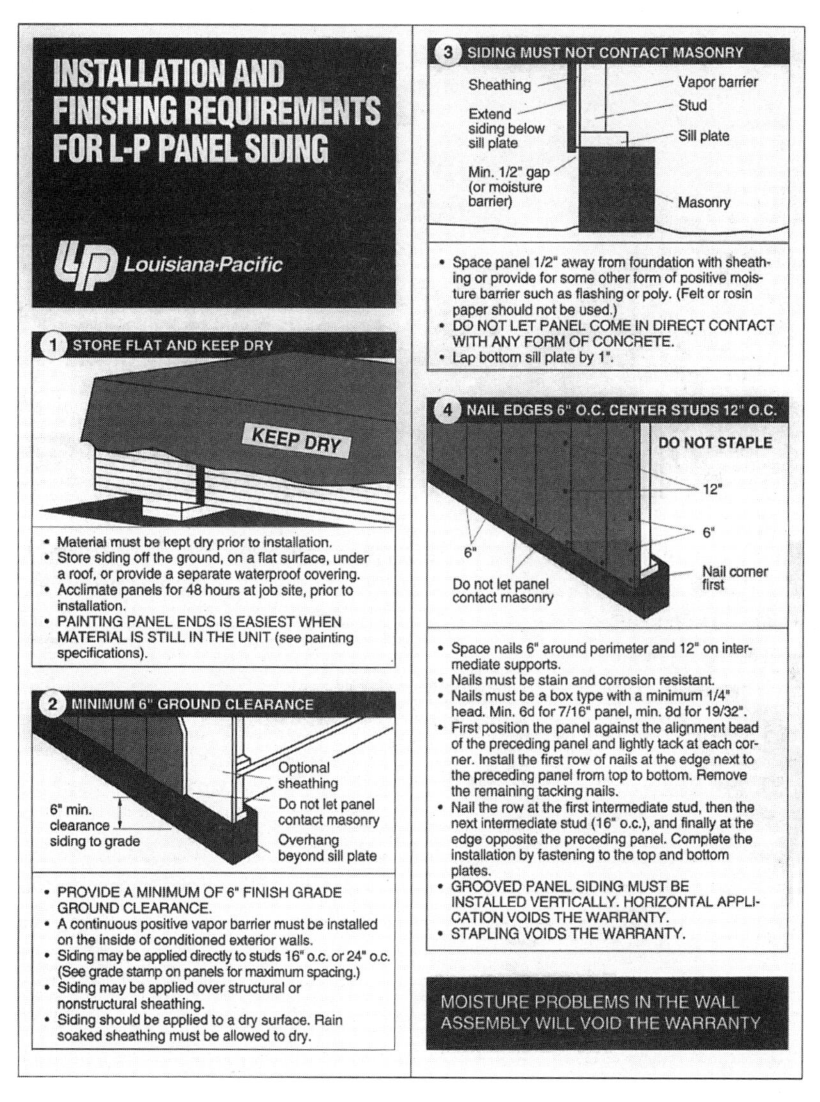

4-9 *An example of part of the installation instructions for OSB sheet siding. You must carefully adhere to these procedures in order to protect the warranty coverage.* Courtesy Louisiana-Pacific

stallation during a time of rising labor costs and lower material prices as good-quality siding lumber became more and more scarce.

But long before plywood and other sheet materials, homes were sided with individual boards. Strip the paint off a classic old home from the 1920s or 1930s, and you're almost sure to find beautifully milled, straight-grained, virtually clear siding boards, which were

4-10 *The makeup of oriented strand board.* Courtesy Louisiana-Pacific

the norm at the time. They were also what lent the distinctive look to many of these home styles, a look that people really began to clamor for again in the 1980s and 1990s. But natural lumber of the quality necessary to meet the appearance, weather-resistance, and durability requirements of siding was in short supply, and often prohibitively expensive.

This situation created another perfect market for engineered lumber and wood composites, and well as nonwood materials. Company after company began introducing manufactured wood siding products in an ever-expanding array of sizes and profiles to meet the demand of builders who wanted to side their moderate- to upper-end homes with individual boards, but still needed to keep them affordable.

During the course of this growth in availability and acceptance, a number of interesting issues emerged. Much of this engineered and manufactured siding material was proving to be so good that manufactures could actually offer a warranty. As an additional sales and marketing tool, they could offer some very nice assurances to builders and homeowners that their product would not warp, split, crack, or otherwise fall apart for the next 20 or 30 years. Imagine trying to offer that assurance if you were selling, for example, tight-knot cedar.

Most manufactured wood siding products, correctly installed and weather-protected, are extremely durable and long-lasting. They arrive at the job site having been made to very strict moisture contents and manufactured from compressed and resin-coated wood fibers instead of solid lumber, so the dimensional stability is excellent. Most are manufactured in 16-foot lengths, which simplifies installation and reduces or eliminates many of the end-to-end joints found in natural wood siding. These siding boards are less expensive, lighter, more uniform, and have a lot less of an impact on forest timber reserves.

Hardboard

Hardboard is seeing a lot of use in the area of board siding as well as sheet use. In fact, it actually seems to perform better as a siding board than as a sheet, apparently due to the better dimensional stability that results from the long, narrow configuration of boards than from the larger square-foot surface area of sheets.

Hardboard board siding is available in a number of different profiles, including individual boards for use as lap siding and multiple-board profiles, in which a single manufactured panel has the face appearance of two, three, or even four individual boards. Most of these multiboard profiles have shiplapped edges to increase weather resistance and also to act as a self-alignment feature, which speeds installation and keeps the boards uniform. Some of the lap siding boards also have splines, rabbets, or other indexing devices that are molded or inserted into the lower rear face of the board to simplify its alignment with the board below.

The manufacturing process for board siding is essentially the same as for sheets. Wood fiber is blended with resins and pressed over plates under heat and pressure. Here again, the plates are molded using real wood as a pattern, and the fine texture of the wood fiber used in the manufacturing process presses up into a very detailed and realistic natural wood-grain appearance.

All hardboard siding is susceptible to damage from moisture, which can cause the wood fibers to swell and distort, and eventually come completely apart. For this reason, all hardboard comes from the manufacturer with a factory-applied primer, usually a very hard, baked finish that seals the wood's surface face and edges. The back is usually left unprimed. It's extremely important to exactly follow the manufacturer's directions for painting, and to caulk or otherwise seal all joints, gaps, and penetrations with appropriate material. At least three hardboard siding manufacturers—ABTCO, Masonite, and Weyerhaeuser—also offer their siding with a factory-applied finish color coat. These finishes are applied and then cured under very controlled factory conditions, so the resulting paint job is smooth, hard, and free of surface defects. You might experience occasional minor color differences, however, especially if the shipment comes from a couple of different manufacturing runs, which can result in a finished overall house that doesn't quite have the consistency and even appearance you can achieve with a site-applied paint job.

If possible, try to see a completed house that's been sided with prepainted material before deciding about whether to use factory-painted or just factory-primed material. If you opt for the factory-painted material, be certain to specify to your supplier that all the material come from the same dye lot. If that is not possible, don't intermix the boards from one color run with those from another; do entire walls with the same color-run material, and save the other color-run material for separate walls or sections of the house.

ABTCO and Masonite, to name just two manufacturers, offer a complete line of accessories for their siding products, all of which simplify the installation and greatly enhance the uniformity of the finished job. These include full-length, metal, inside and outside corners; individual metal corners for use with lap siding where full-length outside corner boards are not desired; joint moldings that match the exact configuration of the siding, for sealing butt joints; metal starter strips for obtaining a uniform angle of bevel on the bottom lap siding board; J-trim, also metal, for use against windows, doors, and other vertical or horizontal surfaces that the siding butts up against; soffit panels, both solid or perforated for ventilation; facia covers, if you want a color-matched metal cover to put over your wood facias; roll flashing, for making your own trim pieces and flashings in nonstandard or unusual areas; and a full selection of color-matched paint, nails, and caulking.

Sizes

Standard thicknesses for virtually all of these products is 7⁄16 or ½ inch, with lengths ranging from as short as 8 or 9 feet up to the relatively standard length of 16 feet. Widths vary greatly with the product and the intended finished look of the installation, and include 4, 6, 7, 9½, 12, and 16 inches, with exposures ranging from 4 up to about 15 inches.

Surface textures

Here again, there are several surface textures available with hardboard siding, all designed to closely resemble other natural wood materials. They include rough-sawn cedar, which has a fairly pronounced grain; cross-sawn cedar, which has a pronounced grain and then saw marks perpendicular to the grain, as though the board was milled and then run back through a band saw for final sizing; cedar shake, which is molded to resemble individual shakes or shingles; pecky cypress, with the appearance of worm holes and other natural defects common to cypress; pine, which has a little softer grain texture than the cedar; stucco, made to resemble hand-textured stucco (actually quite realistic as long as the joints in the panels are covered); and smooth, which is just that.

Coverage

When installing a 4 × 8-foot sheet of panel siding, you know you'll be covering 32 square feet of wall area. Those calculations are not quite as easy to make when using lap or shiplapped siding boards, where you need to deduct out the laps from the actual area of coverage, and also to allow a little bit for cutting waste. Most manufacturers are good about providing you with coverage information for each of their products, so ask for those numbers before working up your cost estimates or material lists.

For example, one manufacturer notes that for their 6-inch lap siding (4 13⁄16-inch exposure), you'll need approximately 1270 square feet of material to cover 1000 square feet of wall area. For their 8-inch siding (6 7⁄8-inch exposure), that number drops to 1180 square feet of siding needed to cover the same 1000 square feet of wall area. The cutting waste allowance is typically 5 percent.

Thermal performance

As is to be expected, siding adds a little bit to the overall heat-loss resistance of the wall structure, but not a lot. The average thermal performance for 7⁄16-inch siding is R-0.45; for ½-inch siding it's about R-0.58.

Oriented strand board

Oriented strand board (OSB) is the basis for Louisiana-Pacific's Inner-Seal siding, mentioned earlier in the chapter (Fig. 4-11). The OSB used in this product is essentially the same as the OSB sheets used for sheathing and other construction applications (see Chapter 3). Here again, the OSB substrate is coated with an embossed and primed resin coating (Fig. 4-12). See the section *Oriented strand board* earlier in this chapter.

OSB siding is intended for use as a lap siding (Fig. 4-13). The edges are slightly beveled back to lessen edge exposure to weathering and improve water runoff properties. The edges and ends of the boards are factory-primed to help prevent moisture from entering the board.

Sizes

OSB siding is commonly manufactured in a 7⁄16-inch thickness, and can be applied directly to the studs on up to 24-inch on-center spacing, or over sheathed walls with 16- or 24-inch on-center spacing (nonsheathed walls must be laterally braced for rigidity according to local codes). Common widths are 6, 8, 9½, and 12 inches. (These are

4-11 *Louisiana-Pacific's popular Inner-Seal brand of OSB lap siding.*

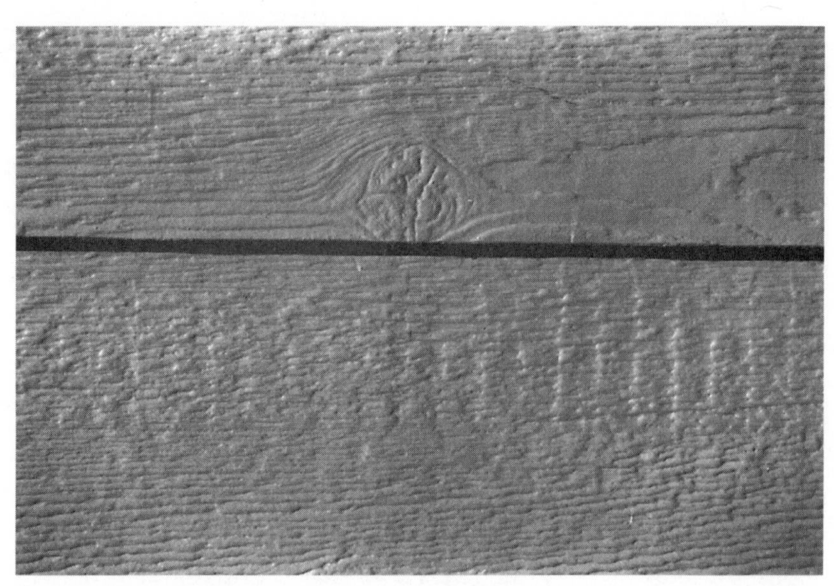

4-12 *A close-up of the realistic wood-grain texture that's pressed into the outer resin layer of Inner-Seal OSB siding.*

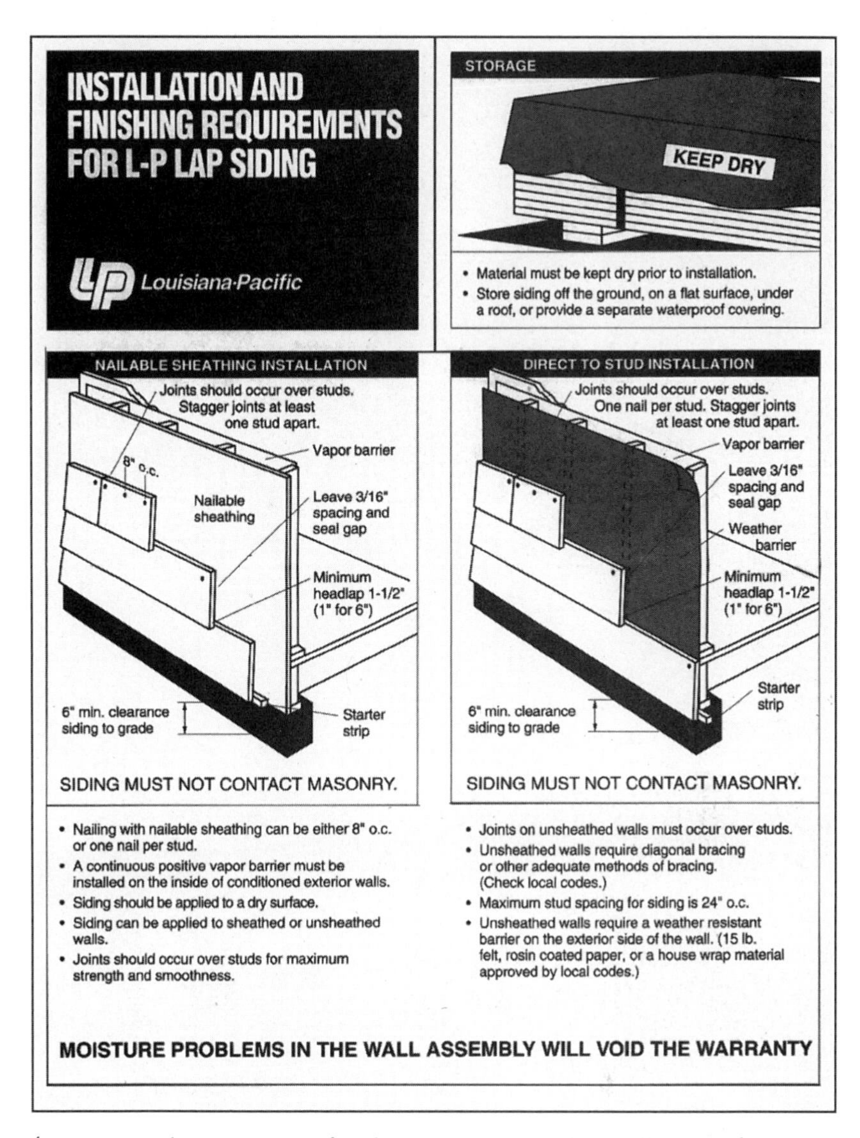

4-13 *Partial instructions for the protection and installation of Inner-Seal OSB lap siding material.* Courtesy Louisiana-Pacific

all nominal sizes; the actual size is about ⅛ inch less. The normal weather exposure is about 2 inches less than nominal dimensions.) The length of the boards is a standard of 16 feet, while the weight varies with the width of the board. The 6-inch boards, for example, weigh 15 pounds each, and the 12-inch boards weigh about 23¾ pounds each.

5

A selection of other wood-based and nonwood products

In addition to the siding products discussed in the previous chapter, there is a whole spectrum of other wood-based products available for everything from doors to trim to furniture. Most are manufactured from waste wood scraps and fibers, have a very low formaldehyde content (a health problem in early generations of manufactured wood products), are economical and easy to use, and, by virtue of their ever increasing market share, are easy to come by.

But wood-based products are certainly not the only things out there that warrant your consideration when siding or trimming a house, building a deck, installing a roof, or completing any number of other applications around the construction site. The same scarcities of good-quality, affordable, solid sawn lumber that have helped the engineered wood industry evolve have also given a boost to manufacturers of a wide variety of other, nonwood products.

This chapter covers, in alphabetical order, several new products that are worth considering for your next construction project, both wood-based (having some wood fiber content) and nonwood (having little or no wood fiber content). They includes board and panel siding, specialty trim materials, underlayment boards, doors, windows, and a few of the other diverse, new, and innovative manufactured products—as well a couple of not so new ones—that are helping change the way we build.

Aluminum shingles

As an alternative roofing material to wood or composition, you might want to check out some of the latest generation of aluminum shingles now on the market (Fig. 5-1). While still somewhat on the expensive side, they're an improvement over the first generation of this product and are a great solution where appearance, long life, and superior fire protection are key elements in the design and construction of a residential or commercial structure.

5-1 *Aluminum roofing such as Rustic Shingles creates a beautiful, durable, and fire-resistant roof.* Courtesy Classic Products, Inc.

Most aluminum shingles use an interlocking design that locks each shingle to the others around it on at least two sides, as well as attachment tabs that let you nail it to the roof sheathing. The interlocking features make for superior wind uplift resistance while still allowing for the metal's natural expansion and contraction that occurs with temperature changes. The shingles can be applied over new sheathing such as plywood or OSB, or they can also be installed right over old composition roofing, depending on the condition of what's existing.

The basic shingle begins with an aluminum alloy sheet that is cut and formed into individual shingles, designed with a natural woodgrain pattern that closely resembles wood shakes. The aluminum is cleaned and rinsed, then coated with a chromate solution for weather resistance. One or more coats of primer and baked-on enamel are applied to the top surface, and a clear coating that helps equalize temperature differentials is also applied to the underside of the shingle. Several colors are typically available, mostly in earth tones, along with red, green, and blue.

Fiber cement siding

Siding boards and panels manufactured of fiber cement can be used in any residential or commercial application where wood or wood-composite sidings would normally be used. They're an especially good alternative siding material in situations where a high resistance to fire, strong winds, or severe weather is important, or where the ability to withstand possible damage from moisture or termites is crucial.

Fiber cement siding is manufactured from a combination of cellulose fiber material, portland cement, and silica-sand, along with water and other additives. It is formed into relatively lightweight siding sheets or individual siding boards, typically with either a smooth face or an embossed face that, like the wood-product sidings, closely resembles natural milled or milled and cross-sawn cedar. The fiber cement is autoclaved—a process of drying and curing the material using superheated steam under pressure—so it has a very low moisture content that promotes excellent paint adhesion. And to put your mind at ease concerning some of the forerunners of these products, all fiber cement products currently being produced are also asbestos-free.

Installation is the same as for most wood products. You can fasten the siding using hand- or air-driven galvanized nails, wood screws, or self-tapping screws for metal studs. You can use standard sawn lumber or wood-based trim and corner materials, and most manufacturers offer metal inside and outside corners to complete the installation, as well as metal joint covers that are textured to match the siding. Caulking and painting operations use the same materials and techniques as those used for wood.

Cutting is, admittedly, a little more tedious and time-consuming than with wood or wood-based siding. The manufacturers recommend scoring the face side of the board with a special tungsten-tipped scoring knife to a depth of about one-third of the board's thickness, then placing a straightedge against the score line and snapping the board upward along the score. If you are installing a lot of this material, there are also special tools called *guillotine cutters*—essentially a big shearing tool—that you can purchase or sometimes rent from your supplier. You can also cut with tungsten-bladed power saws and jigsaws. Consult the manufacturer for specific details and recommendations, and remember that you *must* wear a respirator and eye protection during power-sawing operations.

Sizes

Material sizes for these products are very much like those for other panel and board siding materials. Standard board thickness is ¼ or 5⁄16 inch, and common widths are 9½ and 12 inches. The standard board length is 14 feet, as opposed to the more common 16 feet found in wood-based siding, which is probably a concession to the greater weight. Panel sizes are a very standard 4 feet wide and 8, 9, or 10 feet long.

Weight and thermal resistance

There are two minor drawbacks here; the weight is relatively high and thermal resistance is relatively low, but neither should be a deciding factor in your decision. The weight for the 5⁄16-inch-thick material, which is the more common thickness, is 2.3 pounds per square foot, as opposed to OSB siding, for example, which weighs about 1.5 pounds per square foot. Therefore a standard piece of 12-inch OSB siding, 7⁄16 inch thick and 16 feet long, weighs 23¾ pounds, while a standard 12-inch fiber cement siding board, 5⁄16 inch thick and 14 feet long, would weigh in at around 32.2 pounds. Thermal resistance is R-0.15 for ¼-inch material, and R-0.18 for the 5⁄16-inch material.

Fiber cement shingles

Another excellent use of the fiber cement technology is in manufacturing roofing shingles. Fiber cement shingles are durable and extremely long-lasting, and they have a Class A fire rating, the highest available, when installed according to the manufacturer's specific recommendations for sheathing and felt.

If you're looking for a product that will duplicate the rich look of traditional cedar shake roofing without the maintenance and fire-danger problems associated with real wood, you can't beat fiber cement. It's the closest thing you'll see to an exact color and texture match, including all the random width, length, and texture variations so common with natural cedar shakes. There is also a cement fiber shingle that is virtually indistinguishable from natural slate, even when viewed close up, but without natural slate's brittleness and weight problems.

Cement fiber shakes are ¼ inch thick, 22 inches long, and come in uniform or random widths of 6, 8, and 12 inches (they look best mixed). The weight is about 400 pounds per square (100 square feet), based on a 10-inch exposure. For the slatelike shingles, the length is 18 inches and the width is 8 inches. The weight for this roofing is also about 400 pounds per square, but with an 8-inch exposure.

Fiber cement underlayment

Fiber cement tile underlayment is a great material to use under ceramic tile for walls, floors, and countertops. In the last few years, it has virtually replaced traditional mortar-based ("mud-set") ceramic tile installations in a wide range of applications, and has boosted the success of do-it-yourself ceramic tile installations.

Fiber cement underlayment is a blend of portland cement, ground sand, cellulose fiber, additives, and water. It is formed into square-edge sheets in ¼-inch and 7⁄16-inch thicknesses, and in 3 × 5-foot and 4 × 8-foot sheets. It can be installed over structurally sound panel-type subfloor materials such as OSB or plywood, over plywood or old laminated countertops, over drywall, or directly to the wall studs (7⁄16-inch-thick materials only).

You can cut the panels using a circular saw equipped with a carborundum wheel, and fasten them with galvanized roofing nails or drywall-type screws. For most installations, it is recommended that you also apply flexible latex or acrylic-modified thinset mortar. Be sure to follow the manufacturer's instructions for installing the underlayment panels in specific applications, and applying the ceramic tile over the underlayment.

Fiber cement underlayment is also a great material to use under a variety of resilient floor coverings such as sheet vinyl or vinyl tile. It is dense and solid, has excellent dimensional stability, and is resistant to heel damage and other impacts. Be sure that the underlayment is rated for use under resilient flooring (some are intended for ceramic tile use only, while others can be used for either), and also check with the flooring manufacturer for compatibility with the specific type of floor covering and adhesive you'll be using.

Fiberglass doors

How about a front door that has all the natural warmth and beauty of natural wood, but with much better thermal performance and exceptionally low maintenance. That's now possible, thanks to the new line of fiberglass doors now on the market.

Fiberglass doors are made from compression-molded fiberglass over an insulating core of solid polyurethane foam. The fiberglass looks very much like natural red oak, and can be stained or painted using conventional materials and methods. The resulting door composition won't warp, split, or crack, has the dense feel of a solid-core wood door, and has approximately four times the insulating value.

There are all sorts of styles available in fiberglass doors, including single doors, double doors, and a full range of sidelights. You can get flush doors, carved doors, panel doors, and doors with all sorts of glass inserts and natural brass highlights. Sizes range from 2 feet, 6 inches to 3 feet for single doors, with sidelights available ranging from 12 to 36 inches, giving you the potential for a 12-foot double door and sidelight entry.

Fiber-reinforced gypsum underlayment

One area where builders and homeowners have continued to have problems is with underlayment beneath sheet vinyl and vinyl tile flooring. The traditional material for this application has long been particleboard, but many manufacturers of the latest generations of vinyl flooring are recommending against it. Plywood is still a very good underlayment product, but the core voids that are prevalent in many grades of plywood can sometimes create problems with this material as well.

One possible solution is a new product called fiber-reinforced gypsum underlayment (Fig. 5-2). This is another of the new "green" (environmentally-friendly) building materials that are making more and more frequent appearances on the construction scene. The panels are a mix of recycled newspaper and gypsum, blended without the addition of formaldehyde resins, which are pressed in hard, smooth-surface panels. There is a ¼-inch-thick solid panel version, and a ⅜-inch layered version that has a fiber-reinforced gypsum face and back over a core of perlite. Both are manufactured in easy-to-handle 4 × 4-foot panels.

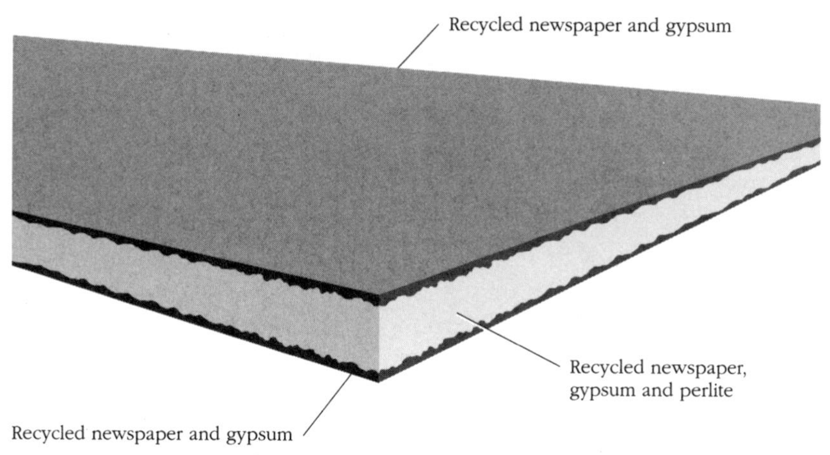

5-2 *Cut-away view showing the makeup of fiber-reinforced gypsum underlayment and wallboard products.* Courtesy Louisiana-Pacific

Fiber-reinforced gypsum underlayment is a good choice for use under vinyl sheet flooring, vinyl tiles, ceramic tile, wood flooring, quarry tiles, and carpet. It is dimensionally stable, has no core voids, and won't delaminate. Before using this product with any specific resilient vinyl flooring, however, be sure to check with the flooring manufacturer to be certain it's an acceptable underlayment to use with their product.

The underlayment sheets can be installed over any solid subfloor material, such as plywood or OSB. Store the underlayment inside, laid flat on a solid surface. If possible, it's best to allow the underlayment panels to acclimate to the temperature and humidity conditions of the job site prior to installation.

Clean the subfloor thoroughly, and patch any dents or other defects. Check the edges for flex, and install additional screws if necessary. Fill the joints in the subfloor panels, and then sand off irregularities in the joints. Lay the underlayment panels at right angles to the direction of the subfloor panels, and place them so the joints in the underlayment overlap the joints in the subfloor panels by a minimum of 6 inches. Also, lay out the underlayment so the joints in the panels are staggered. There should be a gap of approximately ¼ inch between the panels and the walls, and the panel-to-panel joints should butt lightly together. You can cut the panels by scoring them twice with a razor knife and then snapping them along the scored line, or by cutting them with a low-RPM cordless saw.

The manufacturer recommends installing the ¼-inch-thick panels with adhesive and staples, but here again you should check with the manufacturer of the flooring to be certain it's acceptable to install their product over underlayment that's been glued down. Apply construction adhesive in a 6-inch strip around all four edges of the panel, and in a strip down the center. Use a cartridge gun to apply the glue, then trowel it down. You can also apply any approved, premixed adhesive from a bucket, using a notched trowel.

Set the glued panel in place, and fasten it using an 18-gauge chisel-point staple with a ¼-inch crown. Set your air pressure so the staples are set flush with or slightly below the surface of the panel. Place your staples about ½ inch in from the edges of the panel, and space the staples 3 inches apart along the edges and 6 inches on center in the field. For ⅜-inch-thick panels, gluing is not necessary; just fasten the panels in place using staples as previously described.

For remodeling projects, you can also install fiber-reinforced gypsum underlayment panels directly over most types of old flooring, with the exception of cushion-back or foam-back flooring. Prep the old floor as previously described, then fasten the underlayment panels by using adhesive and staples or by nailing them down using ring-shank or screw-shank nails. The nails need to penetrate at least ¾ inch into the subfloor, but not all the way through it.

Fiber-reinforced gypsum wallboard and sheathing

The same material used in the manufacture of fiber-reinforced gypsum underlayment (see the previous section) is also used to manufacture panels for wallboard and wall sheathing use (Fig. 5-3). All of the techniques for cutting, installing, and fastening these panels are the same as for the underlayment panels (Figs. 5-4 and 5-5).

Wallboard and sheathing panels are available in three standard thicknesses: ⅜, ½, and ⅝ inch. The panels are 4 feet wide, and come in standard lengths of 8, 9, 10, and 12 feet.

Hardboard doors and trim

In addition to the siding materials mentioned in the last chapter, hardboard is being used for an increasing number of items around the construction site. And as more people recognize the value of products made from this material, its acceptance will grow and you'll probably see more and more hardboard products appearing both in lumberyards and on job sites.

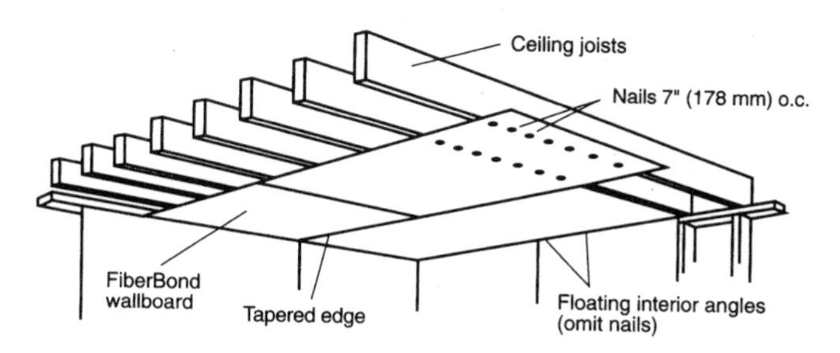

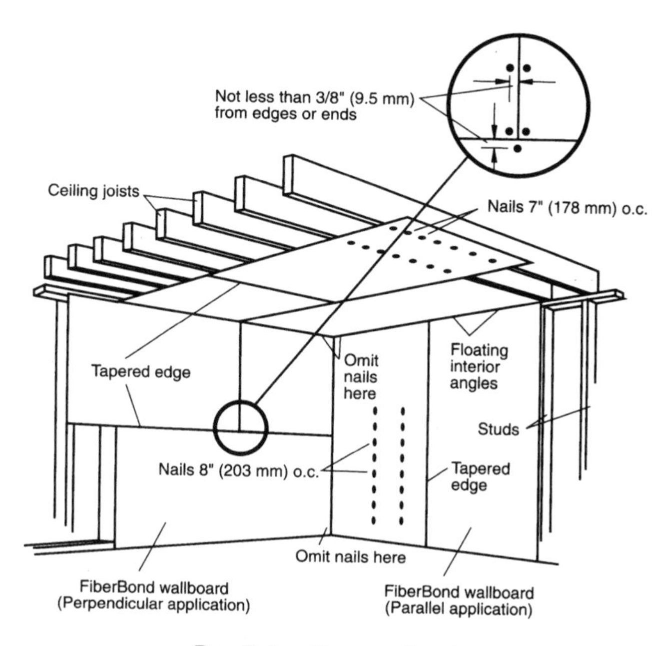

5-3 *Installation instructions for fiber-reinforced gypsum wallboard.* Courtesy Louisiana-Pacific

One very promising use for hardboard is as an exterior trim. With the growing use of lap siding for homes in all segments of the market, the need for compatible trim has grown along with it. Up until fairly recently, only 5⁄4-inch solid sawn stock—which measures a full 1 inch thick—was thick enough for two overlapping layers of 7⁄16-inch-thick lap siding to butt up against at corners and around windows and doors. Full 5⁄4-inch stock is relatively expensive, and since it's almost always used next to prime-coated engineered-lumber lap siding such as Louisiana-Pacific's Inner-Seal and will end up

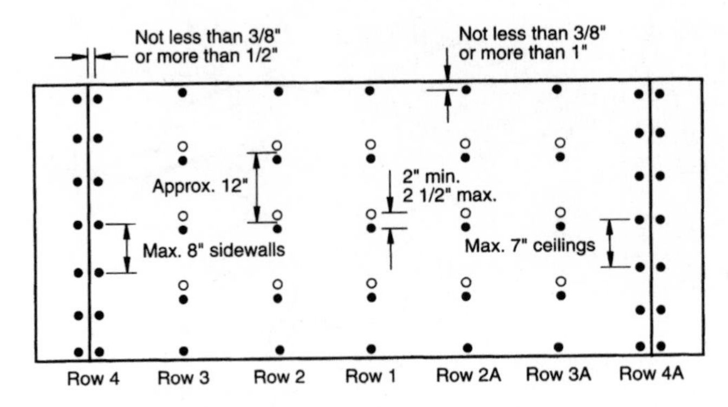

5-4 *Nailing instructions for fiber-reinforced gypsum wallboard.* Courtesy Louisiana-Pacific

painted, there's really no need for expensive clear or Number 1 grade solid wood.

Hardboard trim such as Georgia-Pacific's PrimeTrim is the perfect solution. PrimeTrim is a solid hardboard product with a pressed wood-grain face, resembling the cross-sawn rough cedar surface that's popular on many engineered wood sidings. A smooth version of the board is also available. The board is then factory-primed on the face and both edges, ready for installation and final painting.

Materials such as PrimeTrim have a number of uses, both interior and exterior. In addition to corner boards and window and door trim, it's great for use as facia, soffits, rakes, and band boards, or inside as baseboards or door and window trim. It cuts, drills, sands, and fastens just like solid lumber, but without the splitting and warping. Its dimensional stability is excellent, eliminating the common problem of cracking caulk joints that occur when wet lumber trim boards dry out and shrink. According to the manufacturer's test data (based on tests conducted at Mississippi State University), PrimeTrim tested slightly better in decay resistance than both redwood and cedar—two common trim materials known for weather resistance—and was almost three times as decay-resistant as western pine.

Common sizes for hardboard trim are the same as for solid sawn lumber, ranging in width from nominal 4 × to nominal 12 × (the actual sizes are 3½ to 11¼ inches, the same as solid lumber). The standard length is 16 feet, and two thicknesses are available: a nominal 4/4-inch material, which actually measures ⅝ inch, and a nominal 5/4-inch material that measures a full 1 inch. All are factory primed with an

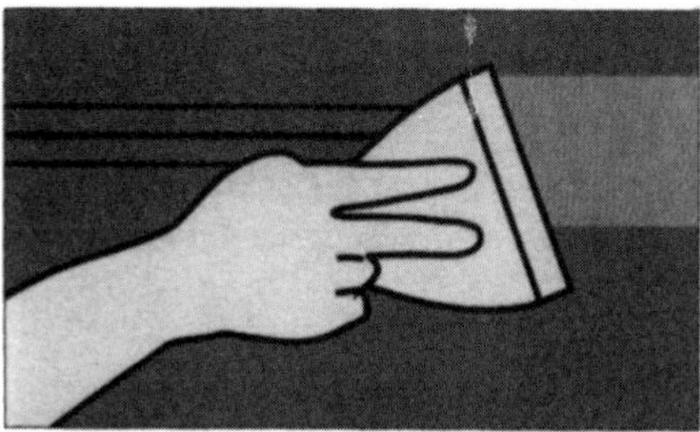

Apply taping compound into recess formed by tapered edges.

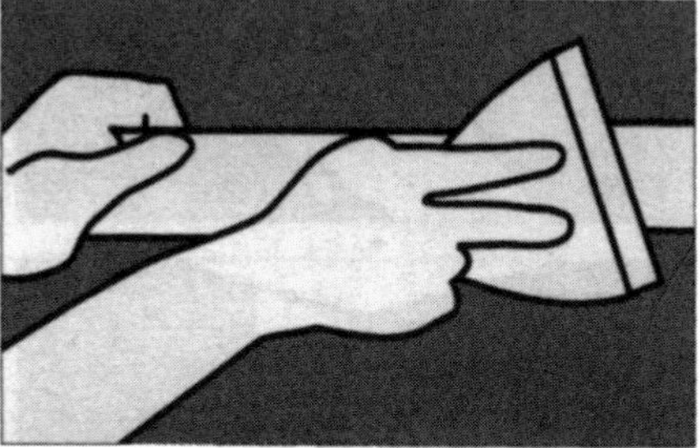

Joint reinforcement tape is embedded in taping compound.

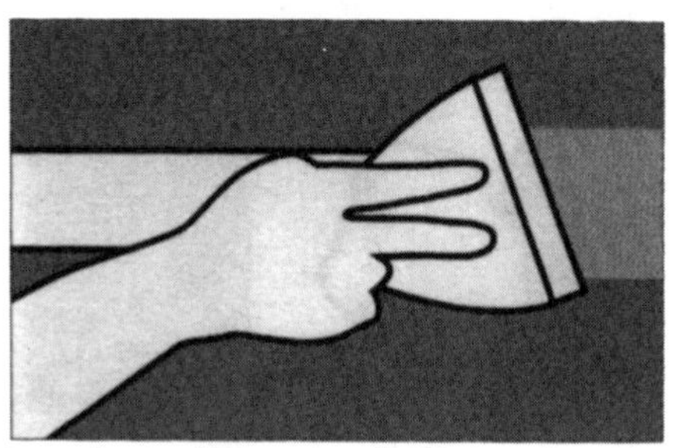

Immediately after tape is embedded, apply a skim coat of taping joint compound.

Finishing coats of ready-mix joint compound are applied.

5-5 *Finishing instructions for fiber-reinforced gypsum wallboard.* Courtesy Louisiana-Pacific

oven-baked primer material, and are manufactured without urea formaldehyde resins.

Another excellent use for hardboard is in the manufacture of doors (Fig. 5-6). Hardboard doors represent one of the fastest growing segments of the engineered wood industry, and have proven to be an exceptional choice for interior doors in homes of all price

5-6 *A typical six-panel painted hardboard door.*

ranges. Hardboard doors are dimensionally stable; are moisture-, insect-, and warp-resistant; and offer the denser feel and some of the noise-blocking performance of solid-core doors at the price of a flimsier hollow-core door.

Hardboard doors are typically constructed with an interior frame of dry, finger-jointed lumber stock, over which the face layers of ⅛-inch molded hardboard are attached. The hardboard is pressed into any of a variety of door configurations, including two-panel, three-panel, four-panel, and six-panel styles. A lot of attention is given to the details of the molds and the pressing process, and the finished doors have all the joints and miters that would be found in a traditionally constructed, solid wood-panel door. Both light wood grain and smooth surface textures are available.

Hardboard doors are available in sizes ranging from 1 foot to 3 feet for swinging doors or pocket doors, and in various combinations up to 8 feet when configured as bifolds. They come with a factory-applied primer and paint out very nicely with either latex or oil-based paints. According to the manufacturer of at least one hardboard door—Masonite Corporation—these doors can also be stained to simulate the look of real wood. The process involves applying a solid color pigment or wiping stain, which is allowed to dry almost completely and then partially wiped off using a rag that's been wet with some additional stain. A video and written instructions on how to accomplish this are available from the manufacturer.

Insulated structural building panels

One of the interesting building products aimed at reducing energy costs is the insulated structural building panel. It consists of a thick layer of EPS (expanded polystyrene) insulation that is sandwiched between structural panels of OSB, creating fully insulated, ready-to-install wall sections. The panels range in size from 4 × 8 feet up to 8 × 24 feet, and in thicknesses from 4½ inches to 12¼ inches, yielding R values in the R-15 to R-48 range.

The panels are made at the factory complete with window and door openings and a chase for the electrical wiring. The panels are delivered to the job site by truck, and erected with a crane. Each panel interlocks with the next one by way of a wooden spline, quickly forming the complete exterior shell of the house. Depending on the manufacturer and the details of both the face panels and the insulation core, insulated structural building panels are suitable for roof and floor applications as well.

Medium-density fiberboard (MDF) molding

In these days of high wood costs and an increased difficulty in finding premium-quality finish materials, moldings manufactured from medium-density fiberboard (MDF) offer a great alternative (Fig. 5-7). MDF is a formed and pressed wood fiber product with excellent dimensional sta-

5-7 *Factory-primed medium-density fiberboard (MDF) Colonial-style molding (left) compared to the same style molding in solid hemlock.*

bility, and it can be formed into a variety of surface configurations that are clean and crisp. It is manufactured from approximately 90 percent preconsumer wood waste, primarily from waste generated by the manufacture of other wood products, so it's another of the new generation of environmentally friendly products that uses no new wood stock. As with other forms of manufactured wood, there is no grain pattern to this material (it looks a little like a piece of tan paper). Most types come preprimed from the manufacturer, and all are intended for painting.

MDF moldings are available in most of the standard molding patterns (Fig. 5-8), including casing, base, chair rail, and crown moldings, as well as solid, flat, lumber stock. You can cut, rout, drill, and sand it with regular woodworking tools (Fig. 5-9), though carbide-tipped

Profile	SKU	Size	Description	
CASING	114325	1/2 × 1 5/8	BEVEL CASING	#108
	114326	1/2 × 1 5/8	BEVEL CASING	#108
	113919	1/2 × 1 5/8	BEVEL CASING	#108
	112844	9/16 × 2 1/4	COLONIAL CAS	#356
	114327	9/16 × 2 1/4	COLONIAL CAS	#356
	111707	9/16 × 2 1/4	COLONIAL CAS	#356
	116588	1/2 × 2 1/8	COL CAS LP356	
	116589	1/2 × 2 1/8	COL CAS LP356	
	116590	1/2 × 2 1/8	COL CAS LP356	
	114328	9/16 × 2 1/4	BEVEL CASING	#324
	114329	9/16 × 2 1/4	BEVEL CASING	#324
	111708	9/16 × 2 1/4	BEVEL CASING	#324
	114330	9/16 × 2 1/4	MOLDING	#348
	114331	9/16 × 2 1/4	MOLDING	#348
	113404	9/16 × 2 1/4	MOLDING	#348
	114332	9/16 × 2 1/4	DBL BULLNOSE	#474
	114333	9/16 × 2 1/4	DBL BULLNOSE	#474
	113920	9/16 × 2 1/4	DBL BULLNOSE	#474
	114334	5/8 × 3 1/8	COLONIAL CAS	#444
	114335	5/8 × 3 1/8	COLONIAL CAS	#444
	113570	5/8 × 3 1/8	COLONIAL CAS	#444
	114345	9/16 × 3 1/8	FLUTED CAS	#340
	114346	9/16 × 3 1/8	FLUTED CAS	#340
	113571	9/16 × 3 1/8	FLUTED CAS	#340
BASE	114336	9/16 × 2 1/4	MOLDED	#625
	114337	9/16 × 2 1/4	MOLDED	#625
	113073	9/16 × 2 1/4	MOLDED	#625
	114339	9/16 × 3 1/4	MOLDED	#623
	114340	9/16 × 3 1/4	MOLDED	#623
	114338	9/16 × 3 1/4	MOLDED	#623
	115163	1/2 × 4 1/8	MOLDED	#310
	115164	1/2 × 4 1/8	MOLDED	#310
	115165	1/2 × 4 1/8	MOLDED	#310
CROWN	114341	5/8 × 3 1/4	CROWN	#51
	114342	5/8 × 3 1/4	CROWN	#51
	113405	5/8 × 3 1/4	CROWN	#51
CHAIR RAIL	114343	5/8 × 3	CHAIR RAIL	#297
	114344	5/8 × 3	CHAIR RAIL	#297
	113499	5/8 × 3	CHAIR RAIL	#297

5-8 *An example of some of the molding patterns and sizes available in MDF.*

5-9 *Detail of an engineered wood window sill. The sill itself was milled from a sheet of ¾-inch MDF, and the trim is standard Colonial MDF.*

blades work best, and install it using screws or hand- or air-driven fasteners. Small ring shank nails are recommended by most manufacturers, and smaller-diameter nails set into the wood better without raising the surrounding surface. Coated chisel-point staples work well also, with an air pressure of around 100 to 110 psi being necessary to penetrate the denser surface material of the molding and properly set the staple's crown below the surface.

For screw installations, use a long, slender screw with a relatively aggressive thread pattern, such as a drywall screw. Drill pilot holes first, and then countersink the face of the board to accept the screw head; trying to "muscle" the screw all the way in without first drilling and countersinking the board will result in a stripped screw or a raised ring of material around the screw head.

MDF is relatively porous in its cut section (below the prime coat), so it glues up well. You can use epoxy resins and adhesives, hot melt adhesive, PVA adhesive, or standard yellow carpenter's glue. This material also paints quite well, having a smooth primed surface and no grain to worry about. Fill voids and fastener holes with spackling compound or a good-quality latex caulking material. Lightly sand the surface with 150-grit sandpaper, then paint with either latex or oil-based paints, applied by brush, roller, or spray.

Polymer composite trim

Have you ever spent hours kerfing the back of a piece of molding to get it to curve and follow a round wall? Or soaked and clamped and bent a piece of trim to get it to conform to a half-round window? If that's not your idea of a fun way to trim out the difficult areas in a custom house, you might want to try one of the polymer composite trims now on the market.

A polymer composite is a mixture of various polymer resins and additives. It's actually a rather complicated chemical process, but the result is a material that can be formed and solidified into virtually any shape. By altering the chemical engineering that goes into the mixture, the resulting composite can be made in such a way that it remains flexible. The final result is a tough, flexible, highly elastic composite material that can be formed into dozens of standard molding patterns.

What this means to you is that you can form a piece of molding in your hands to fit any of those tough areas you have to contend with: curved walls, curved ceilings, arched windows and doors, spiral and circular stairs, curved cabinets and furniture, wall niches—you name it.

You work polymer composite moldings just like wood: cut them to length using a conventional hand, circular, or miter saw, or rip them to width using a table saw. If you're making a lot of cuts, you might need to spray the blade occasionally with silicone lubricant to keep the material from sticking and possibly gumming up the blades. Install using construction adhesive and nails, or specially formulated fast-set adhesives that are compatible with the polymer materials. You'll need to consult with the manufacturer of the product you're using to get specific recommendations for gluing and fastening.

Flexible moldings are usually available through door and window companies, or check with a large lumberyard or home center. You'll find virtually any type of molding you're looking for, including base moldings, base caps, crown moldings, chair rails, door and window casings, cove moldings, and wainscot caps, as well as a wide selection of flexible flat stock. Sizes range from as small as 5⁄16 inch to up over 8 inches, and in various lengths.

Since the polymer material begins life as a liquid, it will follow the finest contours of whatever form it is placed in. As a result, you can find moldings with things like embossed leaves, egg-and-dart, dentils, or any of a wide variety of other intricate architectural patterns. The material is also made to duplicate a variety of different wood grains, including pine, fir, and oak, or simply smooth.

You can paint polymer composite moldings, or stain them for surprisingly realistic results. If you have natural oak doors, for example, you can purchase an oak-grained flexible polymer molding and stain it with conventional stain materials to create a virtually exact match for the door. Here again, refer to the manufacturer's specific instructions for what stains and finishes to use and how to apply them to achieve specific results.

Polystyrene building elements

This is a very new and quite unique product, and *polystyrene building elements* is about the closest generic name you'll see. The brand name is Ener-Grid, and to my knowledge it is currently the only such product on the market.

Ener-Grid is composed of recycled polystyrene, cement, and other additives, which are blended together and then formed into 10-foot-long building blocks, called *elements* by the manufacturer. Each element has both horizontal and vertical cavities on 15-inch centers, and has the appearance of several standard concrete blocks that have been glued together. The elements are stacked on top of each other, and sealed with expandable foam sealant. Rebar is inserted into the cavities, and the cavities are then filled with poured concrete, similar to standard concrete block construction. The exterior is typically covered with a finish coat of stucco, while the inside can be finished with plaster, drywall, or other finish wall coverings.

Ener-Grid is extremely porous, which makes it lightweight (21 pounds per cubic foot, as opposed to 100 to 150 pounds per cubic foot for standard concrete block) and also easy to cut with a hand saw or chainsaw. The porous composition is the result of tiny air pockets in the solidified concrete/polystyrene matrix, which also makes this a very energy-efficient material. The manufacturer claims a tested R value of 36.44 for the Ener-Grid element when filled with concrete (as opposed to R-19 to R-23 for a standard framed and insulated wall section), and the addition of finish coats of plaster and stucco pushes that to around R-40.

Another Ener-Grid advantage is its fire-resistance. The material currently has a UL-tested fire rating of two hours, and the manufacturer claims that further testing currently underway will soon yield a four-hour rating. It has a sound insulation rating of 53 decibels.

Ener-Grid comes in a variety of standard sizes, from 8 × 15 inches to 12 × 30 inches. All elements are 10 feet long.

Vinyl fencing

Having seen first-hand the effects of climatic conditions on a fence constructed of wet, poor-quality solid lumber, I think this is truly the fencing product of the future. Vinyl fencing products are made from solid, first-quality PVC or other vinyl resins, and are totally immune to rot, moisture, insects, heat, and just about anything mother nature can dish out. Early vinyl products were susceptible to damage from ultraviolet light, but new generations of vinyl additives and manufacturing techniques seem to have that problem in hand as well.

You can find vinyl fencing in just about any style you'll find regular wood fencing in. This includes one-, two-, three-, and four-rail fencing (Fig. 5-10); picket fences; slats; 3-, 4-, 5-, and 6-foot fence boards; and lattice tops. They are used with vinyl posts (Fig. 5-11) or with standard wood or metal posts. Installation is simplified by the vinyl material's light weight and uniform size, and only standard woodworking tools are needed to work with the stuff.

You'll find vinyl fencing that's suitable for even the most demanding installations. Many ranches have switched to vinyl because of its

5-10 *A three-rail fence of sturdy, low-maintenance vinyl, an increasingly popular alternative to wood.*

5-11 *Vinyl rails, which are the same size as standard 2 × 6 lumber, interlock with 4 × 4 hollow vinyl posts.*

superior weather resistance and virtual lack of maintenance, and they report that it is durable and impact-resistant, even when used for cattle fencing, horse enclosures, or in other tough livestock applications.

Standard colors are white, almond, and gray, in a wide variety of sizes and configurations. Consult with your fencing supplier for complete catalogs of these products, since new ones are becoming available all the time.

Vinyl moldings and architectural products

It the world of architecture, as in so many areas of our lives, fashions and trends come into vogue and then go out again. True to that inevitable cycling, the 1980s and 1990s have seen a resurgence of interest in "traditional" forms of architecture, with products like lap siding and gridded windows coming back into style. Colonial styling, Greek revival, and steep-pitched Victorian housing styles are popping up all over, pushing aside many of the lower roofed, more modestly adorned styles common to the sixties and seventies.

With the revived interest in these styles has come a fresh demand for more detailed moldings and trim pieces, both inside and outside the home. But once again, the high cost of top-quality, molding-grade, natural wood has encouraged the search for alternatives. One of these is products manufactured from PVC (polyvinyl chloride) resin or other types of vinyl resin.

Vinyl moldings are smooth and solid and, since they begin as a liquid, they can be poured into virtually any type of mold to produce edges and details that are crisp and clean. Vinyl is waterproof, so these are great moldings for use in exterior applications such as patios, railings, and cornice trim. Most are manufactured in white, with some other colors also available, depending on the manufacturer and the type of product. Most can also be painted.

With a little research, you can find virtually anything you need to decorate the inside or outside of any style of residential or commercial building. Vinyl moldings have become a major part of the molding and trim market, especially for exterior use, but a lot of builders are not really familiar with these products. For that reason, I've provided a list of several of the commonly available types of moldings and other available products, in alphabetical order. For details on specific moldings, talk to the people at your local lumberyard or home center, or check with the sources listed in Appendix E at the back of the book.

Arches and keystones. Arches are available in every shape and size, for use over and around doors, windows, openings, niches, or just about any other indoor or outdoor application you can think of, in any of a variety of sizes for single or multiple openings. There are many styles of face trim to choose from as well, including flat, beveled, and several simple and ornately designed molded profiles. Matching or contrasting keystones—the traditional wedge-shaped block at the peak of the arch—are also available in a variety of sizes and styles. Keystones are available separately for use with other arch or half-round applications; when used in conjunction with wood or vinyl trim, keystones are a perfect finishing touch at the top of a half-round window.

Balustrades. If you desire a classic look to your patio, deck, entrance path, or any other interior or exterior application that calls for a railing, vinyl balustrades are the answer. You'll find balusters in several sizes and styles, along with the matching top and bottom cap rails that finish off the installation. There are also riser fittings that are designed to be inserted into the top and bottom rails, allowing the flat-topped balusters to be used along stairways and ramps.

Columns. Columns are a very popular item in vinyl, given their strength and the smooth, clean, easy-care surface that's ideal for exterior applications. Posts and columns come in a wide variety of sizes and styles: square or round, smooth or fluted. Some types are designed for structural support; others are ornamental covers designed to be placed over wood or steel support columns. Columns come in quite an extensive range of diameters, lengths, and configurations, so you'll easily find something to match the architecture of a Colonial plantation or a Greek revival mansion.

Column caps. There are dozens of column and pilaster caps, capitals, column bases, and other products designed to finish off the installation at the top and bottom of the column or pilaster. Once again, these range from simple to ornate.

Corbels. Corbels see a lot of use in architecture, for everything from supporting a shelf to displaying a piece of artwork to forming the structural supports for a massive assembly of exterior roof moldings. You'll find a selection of vinyl resin and fiberglass-reinforced vinyl corbels that encompasses everything from simple scrolls to cherubs and angels.

Cornices. Cornice moldings are available in a variety of sizes, styles, thicknesses, and molding patterns.

Crossheads. A crosshead is an assembly of moldings that sit above a door or window, sometimes alone but most commonly supported by casings or pilasters. In traditional wood trim it is necessary to build a crosshead up from several pieces, but in vinyl you can find a half-dozen or so different configurations in solid, one-piece units.

The size range will accommodate openings from about 2½ feet to around 7 or 8 feet.

Fixture mounts. With the growing use of lap siding, fixture mounts are an important accessory for installing lights, electrical boxes, plumbing, and other items that have to penetrate through the outside wall. Vinyl fixture mounts attach over the siding and match the laps, allowing you to place them wherever it's most convenient after the siding has been installed. There are also flat fixture mounts that are installed before the siding goes on.

Gingerbread. Gingerbread is the term given to all the intricate moldings that so often decorate Victorian-style houses. There is scrollwork for the gable-end barge or rake rafters, different types of brackets, triangular gable moldings and trim, gable posts, scallops, and several other parts.

Lattice. Vinyl lattice panels are relatively new on the market. They look identical to wood lattice, even down to the wood grain, and are manufactured in 4 × 8-foot sheets. The big difference is that the sheets are one solid piece of vinyl, rather than individual slats nailed or stapled together. Vinyl lattice panels won't separate, crack, rot, or be affected by insects or moisture. Standard colors are white, gray, and brown.

Louvers. There are a lot of louver assemblies available for gable-end or dormer use, all of which provide attic ventilation while adding a nice decorative touch to the house. There are dozens of sizes and styles available, including round, half-round, quarter-round, arch-top, octagonal, oval, and triangular, as well as square and rectangular.

Lumber. At least one company offers 100 percent PVC lumber for the construction of decks (Fig. 5-12), deck railings, and docks (Fig. 5-13). Vinyl decking and docks offer the obvious advantages of being weather-proof, color-fast, easy to clean, and virtually maintenance-free, as well as not being prone to splintering and cracking like wood.

Mantels. Mantels are another very popular item, and you'll find both standard straight mantels as well as complete fireplace surround assemblies, from very simple to extremely ornate.

Moldings. There are, of course, lots of moldings, including all the classic types such as crown, chair rail, cove, shoe, and brickmold. Sizes range from as small as 1½ inches to over 16 inches, and in lengths ranging from 7 feet (great for doors) to 12 feet or more. Flat moldings and flat trim in sizes from ¾ inch to 6 inches or more are also available.

5-12 *Pure PVC vinyl is used in the construction of this beautiful, low-maintenance deck and rail system.* Courtesy Thermal Industries, Inc.

5-13 *Vinyl is also used in the construction of this dock.* Courtesy Thermal Industries, Inc.

Outdoor accessories. These are a collection of fun and interesting accessories that allow you to add the perfect finishing touches to any type or style of house you're building. There are urns, statues, fountains, and just about anything else you can image to turn an ordinary house into a country villa.

Pediments. Pediments are decorative trim pieces that sit over the top of a door or window, most commonly in triangular or half-round styles. There are many variations of this, from plain to ornate, and in sizes that fit doors or door combinations up to 8 or 9 feet. Pediments can be used alone, or can be placed on pilasters for an even wider, grander look.

Pilasters. A pilaster is a half-column designed to sit vertically against a wall to support a load. Decorative pilasters such as those described here commonly support a pediment or other decorative finish molding over the top of an opening. As with the columns, you'll find vinyl pilasters in smooth and fluted styles, and in half-round or half-square cross-sections.

Plinth and corner blocks. You can use these decorative blocks at the bottom of side-leg door casings, at the top of the casing, or pretty much wherever you have a meeting of two molding pieces. They add depth and decoration to any type of window or door trim, and also eliminate having to miter the corners where the top and side casings come together.

Shutters. Shutters are another building product that is found in large numbers and a wide variety in vinyl. Both plain and simulated louvered styles are available, in sizes to fit most windows and colors to fit most houses.

Vinyl siding

Long derided on the construction site as something of a "cheap" material, vinyl siding shouldn't be overlooked in this age of higher costs and lower-quality wood siding materials. Manufactured from PVC resins, vinyl siding is virtually immune to any of the problems normally associated with wood siding, including warping, splitting, rot, termite damage, and moisture problems. It is also the closest thing you'll find to a maintenance-free siding; the colors are part of the vinyl itself, and a soft scrub brush and mild detergent will replace your paint brush and roller.

Sizes and styles

Because of their interlocking construction, all vinyl siding products currently on the market are in board form only; there are no panel siding versions being made at this time. There are lots of other choices, though, in both horizontal (the most common type) as well as vertical styles. Sizes include: 3, 4, 4½, 5, 6, 6½, and 8-inch horizontal, as well as 4- and 5-inch vertical, almost all of which are manufactured in 12-foot or 12-foot, 6-inch lengths.

Accessories

As might be expected, there are also a wide variety of accessories available to complete the installation of vinyl siding. These include soffit panels, inside and outside trim, door and window trim, vents, fixture blocks, and various flashings. Before beginning any vinyl siding installation, be sure to consult with the specific manufacturer's catalog to select trim and accessory pieces that are compatible with the siding you've selected.

Colors

Permanent color is and always has been one of the main selling points for vinyl siding, and this is now truer than ever. There is a very wide selection of contemporary and traditional colors now available that are subtler than the vinyl colors of the past, including some that are blended to appear slightly weathered and others that have a more natural wood look. All trim pieces can be purchased in colors that match or contrast with the colors of the vinyl siding.

Vinyl windows

No mention of vinyl products would be complete without including vinyl windows, which have taken over a huge share of the new and replacement window market (Fig. 5-14). Vinyl windows are in many ways the ideal window product. They are weather- and insect-proof, they have thermal performance characteristics that equal wood, they have the same low-maintenance features of aluminum, and they are economical enough for starter houses while still more than attractive enough for use in upper-end homes of any style and price category.

Vinyl windows are manufactured in all traditional window configurations (Fig. 5-15), including single hung (opening up from the bottom), double hung (opening both up from the bottom and down from the top), horizontal sliding, fixed, casement (the sash opens out

5-14 *A beautiful, traditional-style home using gridded vinyl windows.*

5-15 *A half-round shape with a sliding window below is just one of dozens of window styles and combinations available from vinyl window manufacturers.*

vertically), and awning (the sash opens out horizontally). They are also available as round, half-round, quarter-round, octagonal, triangular, and other geometric shapes.

You can also order vinyl windows in combination styles, allowing you to combine two or more styles together into a single unit. For example, you could order a sliding window with a half-round fixed window on top, a single-hung window with a triangular fixed window on top, or a fixed window with an awning below and a quarter-round window above—the combinations are virtually unlimited. The factory will assemble the windows with a matching mullion bar, so even though it's actually more than one window, it will appear to be a single unit.

Another very popular option with vinyl windows is grids, which simulate the look of traditional individual pane windows. The grids are simply matching vinyl or aluminum bars placed between the two panes of glass during the manufacture of the window. They are not removable, and since they are the same color as the window frame, they are also totally maintenance free. The window glass is one piece on either side of the grid.

With the popularity of grids has also come a greater amount of choice. When first introduced, grids were available only in straight and diamond patterns, the two most common configurations. Recently, however, several manufacturers have begun offering other configurations. The craftsman look of the 1940s, with single grids on the top, bottom, and each side, is a popular look, and there are also variations on the diamond pattern, combining straight grids with smaller diamonds. You are also no longer limited to grid colors that match the color of the window. Brass is being used quite a bit, as well as grids in other colors.

Most vinyl windows are sold with double insulated glass, two panes of glass with an air space in between for increased thermal performance. In some parts of the country, double insulated glass still does not perform well enough to meet the new energy standards set forth in some of the building codes. To meet those higher energy codes, you can order your vinyl windows with Low-E glass (a low-emissivity coating that reduces heat loss) [Fig. 5-16]; argon-filled glass (argon is a safe, inert gas that is pumped into the air space in place of standard air, again to lower heat loss); or a combination of both. Some companies offer triple glazing for even better thermal performance; on the other side of the coin, you can also order single-glazed windows for use in interior areas, shops and barns, and other areas where heat loss is not an important factor.

5-16 *A stamp on the metal divider between the panes of glass in an insulated vinyl window indicates that the glass has a Low-E coating for improved thermal performance. Note the grid attachment to the right of the Low-E stamp.*

As with any vinyl product, if your house calls for natural wood windows, this is not the product for you. Vinyl is vinyl, and it's available only in a limited number of colors, usually white, almond, and sometimes gray. A couple of manufacturers produced a dark brown window a few years back that was intended to compete with bronze aluminum, then the industry leader in windows, but the dark color faded in intense sunlight. There are rumors of new colors coming on the market, and it's very likely that, given the popularity of these excellent windows, it won't be long before your color choice will increase dramatically.

Wood-polymer lumber

Taking the concept of engineered lumber, recycling, and environmental concerns a step further, there is now a building product on the market that is manufactured almost completely from waste materials. Called Trex by its manufacturer, Mobil Chemical Company, it is a composition product made from 50 percent thermoplastics and 50 percent waste wood fiber, with a very small amount (0.4 percent) of iron oxide added as a pigment. The plastics are obtained primarily from recycled grocery bags and stretch film, and the wood fiber comes mostly from used pallets and sawdust from the furniture-making industry. No new wood is used in the manufacturing process, and the finished product is itself fully recyclable.

The plastic/wood mixture is extruded in the manufacturing process into solid pieces of lumber that are consistent with the standard nominal dimensions of solid sawn lumber (see later in this section). The surface of the material is smooth with slightly rounded edges and square-cut ends. Two colors are available: natural, which weathers to a driftwood gray color in about 6 to 12 weeks after exposure, or brown, which the manufacturer claims is colorfast.

Being a manufactured product, the moisture content is very low and very consistent, making for excellent dimensional stability. The manufacturer offers a limited warranty against many common lumber defects, including splintering, splitting, checking, rot, decay, and even termites.

Trex is not a substitute for structural lumber, and the manufacturer makes that very clear up front. It is not intended to replace wall framing, beams, or posts, nor is it to be used in other load-bearing applications. There are, however, any number of other uses for which wood-polymer lumber is ideal. It's perfect for decks and deck railings, since it's fully resistant to rot, weathering, and splintering. Other uses include walkways, pool surrounds, benches, outdoor furniture, docks, planters, fencing, and landscaping applications.

Trex is most commonly used in combination with structural supports of solid sawn lumber or engineered lumber. A fence, for example, might be constructed using metal or pressure-treated 4 × 4 posts, and then the Trex is attached to the posts. Deck rails are another good example, where you might use natural wood posts with wood-polymer lumber rails and pickets.

Trex is an excellent product for use in decks. It is smooth, yet maintains good traction even when wet. It requires no maintenance, and will not split or crack in later years and become a hazard to bare feet. Being a nonstructural product, when used for decking the manufacturer has specific guidelines for structural support spacing. For example, 2× Trex can be installed over joists spaced up to 20 inches center-to-center for residential decks or light-duty docks, or over 16-inch spacing for commercial decks, boardwalks, and marinas.

That spacing is for decking that is run perpendicular to joist supports, and needs to be modified if you are running the deck boards at an angle. For example, if you are running boards at a 45° angle to the joists, you would need to modify the center-to-center spacing by a factor of 0.7. For example, if you normally use joist spacing of 16 inches on center for perpendicular decking applications, a deck running at 45° would require joist spacing of 11.2 inches center to center (16 × 0.7 inches). 2× decking can also overhang deck supports by up to 6 inches.

Trex weighs more than the equivalent size of natural wood, and is also more flexible. Exercise care in both lifting and carrying the material, as well as stacking it prior to installation. As to the workability of the product, the manufacturer claims that it:

"Trex can be sawed, routed, sanded, nailed, drilled, and turned on a lathe." Carbide blades are recommended, as well as a relatively slow cutting speed. Use saw blades with 18 to 20 teeth to keep both the blade and the material cooler. When drilling large or deep holes, periodically remove the bit and clear drilling debris from the hole, and do not reverse the drill when backing out of the hole. When routing, use carbide-tipped router bits. Sanding and planing operations should be the same as with natural wood.

"Holds fasteners tighter than wood." Mobil recommends galvanized or stainless steel nails or staples. Ring-shank or spiral nails are suggested in place of common nails. Nails can be installed by hand, and the manufacturer states that no more force is required for nailing than with most common woods. You can also use an air-driven nail gun with a pressure of at least 110 psi. You can use screws or lag bolts; predrill the material with a hole approximately three-fourths the diameter of the screw or bolt you're using. For adhesive applications, use PVC cement or two-part epoxy for exterior applications. Do not use any adhesive that lists chemicals harmful to polyethylene.

"Readily accepts paint and stain." Two different colors, natural and brown, are available from the manufacturer, and no other finishing is necessary for weather resistance. However, you can paint Trex as you would natural wood. No special preparation or priming is required, but the material should be clean and dry. Latex paint is preferred. You can also apply wood stains and even polyurethane, but for the best color penetration and uniformity, it is recommended that you wait until after the natural fading process is complete.

A number of standard lumber sizes are available, including 1 × 6, 5/4 × 6, 2 × 4, 2 × 6, 2 × 8, 2 × 10, 3 × 6, 4 × 4, 4 × 6, 6 × 6, and 6 × 8 (as with solid lumber, all sizes are nominal; the actual sizes are the same as for solid sawn lumber). Trex also comes in 2½, 3½, 4, and 6-inch rounds (actual size), in lengths of 8, 12, and 16 feet. Other sizes should become available as the product gains acceptance. As of this writing, the prices for wood-polymer lumber are a little higher than standard grades of lumber (Number 1 and Number 2 grades of fir or cedar, for example), but lower than clear grades of the same material. Again, as with engineered wood, the prices for the manufactured lumber should stay relatively consistent against the more noticeable fluctuations in sawn lumber prices, and will probably come down some as the products become more accepted.

6

The engineered lumber manufacturing process

I could say a lot about the need for engineered lumber products, about their strength, their performance, and their stability advantages over conventional solid sawn lumber products. I could argue the environmental issues, and discuss whether or not we need these products. But when it comes to the debate over whether or not engineered lumber will be a part of the construction scene in the foreseeable future, you need only look at the amount of time, effort, and, most of all, money that has been poured into the development of the technology and the manufacturing plants to know that this is no flash in the pan.

Companies such as Trus Joist MacMillan have worked for decades on the research and development for these products, having invested countless millions of dollars in the industry. Their faith in this technology is paying off in terms of consumer acceptance and steady—at times phenomenal—growth. Taking a tour through one of their plants, such as the one I took in Eugene, Oregon, speaks volumes about supply and demand.

When Trus Joist first converted an old plywood plant in Eugene over to the manufacture of LVL and I-joists, the plant was putting out about 75 feet of I-joist per minute off their line, with a total output of around 2500 linear feet per week. In just a few short years, that production—as of May, 1996—was up to 350 feet per minute, with a total weekly production of 1.9 million linear feet. This output is from just one of the company's 15 plants currently operating in the United States and Canada to produce LVL, PSL, LSL, and I-joists, and this is just from one company out of the approximately half dozen now active in the engineered lumber industry.

Statistics are always interesting, but the real point is that manufacturing on this scale would not be occurring for an experimental product with an uncertain future. Take one look at the 22 rail cars and 120 semi truck loads of material leaving just one factory in the course of just one week, and you'll know for certain that the engineered lumber products you see in the houses of today are a tiny percentage of what you'll be seeing in the years to come.

Laminated veneer lumber (LVL)

Laminated veneer lumber (Fig. 6-1), abbreviated LVL, begins its life in the same way that plywood does, as veneers of wood. The $\frac{1}{10}$- to $\frac{3}{16}$-inch-thick veneers (Fig. 6-2) are peeled off solid logs approximately 9 feet long and of various diameters (Fig. 6-3). Several different species of wood are used, mostly Douglas fir, southern pine, and yellow poplar. The veneers are then clipped to widths of 27 and 54 inches, banded into units, and shipped to the LVL plant. Some manufacturers own veneer plants and produce their own veneers, others purchase veneers from outside suppliers, and some do both.

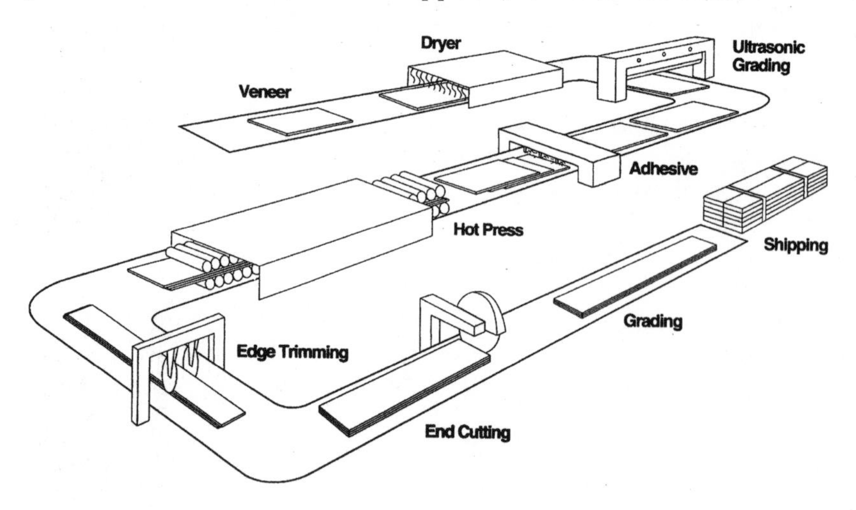

6-1 *The typical manufacturing sequence for laminated veneer lumber (LVL).* Courtesy Trus Joist MacMillan

6-2 *Wood veneer similar to the material used in making plywood is used in the manufacture of LVL.*

6-3 *Cut logs stacked and awaiting peeling at a veneer plant.*

Once the wet, unsorted veneer sheets arrive at the plant, they are stored outside, still bundled, awaiting processing (Fig. 6-4). Forklifts move the wet veneer sheets into the plant, loading them onto conveyor belts that move the sheets into massive, gas-fired, multilevel drying ovens (Fig. 6-5). The veneer passes slowly through the dryers, either

6-4 *Stacks of wet veneer, bundled and delivered from the veneer plant, await processing into LVL members.*

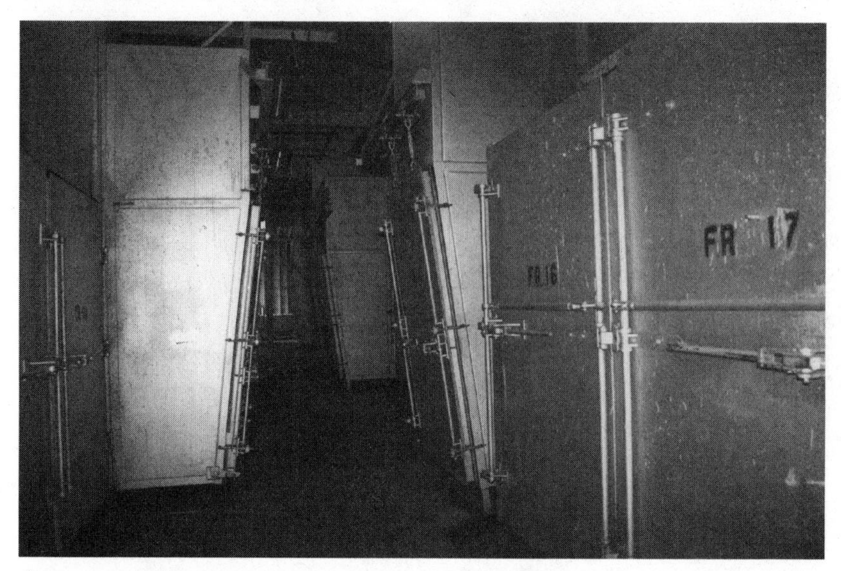

6-5 *Rows of high-capacity gas-fired ovens are used to dry the veneer to a uniform moisture content.*

on a straight belt or as part of a multiloop system, emerging from the far end of the ovens dried to a uniform moisture content (Fig. 6-6).

As the sheets of veneer exit the dryers (Fig. 6-7), the moisture content of each one is checked electronically by a series of automatic

6-6 *Dry veneer sheets exit the ovens once the drying sequence is complete.*

6-7 *As the dry sheets of veneer leave the oven, a conveyor belt moves them to the ultrasonic grading equipment (not shown).*

sensors. If the system detects a sheet that remains too wet, it triggers a small paint head that automatically marks the panel with a shot of paint. That panel is removed by an operator further down the line and set aside, then later inserted back into the wet side of the line of veneers to be processed through the dryers to be dried further.

Next, the dry veneer sheets are graded for strength, with a rating of Grade 1 (the highest) down to Grade 3. Some of the grading is done by having trained operators physically examine the wood, but most is done automatically. Trus Joist, for example, uses an ultrasonic testing device, computer-controlled for speed and accuracy. As a sheet of veneer moves down the line, it passes under a pair of wheels. One wheel transmits a sound wave through the veneer sheet, which is received by the opposite wheel. The computer instantly analyzes how long it takes that sound wave to pass through the veneer, on the principle that the wave will pass more quickly through a sheet with straight grain and minimal defects than it will through one with irregular grain or a lot of knots and other flaws. Based on that evaluation, the computer designates the veneer as Grade 1, 2, or 3, and a paint head automatically marks the sheets with the appropriate grade designation.

Continuing down the line, the veneers are manually sorted according to the computer-generated grade marks (Fig. 6-8), and any veneer sheets that were marked as still being too wet are removed

6-8 *Dry and graded veneers are hand-sorted according to color-coded marks, and stacked in labeled bins (background), awaiting processing.*

from the line at this point and routed back to the beginning of the line for further drying. Short veneers and others with various defects are also removed from the system at this point.

The veneers are next fed into the automatic adhesive spreader in a predetermined pattern. That pattern is governed by two things: the intended thickness of the finished LVL member, which is controlled by the number of plies in the composite, and by its intended final strength characteristics, which is controlled by the grades of veneers used and how they are stacked in relation to one another.

A piece of LVL intended for use in the construction of I-joists, for example, might be nine veneers thick and have Grade 1 veneers at the top and bottom, graduating to Grade 2 and Grade 3 veneers near the center, where the strength requirements are less important. It is the ability to manipulate the strength characteristics of products in this manner to meet specific needs that is really at the heart of the entire engineered lumber concept.

At the adhesive spreader, the veneer sheets are coated with a phenol formaldehyde resin adhesive (Fig. 6-9). The resin application process is automatic, coating both sides of each veneer except for the top and bottom layers, which are coated on one side only. The

6-9 *A cut veneer sheet exits the glue head, coated with resin on both sides. The gluing process is fully automated, and the machine coats one or both sides of the veneer panel according to the preprogrammed number of layers in the LVL.*

veneers are stacked on top of one another in the prearranged pattern of grades and number of veneer layers, with the ends of each veneer in each layer overlapping to ensure good adhesion—end-to-end butt joints of the veneer could weaken the layers and result in possible failure at that point.

The glued-up veneers are then fed into a 100-foot-long continuous press, which uses a combination of heat and gradually increasing pressure to bond the veneer layers to one another. As the resulting pressed and glued LVL product—called a billet—exits the press (Fig. 6-10), it passes between a pair of saws that trim it to width. Finished widths vary with the intended end use of the product.

6-10 *Completed LVL billets exit the press.*

As the fresh billet exits the hot press and passes between the width-trimming saws, other saws cut the billet to its finished length. Since the LVL manufacturing process is a continuous one, of glued veneer sheet upon glued veneer sheet, the resulting billet could in theory be literally endless. The billet is cut to uniform length, however, as determined by customer order, or simply to meet the constraints of storage and shipping—usually 66 feet long.

The last step in the process is final grading and testing, then wrapping. Pieces of LVL are selected at random off the line, usually at the beginning of each new shift of workers, and are tested for the proper strength characteristics. The finished billets are then wrapped and stacked for shipping, or routed to an I-joist assembly line for use in the manufacture of I-joists.

Parallel strand lumber (PSL)

Parallel strand lumber, or PSL (Fig. 6-11), was first introduced in the 1980s, and was the end result of over 20 years of research and development and the investment of over 100 million dollars. As with LVL, this product also uses wood veneer peeled from a variety of wood species, most commonly Douglas fir, hem-fir, southern pine, and yellow poplar. Small-diameter logs are commonly used in the process, which is ideal for second-growth Douglas fir.

In this process, the veneers are again peeled and shipped wet from the veneer plant to the manufacturing facility. There, they are dried to a uniform moisture content that ensures consistency of performance in the completed product. Again, the veneers are carefully checked for moisture as they exit the massive gas-fired dryers, and those showing an excess amount of residual moisture are routed back through the process for additional drying.

The dry veneer sheets move down the line and enter a clipper, which hydraulically clips the veneer into thin strands ranging from 4 to 8 feet in length. From the clipper, the strands move through additional machinery that detects and removes any foreign objects, and also separates out any short or defective strands.

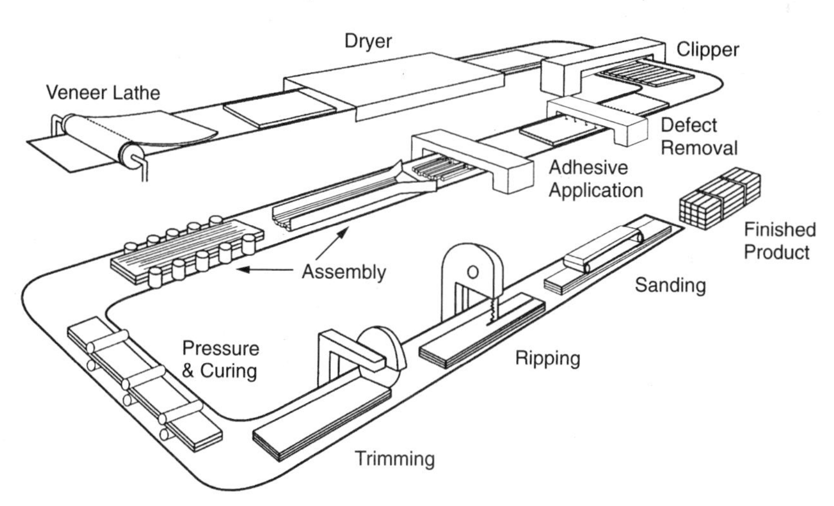

6-11 *The manufacturing process for Parallam parallel strand lumber (PSL).* Courtesy Trus Joist MacMillan

The strands then move through an adhesive applicator, which coats the veneer pieces with adhesive. For this process, the adhesive is a phenol formaldehyde resin compound, mixed with a small amount of wax. The wax promotes a better bond between the wood fibers, and helps the finished material resist moisture.

The long, adhesive-coated strands are aligned parallel with each other, which is done to take advantage of the natural strength of the wood fibers. The strands are formed up into loose billets, and then move into a specially designed microwave press. The resulting pressed billets are 11 × 17 inches, and up to 66 feet in length. There are approximately 3,500 strands of wood compressed into each 8 feet of finished PSL.

The microwave is at the heart of the PSL assembly process, since this is what allows the billets to be made so thick. Plywood, for example, cannot be pressed up in panels this thick because, with the plywood manufacturing process, heat is applied from the outside of the panel to cure the adhesive; the outer veneers of the plywood would burn before a sufficient amount of heat ever reached the inner core. With the PSL microwave technology, however, the energy created by the oven causes the billet to heat from the inside out, just like the microwave oven you use at home, which allows the adhesive to be heated and cured completely without damaging the wood. Of course, this is no ordinary microwave oven. According to the manufacturer, it generates enough microwave energy (400,000 watts) to power four FM radio stations.

After curing, the rough billets continue down the line for final sizing. They are trimmed to length, then ripped to their final width, depending on their intended use. From there they move through automated sanding machinery, which smooths the surfaces and slightly eases the edges. Then the finished product is wrapped and banded as needed for shipment.

Laminated strand lumber (LSL)

The newest of the engineered lumber products is laminated strand lumber, known as LSL (Fig. 6-12). This is the most sophisticated of the new products (Fig. 6-13), using a blend of high technology and very small, fast-growing wood species. LSL, which is the result of 10 years of some of the most intensive research and testing to date in the engineered lumber field, is considered by many to be the wave of the future.

LSL technology uses as much as 76 percent of the log, which gets maximum usage out of the timber source. Small-diameter logs are

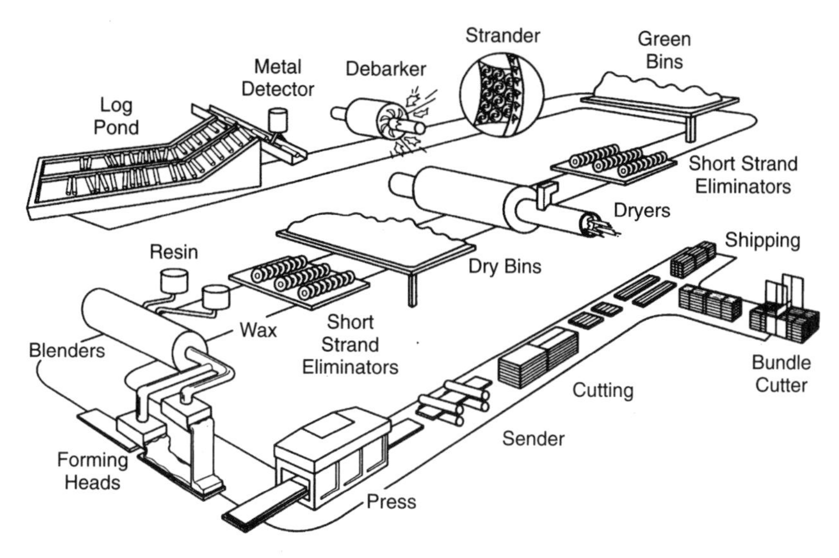

6-12 *The manufacturing process for TimberStrand laminated strand lumber (LSL).* Courtesy Trus Joist MacMillan

used here to their best advantage. The equipment will handle logs of from 4 to 22 inches in diameter, with most of the logs averaging only 9 to 12 inches. Small-diameter lathe-turning technology is also improving in the wake of these products, allowing the log to be turned down to a core as small as 2⅝ inches. LSL technology shows a lot of promise at the moment, and might well turn out to be the first of several generations of engineered lumber products designed specifically for use with tree-farm timber and even "waste" wood.

The process begins with small-diameter logs from fast-growing species such as aspen (Fig. 6-14) and yellow poplar, which are unsuitable for milling into solid sawn lumber for construction use. The logs are typically stored wet in a log pond, as is common with conventional saw-milling operations, and then move through a metal detector and other scanning machinery to remove foreign objects that might damage the manufacturing machinery or injure workers on the line. The logs are sawn to a uniform length of 100 inches, then debarked and cleaned using a water and steam process that softens the wood and readies the log for stranding and further processing. The cleaning bath is performed in a covered tank of 140° water, and lasts about five hours.

The clean, debarked logs are next cut to 8 feet (96 inches) in length, and moved into a machine called a strander. The strander cuts the log into small, thin strips up to 12 inches in length; in fact, the ideal strand size for the LSL manufacturing operation is 12 inches

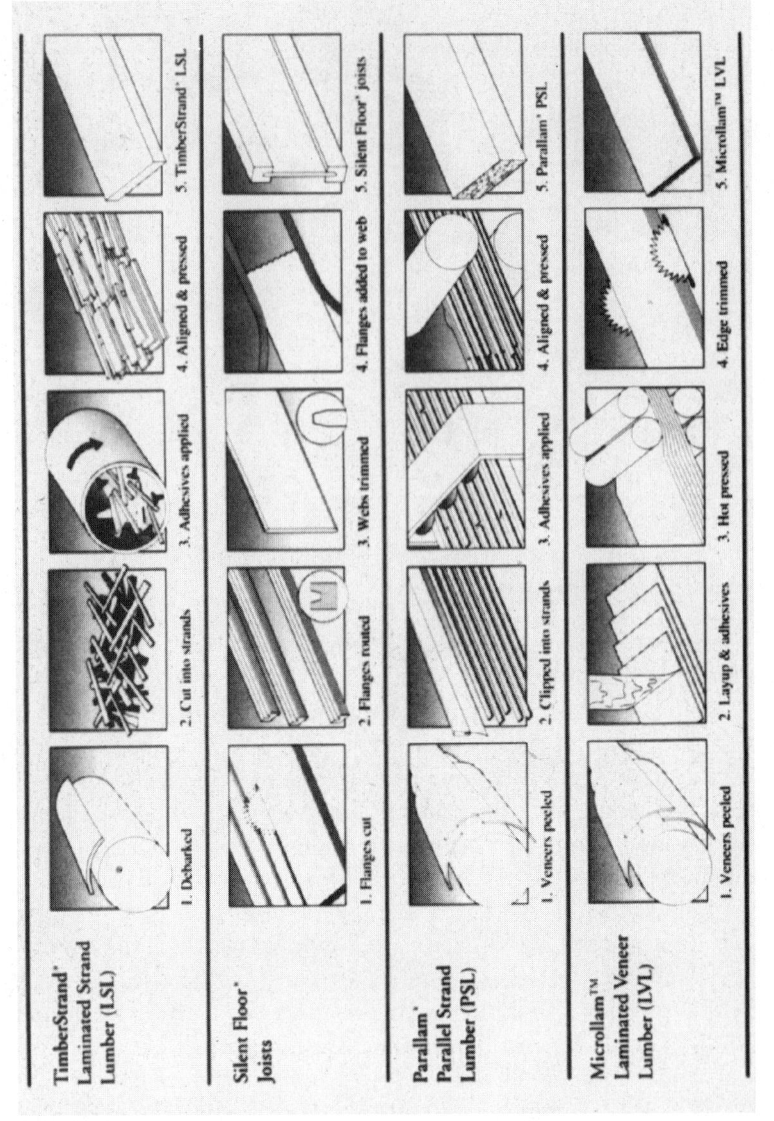

6-13 *Comparing the manufacture of LSL, PSL, and LVL products, as well as the I-joist.* Courtesy Trus Joist MacMillan

long, 1 inch wide, and 1/32 inch thick. The strander is another one of the technological innovations that have come about with the engineered lumber revolution, and is capable of reducing an entire cord of raw logs into strands in just 90 seconds.

The wet or "green" strands of wood move through special sorting machinery that sorts out and eliminates short strands and other pieces unsuitable for manufacturing. The strands, after storage in the log pond and exposure to the long warm cleaning bath, can have a moisture content as

6-14 *Small-diameter, fast-growing trees such as aspen are used in the manufacture of LSL products and other engineered lumber.* Courtesy Trus Joist MacMillan

high as 70 percent, so next they're moved through a dryer that dries them to a uniform moisture content of around 6 percent, which helps ensure a good adhesive bond and consistency in the finished product.

The strands are automatically checked for proper moisture content, and then move through another short-strand eliminator. Short strands of wood will impair the overall structural integrity of the finished LSL member, so the short-strand eliminator is designed to remove strands that are less than 4 inches long.

Further down the line, a mix of polyurethane resin and wax is combined in giant blenders. The dry and sorted strands move through a machine that blends the resin/wax mix with the strands, thoroughly coating them. The coated strands then move through a machine called a forming head, where they are aligned parallel with each other to take advantage of the natural strength of the fibers. The parallel strands are then formed up into loose mats of uniform thickness, depending on the desired thickness of the final pressed mat. Before moving into the press, the loose mat is scanned and weighed by sophisticated automatic machinery to ensure that the finished product will be uniform in size and performance characteristics.

The loose mats then move into a unique press that uses a steam-injection system to press and form the mat into a solid slab of wood. These massive, solid blocks of wood are 8 feet wide, up to 5½ inches thick, and up to 48 feet in length when they exit the steam-injection press. The rough, freshly formed billet next moves through a sander, which is designed to remove the rough edges from the billet and better facilitate the cutting process. The sanding process also ensures uniformity in the thickness of the finished product, which will vary by only a maximum of $^{5}/_{1000}$ inch plus or minus from one piece to the next. The sanded billet is then cut to final width and length, either one of several stock sizes or custom-ordered dimensions, then wrapped and bundled to be shipped.

Oriented strand board (OSB)

Oriented strand board (OSB), is an engineered panel product used in a wide variety of applications on the construction site. The area where it is most commonly seen is for use as a wall and roof sheathing, where its availability, consistency, and lower cost have helped it to all but replace plywood as a sheathing material. OSB has the additional advantage of having a checkered, nonslip surface on one face, which makes it an ideal product for use on a roof. With the growing popularity of I-joists, OSB has found another new home as a web material. It is lightweight, relatively inexpensive, and has the necessary strength and uniformity to make it a great product for this application.

The manufacturing process begins with small-diameter logs of various species, making OSB yet another engineered wood product with tremendous potential for using fast-growing, small-diameter logs that previously had no real structural value. The raw logs are first debarked and cleaned, and then ripped into long thin strips having an average width of ¾ to 1 inch wide, and an average thickness of $^{25}/_{1000}$ to $^{30}/_{1000}$ inch. A clipper shears the strands to a uniform length of between 3¼ and 3½ inches long.

The strands are then processed through a dryer to remove excess moisture. The strands enter the dryer having a moisture content as high as 60 to 70 percent, and come out having as little as 2 to 6 percent moisture. As with the other engineered lumber products, the moisture content is a crucial factor in the performance of the adhesive resins, and in the strength, uniformity, and dimensional stability of the finished panel itself.

The strands are sorted by machine, then coated with a waterproof phenolic resin. The coated strands are laid up on a forming machine into perpendicular layers, most commonly three but, depending on the intended final thickness, sometimes five layers. The first layer is laid up with the strands oriented parallel with the long dimension of the panel, the second layer is laid up with strands parallel to the narrow dimension of the panel (perpendicular to the strands of the first layer), then the top layer is again laid out parallel to the long dimension and perpendicular to layer number two. In thicker sheets having a five-ply construction, layer number four matches layer number two across the width of the board, and layer five is oriented down the length of the board, the same as layers one and three.

The raw coated strands are layered up into a 24-foot-long mat. The mat is pressed under heat of 400° Fahrenheit, as well as pressures of up to 750 psi, forming a solid sheet. The amount of heat and pressure used at this stage are only several of the elements in the manufacturing process that can be varied to affect the strength characteristics of the finished panel. Multibladed carbide-tipped circular saws automatically rip the sheets to final length and width, typically 4 × 8 feet, then the sheets are wrapped for shipment.

I-joists

Engineered lumber I-joists have a basic structural design that has been in use for centuries. The I configuration, a common sight in steel beams, has a flat top and bottom flange for strength in load-induced tension and compression situations, and a tall, relatively thin center web that adds bending strength to the member. This combination of flange and web gives the structural member excellent strength in tension, compression, and bending, without all the weight associated with a solid sawn lumber girder or beam.

While the concept of the I as a structurally solid engineering shape might not be new, applying that concept to wood is only a couple of decades old. The first commercially produced all-wood I-joist was created in 1969, with a top and bottom flange of Machine Stress Rated lumber, typically a standard-dimension 2 × 3 with a plywood web.

As the I-joist concept began to catch on, stable sources for an economical and high-quality flange material began to be a problem. Trus Joist MacMillan took a clue from the plywood it was using as web material, and developed the first engineered wood material—LVL—just for use as a flange material. In addition to using a manufactured wood product that helped conserve dwindling wood supplies, LVL also allowed lamination grades to be mixed during the manufacturing process, creating a flange material that was actually stronger at the top and bottom faces of the material than solid lumber. This second generation of I-joists continued to use plywood as a web material, and it was this configuration that became an increasingly common sight on construction projects from 1971 until the mid-1980s.

The next evolution of the I-joist occurred in 1987, when Trus Joist replaced the plywood web material with another new, engineered, environmentally friendly material called oriented strand board (OSB), though LVL was still the material of choice for the top and bottom flanges. Today, most I-joists on the market use LVL for the top and bottom flange material and OSB for the web (Fig. 6-15). Trus Joist, in fact, has a proprietary OSB product called Performance Plus, that is engineered and manufactured specifically for their line of TJI I-joists.

The manufacturing process for creating I-joists is actually two processes occurring simultaneously (Fig. 6-16), at least in most plants. On one line, LVL material is manufactured and then moved to the I-joist line (Fig. 6-17). The full-size billets of LVL are ripped with gang saws into long, uniform strips 1½ to 2 inches wide for use as flange material. These strips then enter a series of milling machines that rout a square or flat-bottomed V-shaped groove (depending on the manufacturer) along one face of the flange material, and also finger-joint the ends of the strips.

On a parallel manufacturing line, panels of ⅜- or ⅝-inch OSB or plywood (again, depending on the manufacturer) are ripped into uniform panel sizes for use as web material (Fig. 6-18). The panels are shaped along both edges as needed to fit the groove in the flange, and also finger-jointed at the ends (Fig. 6-19). As the line carrying the webs meets up with the line supplying the flanges, the edges and

6-15 *The two basic engineered lumber components of the Trus Joist MacMillan I-joist: a top and bottom flange of laminated veneer lumber, and a center web of oriented strand board.*

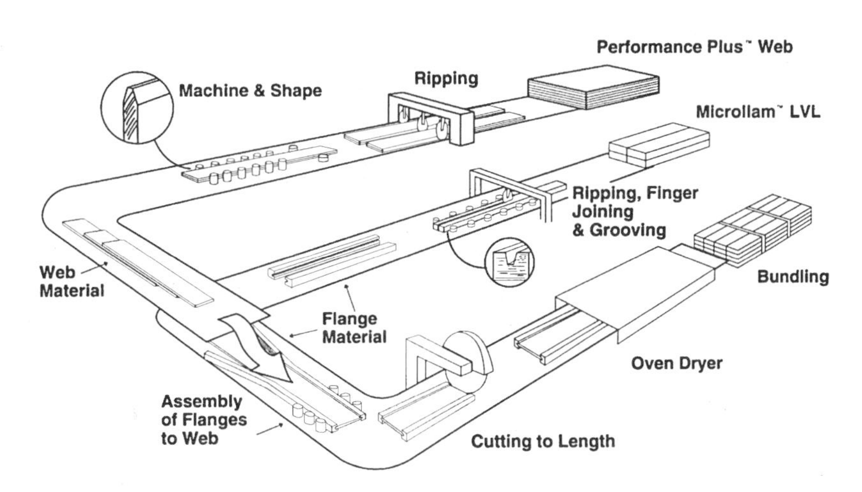

6-16 *The manufacturing process for the TJI I-joist. Notice how the LVL members, manufactured in one part of the building, are combined with the OSB Performance Plus web material, which is brought in from another plant and then placed on a parallel line.* Courtesy Trus Joist MacMillan

6-17 *Billets of LVL move toward gang saws (not shown) and eventual manufacture into I-joists.*

ends of the web panels are coated with adhesive, typically a phenol resorcinol resin. The finger-jointed ends of the flanges are also glued.

In one process that combines the two components, the flanges are pressed onto the edges of the web material. The webs are butted end to end to interlock the adhesive-coated finger joints, and the flanges are also joined end to end as the process moves along. The end result is one long, continuous I-joist in which the web panels are glued to one another, the flange sections are glued to one another, and the webs are glued to the flanges, all under the pressure of high-capacity pressing equipment (Fig. 6-20).

The continuous I-joist is then cut to uniform lengths of up to 60 feet, and the web material is hydraulically punched to create the round knockouts for wiring and plumbing. The cut joist passes through an oven that dries and cures the adhesive (Fig. 6-21), then the joists are bundled together and wrapped for shipment (Figs. 6-22 and 6-23).

6-18 *On a parallel assembly line, OSB sheets are ripped to size for use as the web of the I-joist.*

6-19 *The ends of the cut OSB web sheets are finger-jointed, ready for end-to-end joining.*

6-20 *The completed I-joist, a combination of OSB webs and LVL flanges, exits the press as a continuous billet. Note the outlines of the prepunched web holes, used as needed for wiring or plumbing on the construction site.*

6-21 *After being cut to length, the new I-joist enters the drying and curing ovens.*

6-22 *Completed I-joists are removed from the assembly line with massive overhead lifting gear.*

6-23 *The completed joists are banded and stacked for delivery.*

7

Hangers, connectors, and fasteners

Anyone who has worked in the construction industry in the last two decades or so is sure to be familiar with metal hangers and connectors. From the original metal nailing plates and simple joist hangers (Fig. 7-1) to today's sophisticated skewed (Fig. 7-2) and angled multimember hangers, the use of these strong and convenient hangers

7-1 *Simple joist hangers, used here to connect solid sawn joists to a glulam beam.*

7-2 *A sophisticated skewed top-flange hanger, connecting a roof truss to a glulam support member.*

has supplemented and even replaced many traditional joint-to-joint connection methods.

Hangers and connectors have always been designed around standard sawn lumber sizes. Hangers, for example, were sized to fit 2 × 10 joists or 4 × 6 beams or other common dimensional sizes of lumber. Since these hangers are engineered to meet specific shear, upload, download, and other requirements, having them sized properly for the structural member they will carry is an essential part of their design.

Engineered lumber, however, is different from solid sawn lumber in several important ways. The dimensions are different from the common lumberyard boards you're used to, as is the fiber orientation and the fact that the products are laminations rather than solid pieces. These differences result in different fastener requirements and, in some cases, different installation practices. But here again, the manufacturers have seen the trends and changed with the times. A growing number of hangers and other connectors designed specifically for use with engineered lumber and composite products (Fig. 7-3) have been coming steadily onto the market in the last couple of years, and manufacturers are also changing and adapting traditional sawn-lumber hanger designs to be compatible with the new composites.

Simpson Strong-Tie is probably the most well-known name in the metal hanger industry. Simpson has worked very closely with compa-

7-3 *An example of one of the new generation of metal hangers, designed to fit standard sizes of engineered lumber.* Courtesy Trus Joist MacMillan

nies such as Trus Joist MacMillan and Louisiana-Pacific Corporation to mutually develop engineered lumber/metal hanger systems that work together to speed and simplify installation, while making the most of the strength and span potential of both products. As of this writing, Simpson offers by far the largest selection of modified and specially designed hangers and other connectors for use with engineered lumber systems, composite products, and manufactured trusses. For that reason, I have selected their Composite Wood Products Connectors catalog (C-CWP96) to use in this chapter to familiarize builders with the types of products currently on the market.

In addition to this general catalog of composite wood connectors, Simpson has also developed a series of specific connector selection guides to assist builders and designers in selecting the proper connector to use with I-joists, LVL beams, and other composite wood members manufactured by specific companies. Connector selection guides, which are intended to be used in conjunction with the main composite wood connector catalog, are currently available for use with composite wood products from the following manufacturers: Boise Cascade Timber and Wood Products Division; Georgia-Pacific; Louisiana-Pacific Corporation; Tecton Laminates Corporation; Trus Joist MacMillan; and Willamette Industries, Inc. Connector selection guides for engineered lumber products from other manufacturers will be released soon.

Catalogs, specification manuals, code compliance information, pricing and cost comparison guides, specification software, and many other useful pieces of information on Simpson Strong-Tie products are available from your lumber supplier, or from the manufacturer directly. See Appendix E at the end of the book.

Hanger terms

As mentioned earlier, the concept of engineered lumber can be divided into two basic categories: composite products, which include I-joists, LVL, PSL, LSL, and glulam beams, and manufactured trusses, also commonly known as *plated trusses.* Making this distinction will help you separate and clarify some of the product descriptions that follow. There are also a number of other terms that will help you better understand and use the information in this chapter.

Supporting member. This is the structural member to which the hanger or connector is attached. For example, if a joist hanger is nailed to a glulam beam with the purpose of supporting a ceiling joist, the glulam is considered the supporting member.

Supported member. This is the structural member that is supported by the hanger or connector. In the previous example, that member would be the ceiling joist.

Face-mount hanger. A style of hanger that is attached to the face of the supporting member, using nails or screws driven through the nailing flanges of the hanger. With this style of hanger, the load-carrying capacity is placed primarily on the shear strength of the fasteners.

Top-flange hanger. This style of hanger has an additional attachment flange on top, which partially or completely overlaps the top surface of the supporting member. The hanger is attached to the supporting member with fasteners driven through the top flange, and usually through the face of the hanger as well. Some hangers of this type are also suitable for welding to steel supporting members. Because of the additional load-bearing surface and a better transfer of load, top-flange hangers can, in most cases, carry greater loads than the same size of face-mount hanger.

Seat. The bottom part of the hanger; the portion of the hanger that physically supports the load of the supported member.

Side flange. The two vertical sides of a U-shaped hanger, parallel to each other and perpendicular to the seat. The side flanges are usually provided with nail and/or bolt holes for attachment to the supported member.

Nailing flange. In a face-mount hanger, these flanges are bent out perpendicular to the side flanges, and contain the necessary nail and/or bolt holes used to attach the hanger to the supporting member.

Top flange. In a top-flange hanger, this is the upper portion of the hanger, bent or welded perpendicular to the side and nailing flanges and designed to bear on the upper surface of the supporting member.

Hanger codes and abbreviations

The Simpson Strong-Tie catalog uses a number of codes and abbreviations in designating their hangers, connectors, and other products. This can be a little confusing, and even at times a little contradictory, but for the most part the coding system is consistent from hanger to hanger, and helps with identification and ordering.

To simplify listing and specifying hangers and connectors, a number of abbreviations and letter codes are used, either as prefixes or suffixes to the model number of the hanger. Standard prefixes include:

L (light). This is the lightest gauge of metal, used for relatively light loads.

U (standard face mount). Standard gauge, face-mounted hanger, the most common of the hanger types.

M (medium). Medium-gauge hanger, between light and heavy load applications.

H (heavy). These are heavier-gauge hangers for use with heavy or concentrated loads.

HH (heavy heavy). The strongest of the connector or hanger styles, used for extremely heavy loads, or to connect multiple members.

HG (heavy glulam). Specifically for use with heavy loads involving glulam beams or other composite wood beams.

B (bent). A type of hanger formed from 12- or 14-gauge steel (depending on the hanger size and application) and used in heavier load applications.

There are also some standard suffixes following the hanger model designation, including:

S (double shear nailing, sloped, or skewed). Double shear nailing (Fig. 7-4) is a patented Simpson hanger design that allows you to attach the supported member into the hanger by using nails driven in through the side flanges at an angle from both sides. The S designation is consistent among hangers having the double shear feature, but

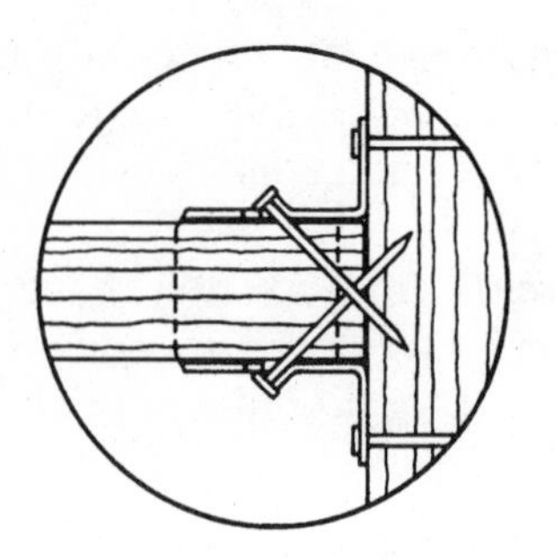

7-4 *Simpson Strong-Tie's patented double shear nailing system, which adds strength to the wood/hanger joint.* Courtesy Simpson Strong-Tie Company, Inc.

it can also be a little confusing, since the S can also refer to a hanger that is sloped or skewed.

T and TF (top flange, truss, or tab). These are hangers having a bent or welded flange on top to bear on and be attached to the top of the supporting member. Here again, the T designation is pretty consistent among top-flange hangers, but T is also used to indicate certain hangers used for plated trusses, and where two T designations are used for a particular hanger, the second T indicates that the hanger has bend tabs.

V (veneer). Specifically for use with veneer composite load-carrying members such as LVL and PSL. The hangers have specific nailing patterns that help prevent splitting the veneer layers.

I (I-joist). Specifically designed for use with I-joists.

PT (plated truss). Specifically designed for use with plated trusses.

L and R (left and right). When ordering skewed hangers, you will also commonly see the suffix L for left-hand skew and R for right-hand skew (the L in this case has a different meaning from the L used as a prefix designation).

PAN

This is an acronym for *positive angle nailing,* which is a patented Simpson Strong-Tie feature (Fig. 7-5). The PAN slot in the side flange of the hanger allows the nail to be driven in only at an angle in relation to the laminated flanges of the I-joist. This allows the nail to penetrate through the various layers of the lamination, rather than driving straight into the joist and causing the laminations to separate.

7-5 *Positive angle nailing (PAN), another patented Simpson feature, helps to align and guide the nail into the LVL flange found on most I-joists, minimizing the chances of splitting the veneer.* Courtesy Simpson Strong-Tie Company, Inc.

Bend tab

Another Simpson hanger feature, bend tabs (Fig. 7-6), are part of some types of joist hangers designed for use with I-joists. After the I-joist has been installed in the hanger, the tabs are bent over the top of the joist's bottom flange, then a nail is driven at a slight angle through the tab and into the wood flange. The bend tab feature helps to lock the I-joist into the hanger, reducing the possibility of squeaks in the floor that result from movement in the joist.

Hangers for composite wood products

This section describes some of the most common types of composite wood hangers, including their features and some common usages.

The letter designations for these hangers are from the Simpson Strong-Tie Composite Wood Products Connectors catalog (C-CWP96). Other Simpson hangers are available in addition to these, and composite wood hangers are also available from other manufacturers.

ITT (I-joist top-flange tab hanger)

This series of hangers (Fig. 7-6) is perhaps the best all-around hanger for use with composite wood I-joists. The design of the hanger's side flanges keeps the top flange of the I-joist from moving laterally, eliminating the need for web stiffeners in most applications. The ITT special bend-tab feature allows you to nail vertically into the top of the I-joist's bottom flange while firmly constraining the joist, which helps reduce the squeaks that result from joist movement. The bend tabs can be left straight up (unbent) for nailing into the web stiffeners (if used).

The seat of the ITT hanger is offset ¼ inch away from the rear plane of the hanger, which allows full structural bearing under the supported member, even if the member is cut slightly short. The top flange and overall length of the hanger are designed so the installed joist sits approximately flush with the top of the supporting member, allowing easier and flatter installation of the subflooring panels. Optional nail holes are also provided to increase the uplift resistance.

The ITTM (I-joist top-flange tab hanger, masonry), shown in Fig. 7-7, is designed for installation into grouted concrete block walls, poured concrete walls, and welded-to-steel members; the MIT (medium I-joist top-flange hanger) and HIT (heavy I-joist top-flange hanger) are two heavier versions of the hanger for use with larger members and larger loads. The MIT and HIT hangers also have the positive angle nailing (PAN) feature, which lessens the risk of splitting the layers in veneered lumber, and also increases the hanger's uplift capacity.

U (face-mount hanger) and HU (heavy face-mount hanger) series

These are all face-mount hangers, designed for use with solid sawn lumber and some I-joist and composite wood applications. They are economical, and have a variety of job site uses. The HU version is heavier, and is intended for applications where increased strength, longevity, and safety are factors of concern. See Fig. 7-8 for an example of a U face-mount hanger.

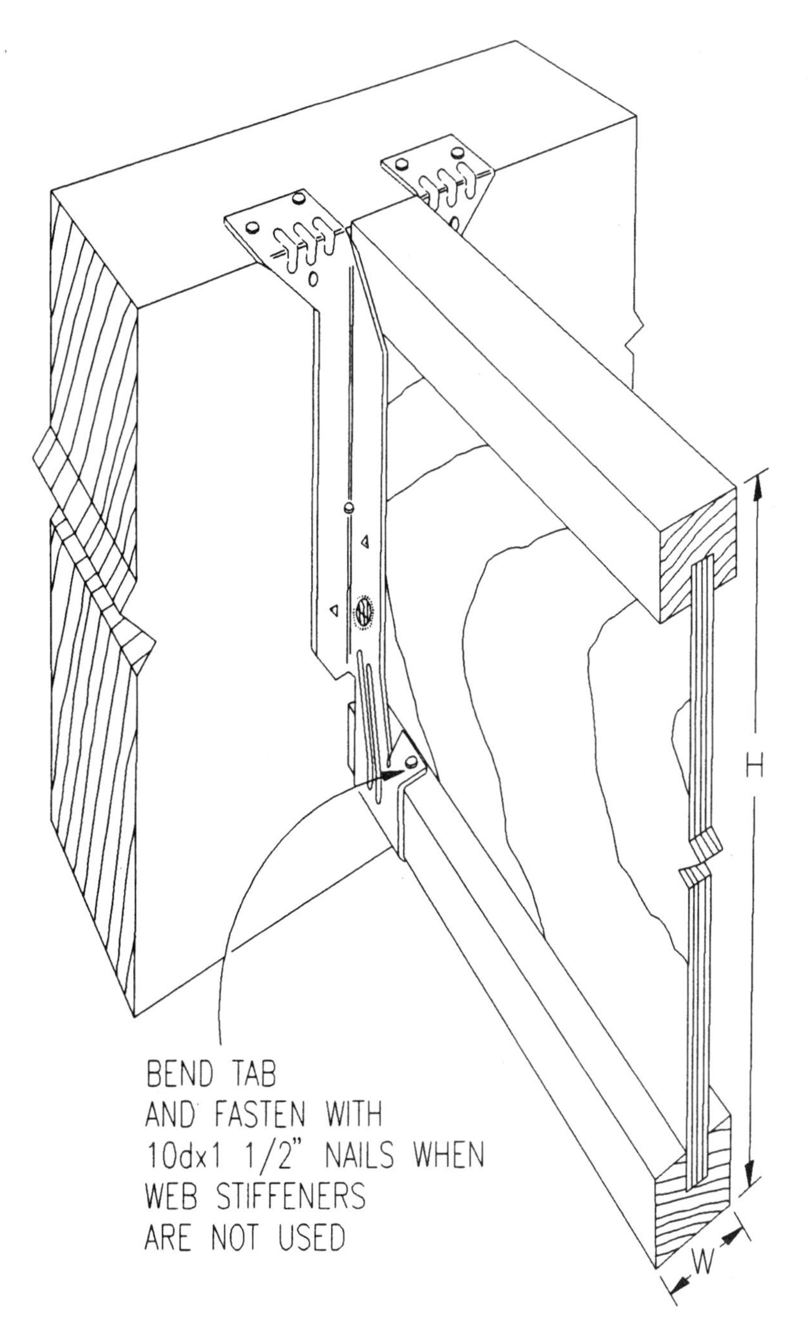

7-6 *ITT top-flange hanger with bend-tab feature, a common hanger for use with I-joists.* Courtesy Simpson Strong-Tie Company, Inc.

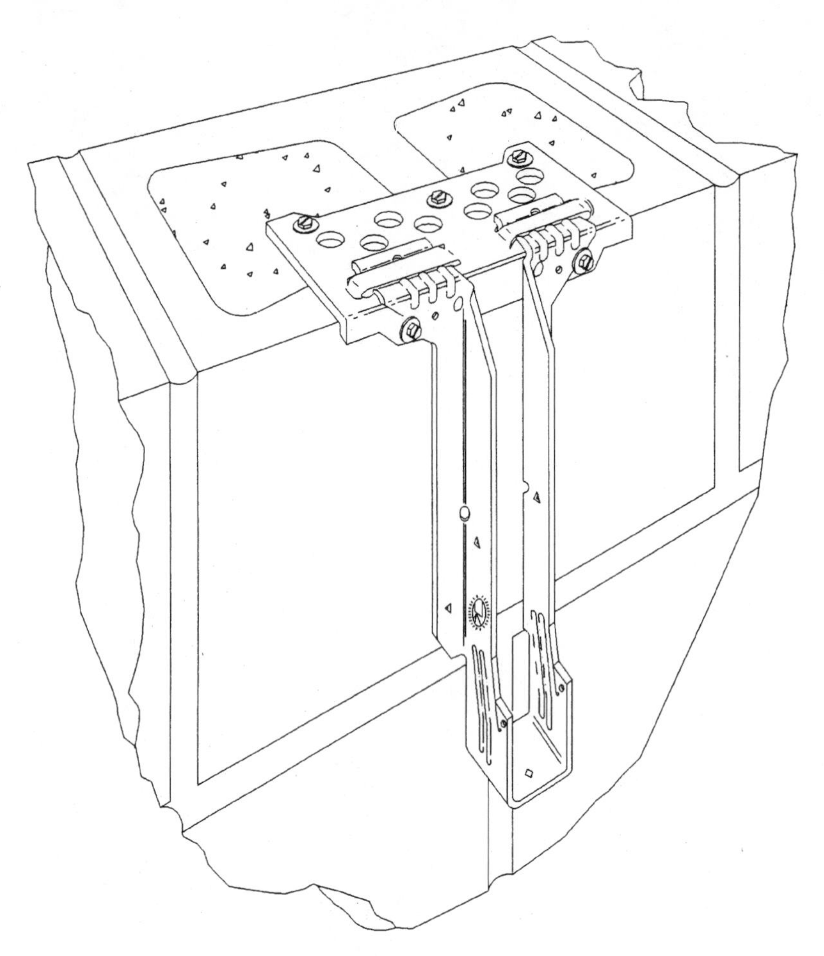

7-7 *The ITTM version of the ITT hanger, for use in masonry construction.* Courtesy Simpson Strong-Tie Company, Inc.

Variations in this series include LUS (light face-mount double shear), HUS (heavy face-mount double shear), HHUS (heavy heavy face-mount double shear), and HGUS (heavy glulam face-mount double shear), all of which can be used with PSL and LVL members and plated truss applications. Double shear nailing cannot be used with I-joists.

IUT (I-joist face-mount tab hanger)

This is similar to the ITT hanger, but is a face-mount instead of a top-flange hanger (Fig. 7-9). While the load capacity of these hangers is a little lower than the ITTs, they are less expensive and allow you to adjust the I-joist up and down to simplify alignment. The design of this hanger

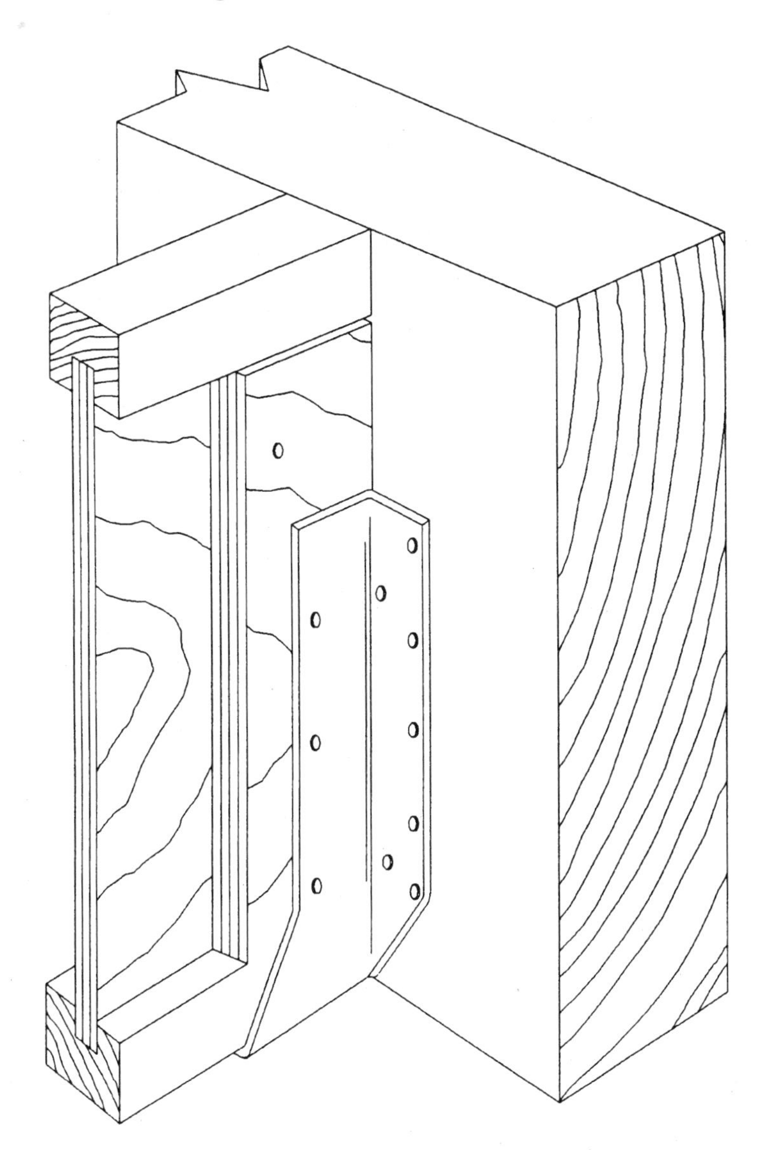

7-8 *A U series hanger, which is a face-mount instead of a top-flange hanger. Note the use of the web stiffener to fill in the web of the I-joist.* Courtesy Simpson Strong-Tie Company, Inc.

restrains lateral movement of the joist, and eliminates the need for web stiffeners in most applications. It also has the bend tab feature (Fig. 7-10). A variation of this is the MIU (medium I-joist face-mount hanger), which is designed for heavier loads without the use of web stiffeners. MIU hangers have the PAN nailing feature instead of bend tabs.

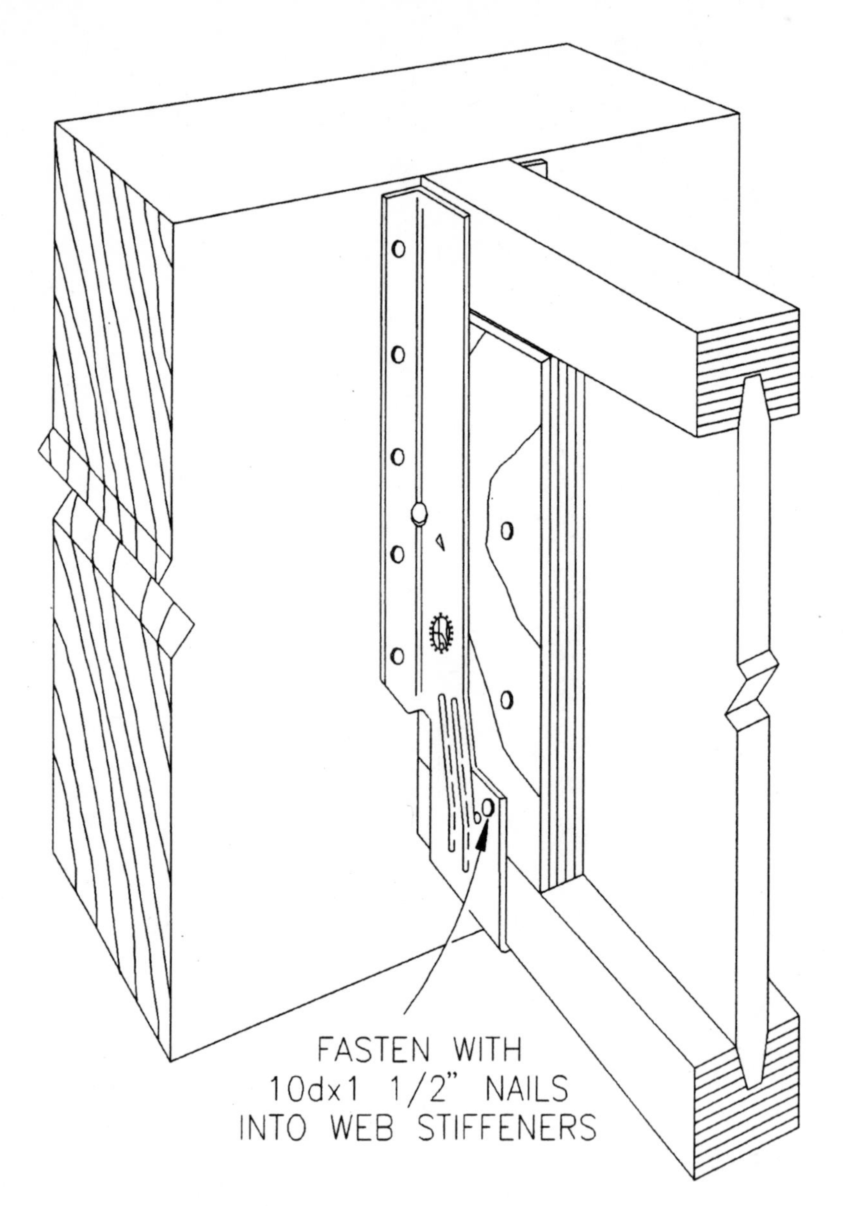

7-9 *IUT face-mount hanger with a bend tab. In this case, the bend tabs are not used since the I-joist has web stiffeners.*
Courtesy Simpson Strong-Tie Company, Inc.

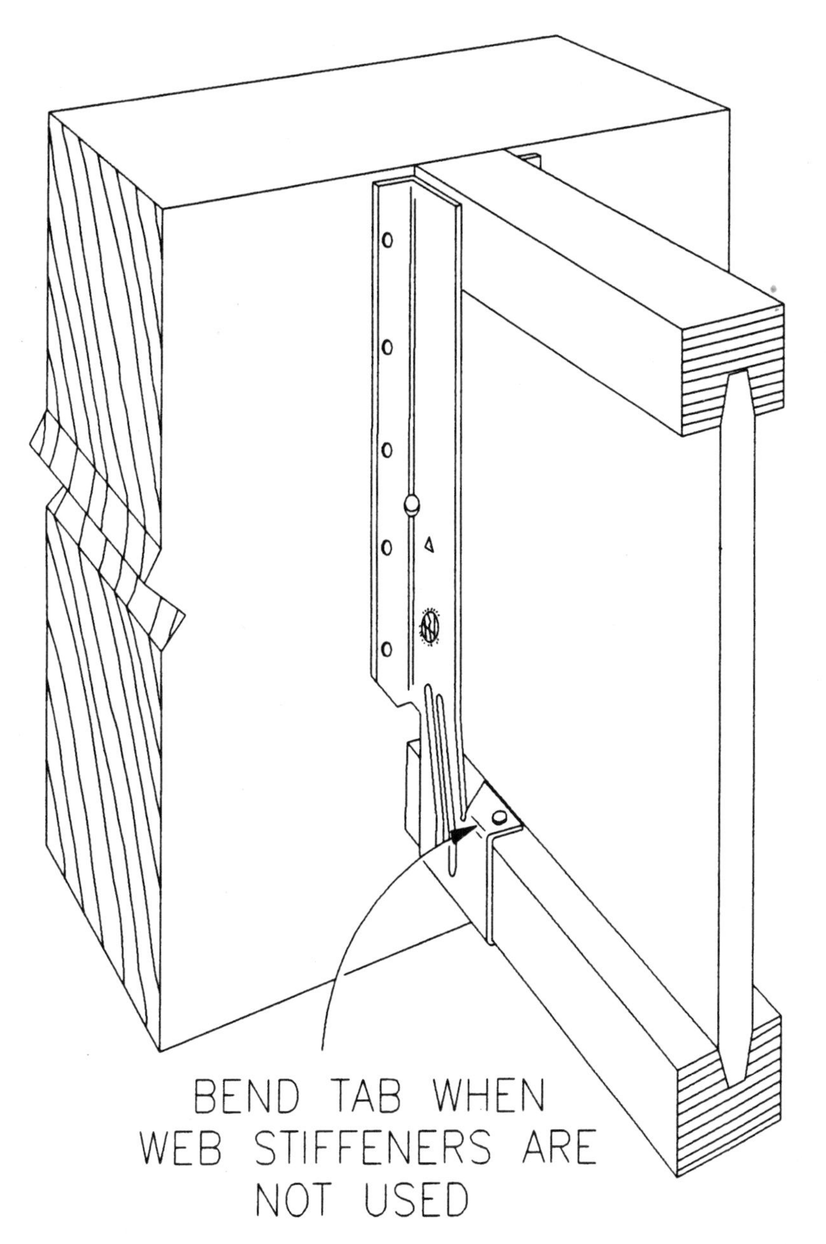

7-10 *IUT used on a joist without web stiffeners, allowing the bend tabs to be used.* Courtesy Simpson Strong-Tie Company, Inc.

THA (truss hanger adjustable) and THAI (top-flange hanger adjustable I-joist)

The THA hanger (Fig. 7-11) has long, perforated nailing flanges that can be field-bent over the top of the supporting member, or left unbent for face-mount applications. This allows for the flexibility of vertically adjusting the hanger while still maintaining the strength of a top-flange hanger. This style of hanger, which also has an offset seat to allow full bearing under members that are cut a little short, can be face-nailed, top-nailed, or both. No bolts are required to obtain full rated load strength, which reduces installation time. Although the S designation is missing on this model, it also has the double shear nail-

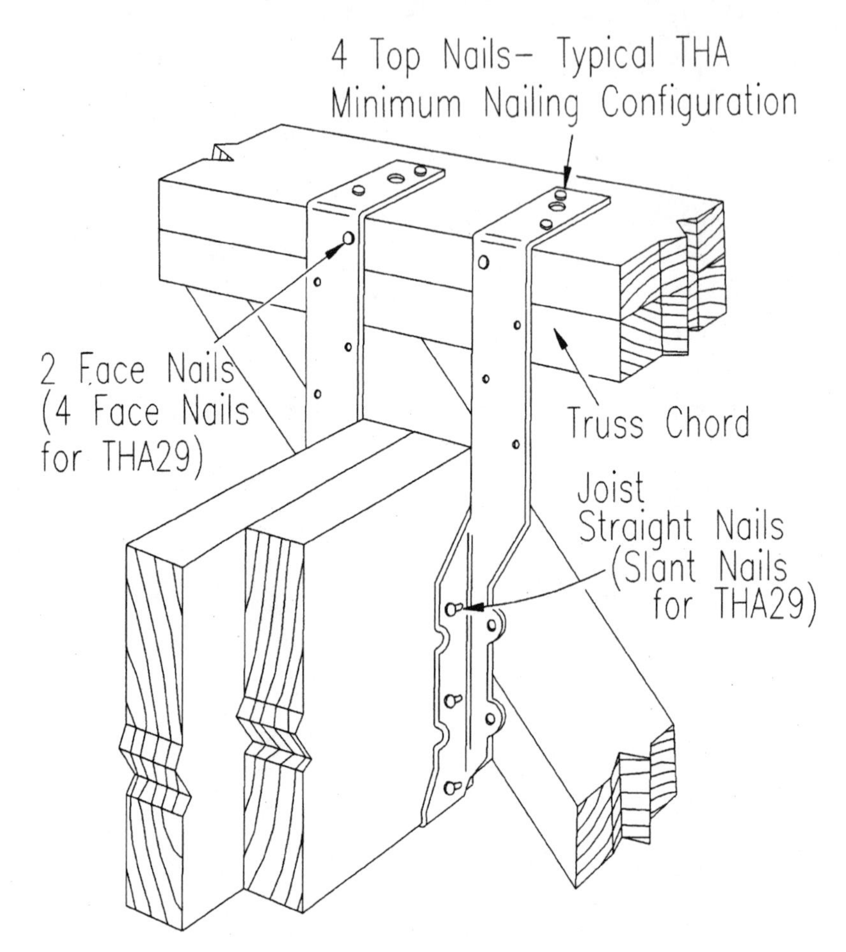

7-11 *The THA style of hanger. The long nailing flanges can be field-bent over the top of the supporting member as needed.*
Courtesy Simpson Strong-Tie Company, Inc.

ing feature. THA hangers are used for hanging plated trusses, as well as with LVL, PSL, and solid sawn lumber.

A variation of the THA hanger is the THAI (truss hanger adjustable I-joist), shown in Fig. 7-12, which is specifically designed for use with

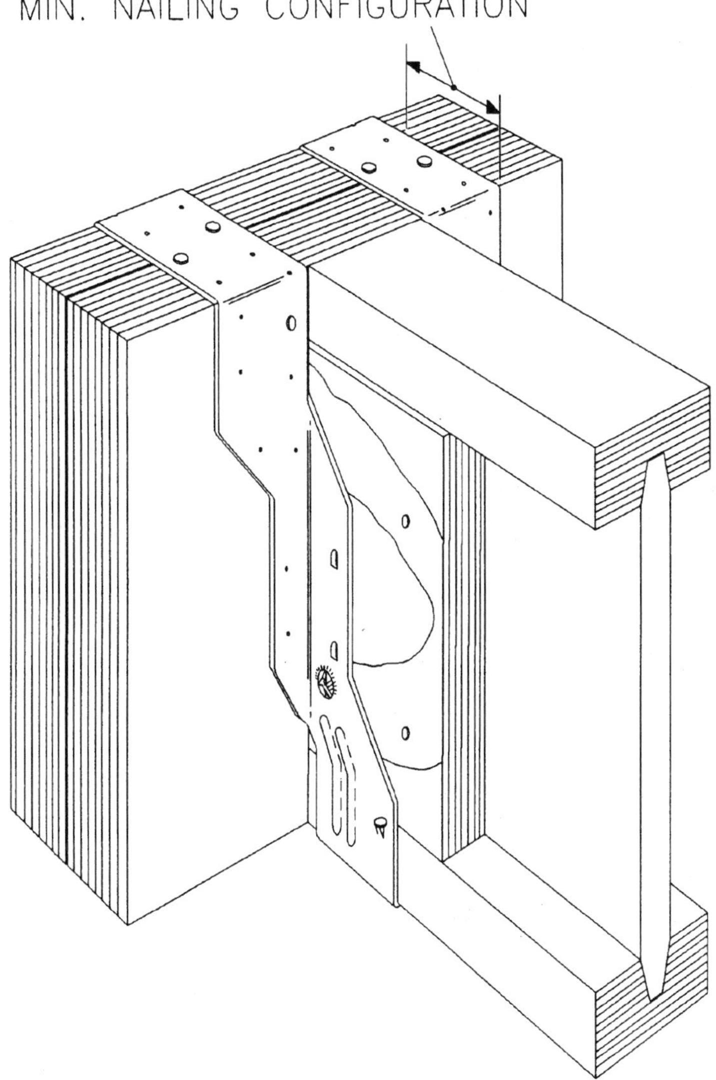

7-12 *The I-joist version of the THA hanger, called a THAI.*
Courtesy Simpson Strong-Tie Company, Inc.

I-joists. It has all the same features as the THA hanger, except that PAN nailing is substituted for the double shear nailing. A special-order version of the THAI, THAI-2 (Fig. 7-13), is designed to accommodate the width and load of a pair of I-joists used in combination with one another.

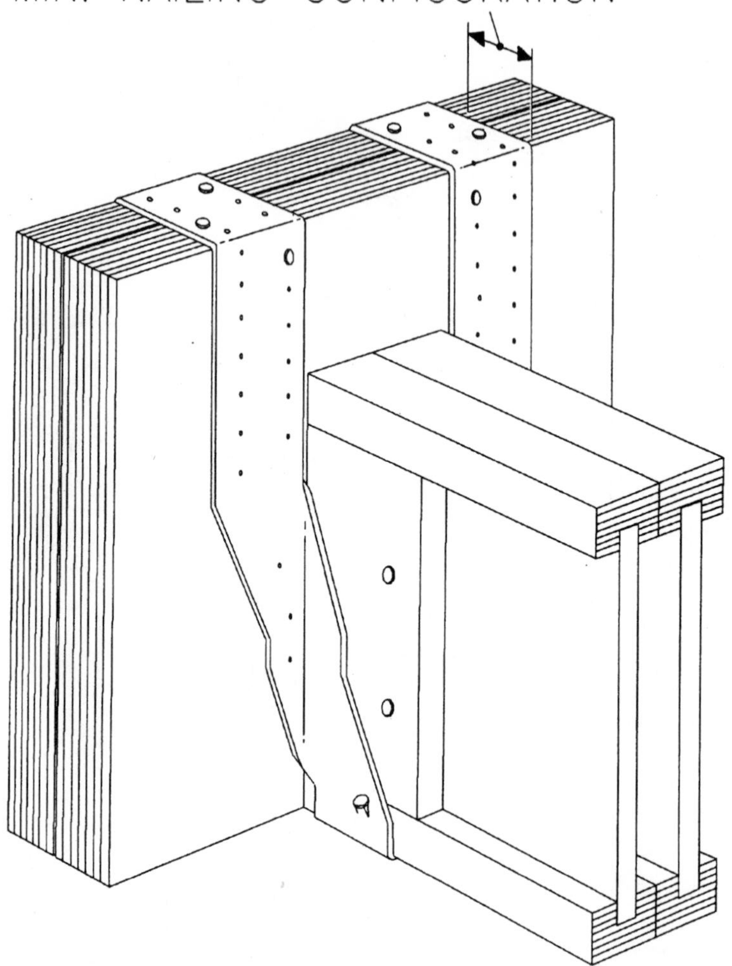

7-13 *Another style of the THA hanger, the THAI-2, is used for supporting multiple I-joists.* Courtesy Simpson Strong-Tie Company, Inc.

LSSU (light sloped/skewed face-nail hanger)

The versatile LSSU hanger (Fig. 7-14) was developed for use in sloped and/or skewed rafter installations, for both I-joist and solid sawn lumber applications. After the rafter is angle-cut, the hanger is attached to the end of it, then the assembled unit is attached to the supporting member. This hanger is adjustable on-site for slopes or skew angles up to and including 45°.

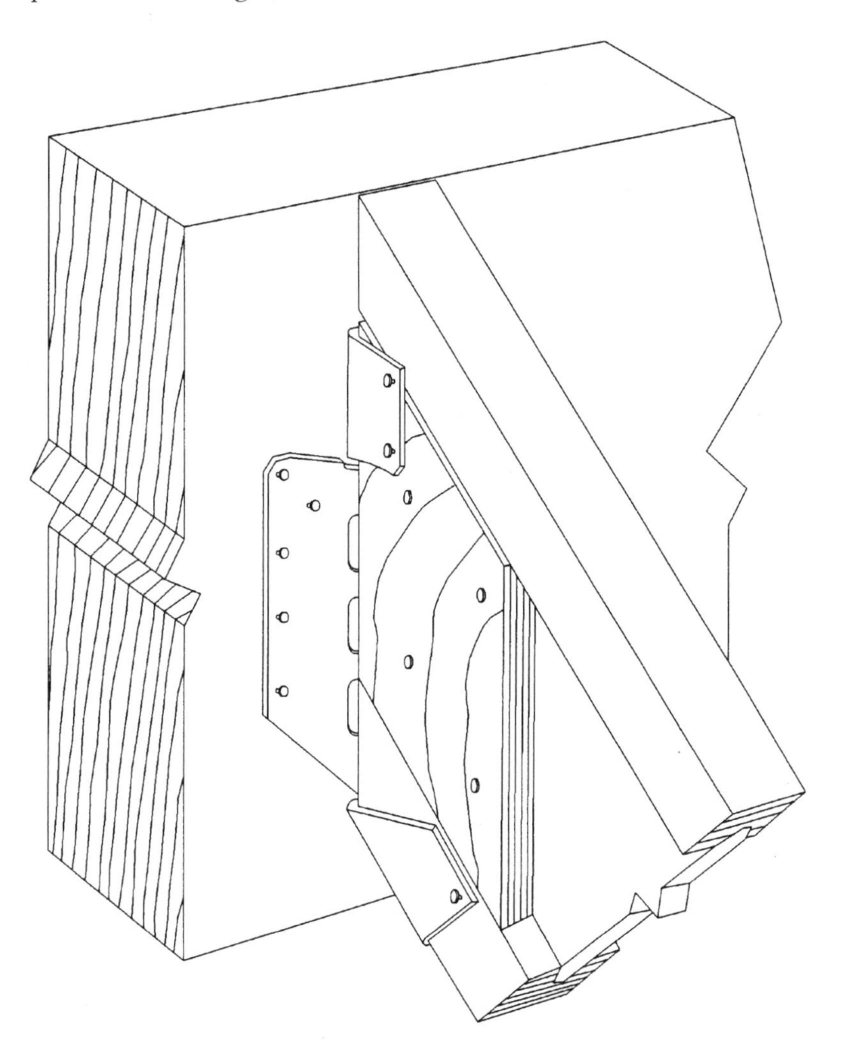

7-14 *This style of hanger, the LSSU, can be used for sloped or skewed installations.* Courtesy Simpson Strong-Tie Company, Inc.

B (bent hanger) series

There are a number of different hangers in the B series, all of which are top-flange hangers intended for use where maximum load values are required (Fig. 7-15). Each style uses a minimum number of fasteners, and is suitable for use with nailer, ledger, and weld-on applications.

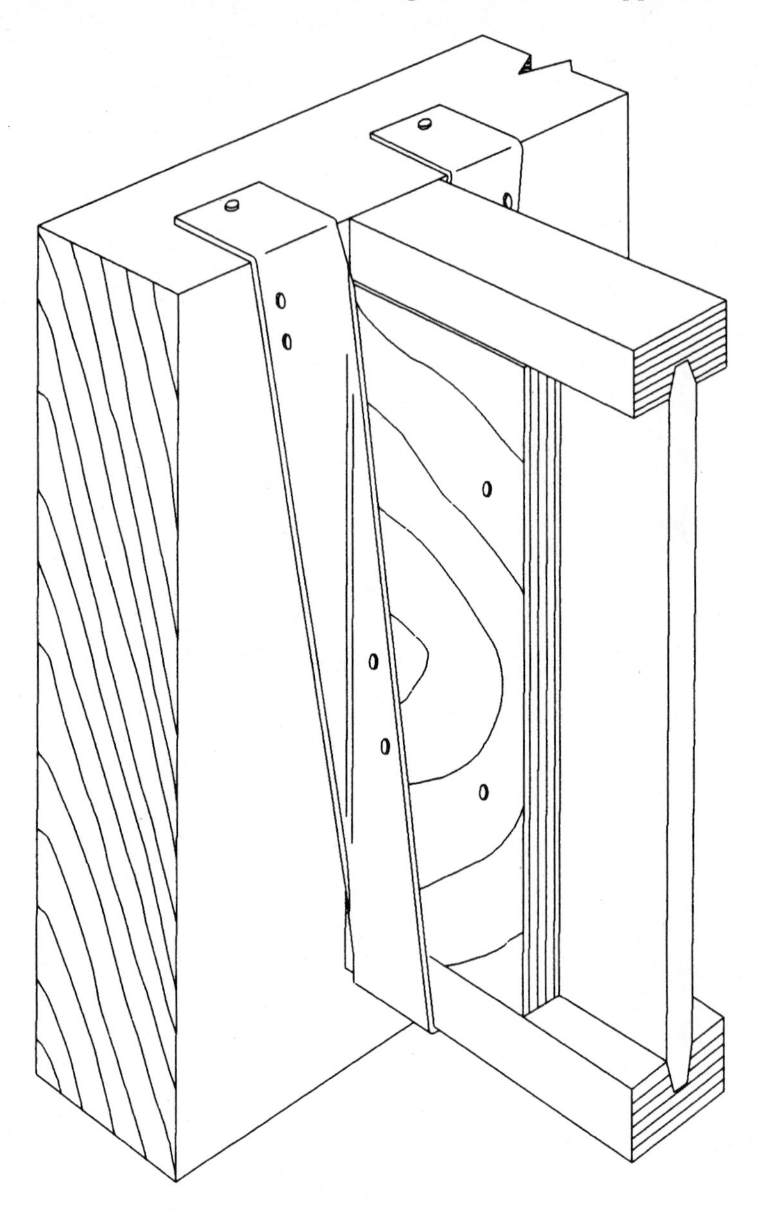

7-15 *A B (bent) hanger, used here with an I-joist and web stiffeners.* Courtesy Simpson Strong-Tie Company, Inc.

A couple of the more important variations within this series include BI (bent I-joist), which is the standard hanger in this series for use with I-joists; LBI (light bent I-joist) and HBI (heavy bent I-joist), which are, respectively, a lighter less-expensive version and a heavier version for larger loads; and LBV (light bent veneer), for use with veneered lumber carrying members such as LVL and PSL. The design of this particular hanger does not contain the upper flange of the I-joist to prevent lateral movement, so web stiffeners are required for all I-joist applications.

W (welded hanger) series

W-series hangers (Fig. 7-16) are all welded top-flange hangers for use in heavy down-load situations, such as supporting large floor beams.

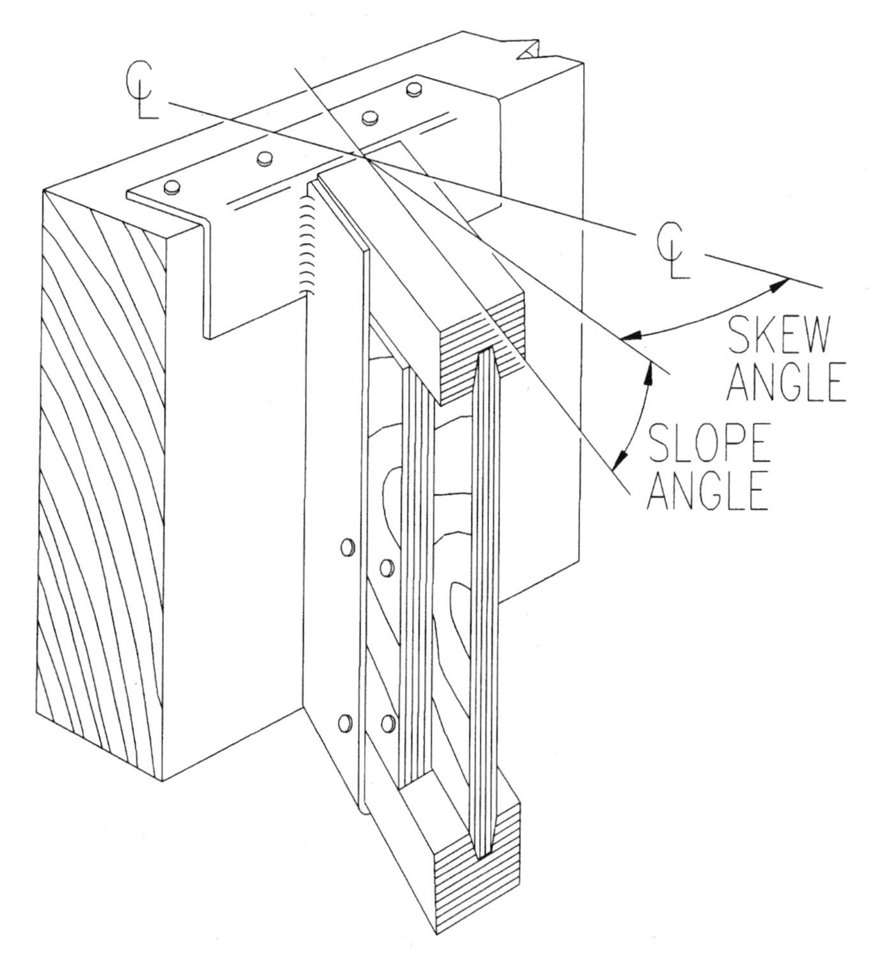

7-16 *W (welded) hangers come in a number of configurations, such as this one for skewed and sloped installation.* Courtesy Simpson Strong-Tie Company, Inc.

They can be used in nail-on, weld-on, and masonry applications, require a minimal number of fasteners, and are available both sloped and skewed. There are several variations within this series for different applications. These include WP (welded purlin) for roof purlin applications; WM (welded masonry, see Fig. 7-17), for installation in grouted concrete blocks; and HW (heavy welded) for very large loads.

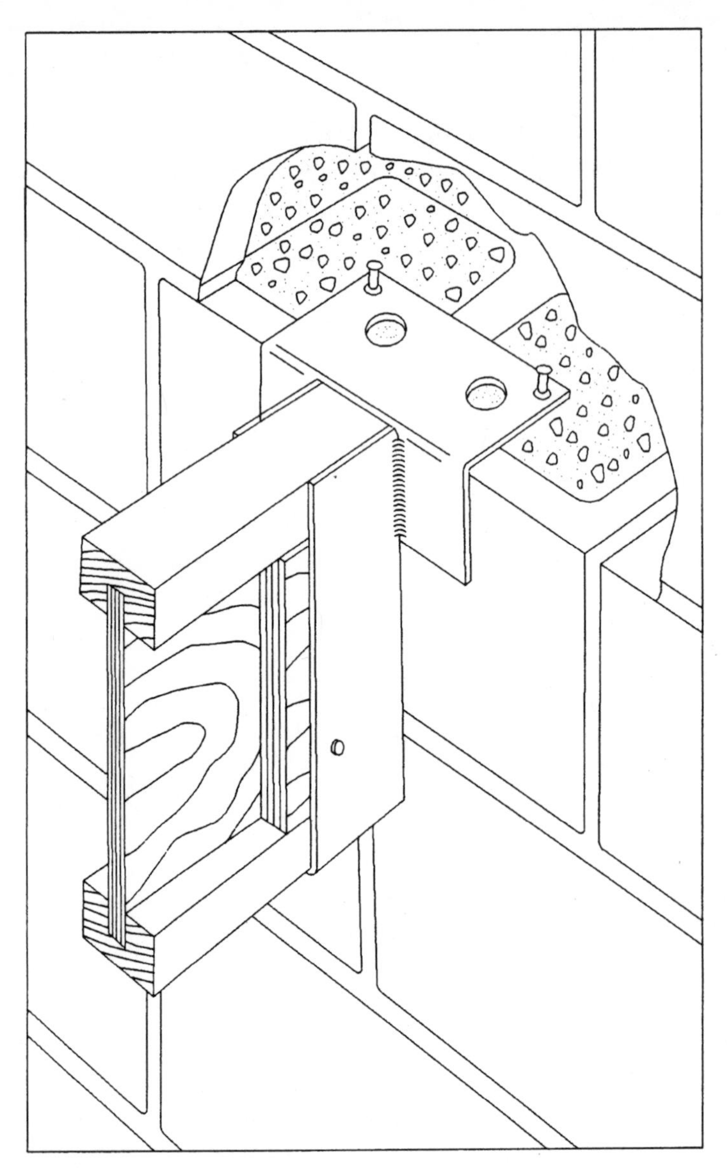

7-17 *The masonry version of the W hanger, called a WM.*
Courtesy Simpson Strong-Tie Company, Inc.

GLTV (glulam timber veneer)

This is a heavy, welded, top-flange hanger specifically designed for use with LVL or PSL members, or with plated trusses (Fig. 7-18). The nail hole locations in the top flange are designed in both size and location to minimize the chance of splitting the veneers in the supporting member. Slope and skew variations are available, and these hangers can be used for nailer, ledger, or weld-on applications. Another variation, HGLTV, is designed for heavier loads.

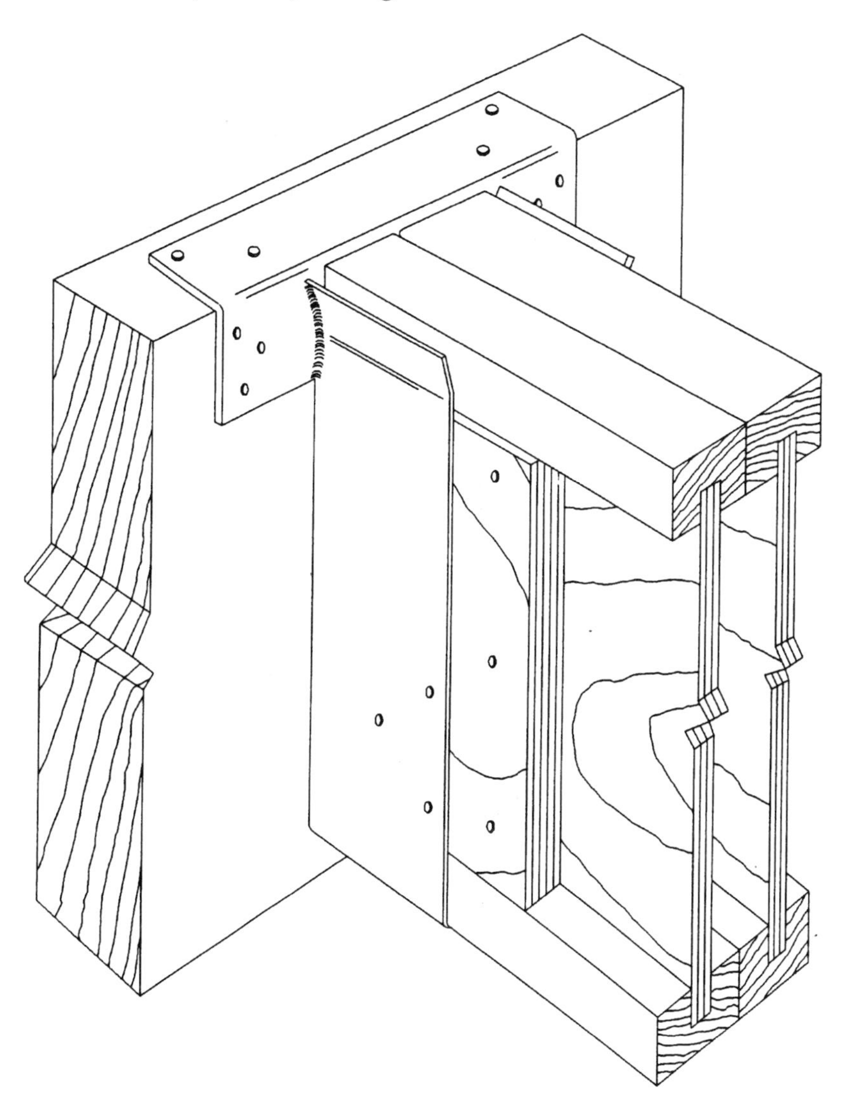

7-18 *A GLTV top-flange hanger for use with multiple I-joists.*
Courtesy Simpson Strong-Tie Company, Inc.

MSC (multiple seat connector hangers)

This is a series of welded, top-flange hangers for use where up to three members come together, as in dormer or hip-roof construction (Fig. 7-19). There are a number of variations and configurations available in this hanger to suit structural members of various sizes, and many different types of joint connection variations. These hangers are also available in sloped and skewed configurations, and the members can be installed without angle or bevel cutting.

SUR, SUL (skewed face-mount hanger, right or left)

Since a number of framing connections involve two members meeting at 45°, SU hangers are available already skewed right or left to that angle. They're available in a variety of sizes for use with solid sawn lumber, I-joists, and composite lumber. A heavier version, the HSUR and HSUL hangers, are also available for supporting larger loads. Both styles of hangers will actually accommodate supported

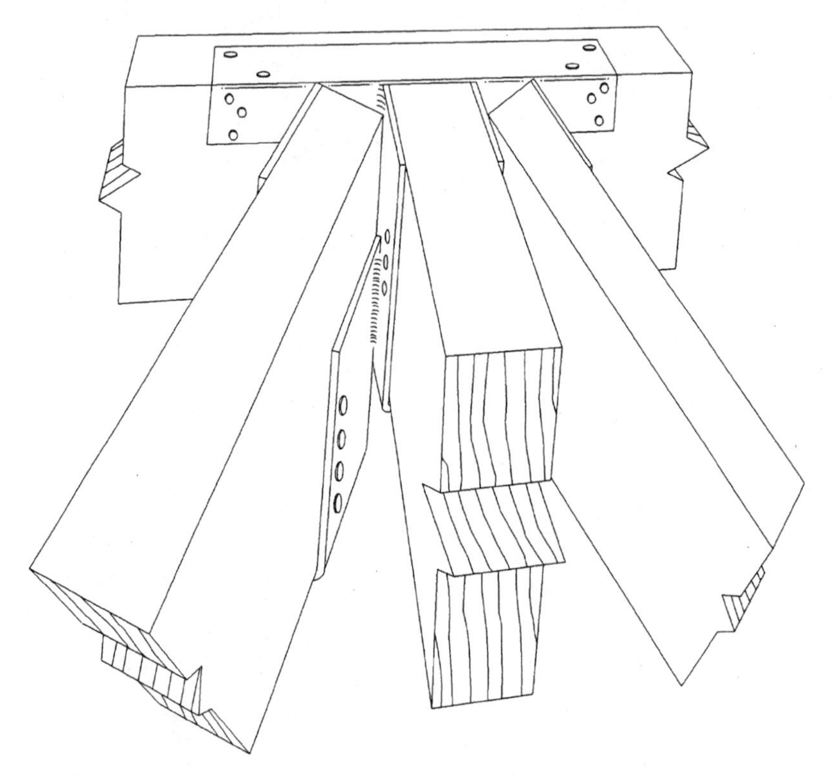

7-19 *This sophisticated hanger, the MSC series, connects multiple members in a variety of configurations.* Courtesy Simpson Strong-Tie Company, Inc.

members within a range of about 40 to 50°, allowing on-site adjustment of framing in problem areas. Web stiffeners are required when using any of these hangers in conjunction with I-joists.

VPA (variable-pitch adjustable connector)

This is the ideal connector for installing I-joists or solid sawn lumber joists on a slope (Fig. 7-20). The connector is used at the point where

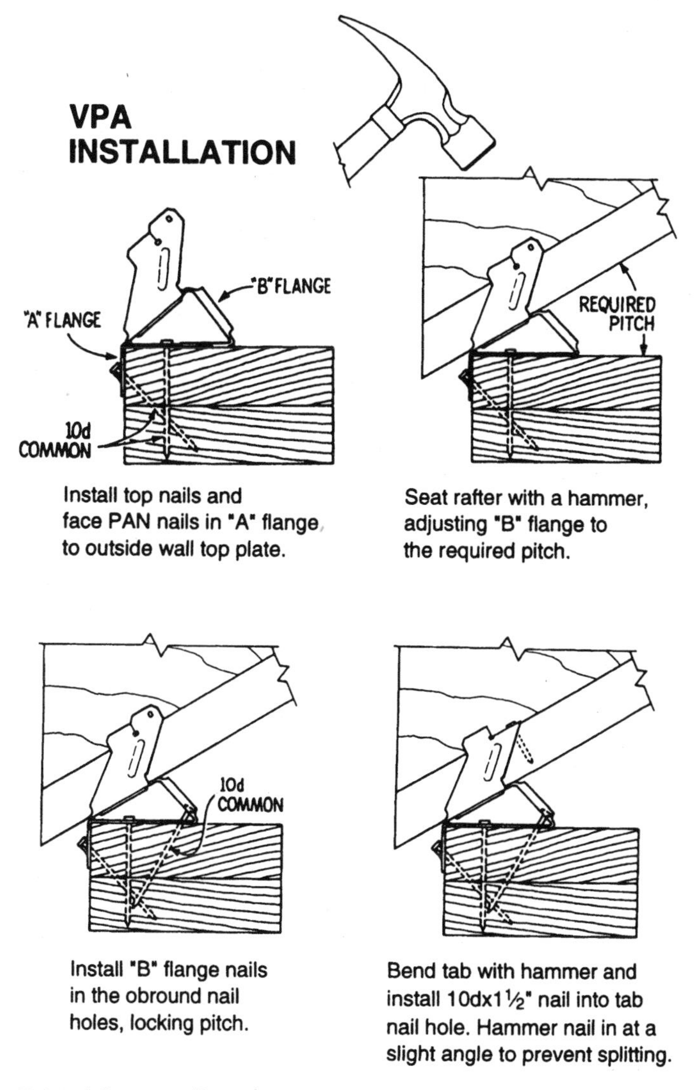

7-20 *The installation procedure for the VPA hanger, which is site-adjusted to accept rafters on a variety of pitches.* Courtesy Simpson Strong-Tie Company, Inc.

the joist makes its intersection with the wall plate, and can be adjusted on-site to accommodate roof pitches from 3 in 12 up to 12 in 12. The hanger is designed with an interlock feature that indicates when the connector is at its maximum allowable pitch.

The VPA connector is ideal for use with I-joists, since it eliminates difficult notching or angled blocking at the joist/plate intersection, and greatly simplifies nailing the joist to the plate without splitting. A bend-tab feature also improves uplift resistance for the I-joist. When used for solid sawn lumber, the VPA eliminates the need for cutting a traditional birdmouth joint in the joist. This allows the joist to maintain its full height, which increases its strength and also increases the amount of insulation that can be placed over the exterior wall plate.

TB (tension bridging)

These are specially engineered metal straps that are perforated with nail holes at each end, and are used in pairs for X-bracing between joists (Fig. 7-21). There are several different sizes available, to accommodate joists from 9½ up to 32 inches high, and with center-to-center spacing of 12 up to 48 inches. In order to be effective, a

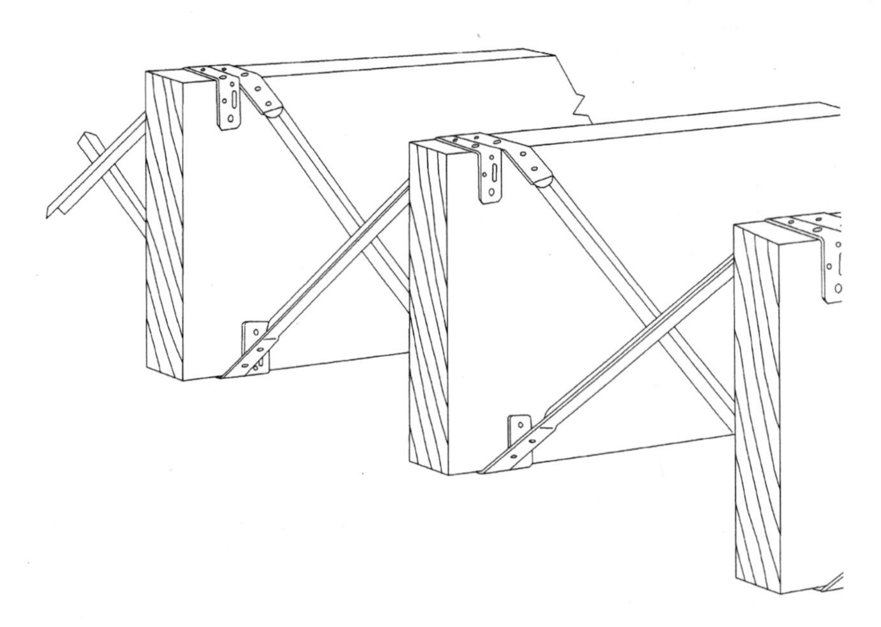

7-21 *Tension bridging between joists.*
Courtesy Simpson Strong-Tie Company, Inc.

minimum of two of the seven nail holes at each end must be fastened to the joist.

Z (Z clips)

Z clips (Fig. 7-22) are used to attach 2 × 4 or 3 × 4 flat blocking in between joists (solid lumber, veneer lumber, or I-joists) or between plated trusses. The clips greatly simplify and strengthen the installation of the blocks, which are used to support the edges of sheathing panels.

MTS (twist strap)

A twist strap (Fig. 7-23) is a versatile, site-bent, perforated metal strap. It is designed for use at a variety of framing intersection and connection points to strengthen the connection, while providing greatly increased uplift resistance. Twist straps feature a staggered nailing pattern to reduce the likelihood of splitting the wood, and can be fastened to either wood or masonry framing.

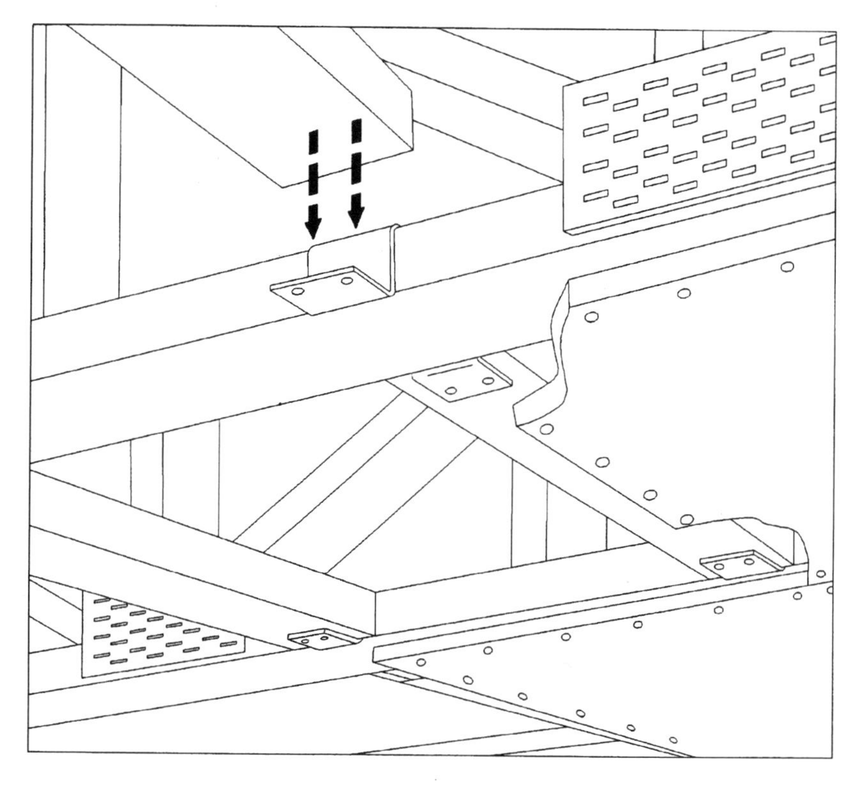

7-22 *Z hangers, for use with ceiling blocking.* Courtesy Simpson Strong-Tie Company, Inc.

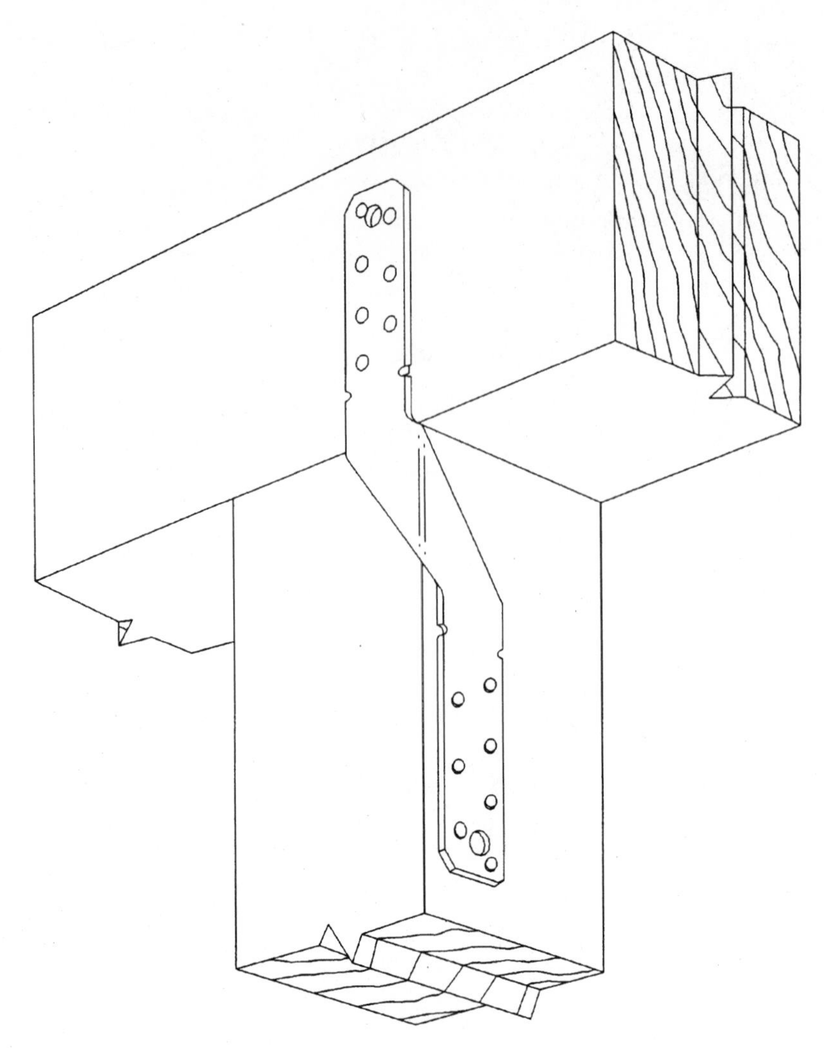

7-23 *An example of an MTS twist strap, another way to reinforce post-to-beam connections.* Courtesy Simpson Strong-Tie Company, Inc.

LTTI (tension tie)

This is a perforated metal strap designed specifically for the purpose of attaching wood-chord open-web trusses to concrete or masonry walls (Fig. 7-24). The strap is attached to the upper truss chord using nails, then bolted through one end into the wall.

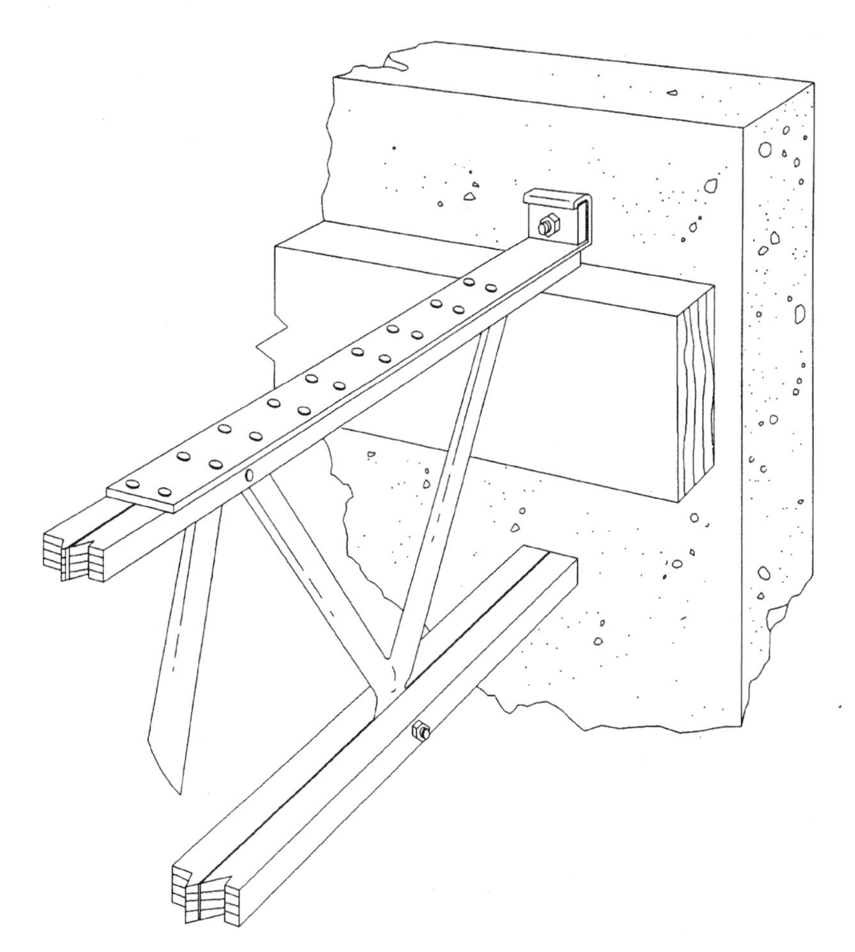

7-24 *A tension-tie anchor, used to connect open-web trusses to concrete. Note the use of the wood nailer to support the truss.*
Courtesy Simpson Strong-Tie Company, Inc.

CSC (ceiling support clip) and FSS (furring stabilizer strap)

These two parts (Fig. 7-25) combine to provide an easy method of installing metal furring channels, while providing a 1-inch space between the channel and the joist for installing Thermafiber insulation. The CSC supports the channel, while the FSS prevents the channel from rotating during installation.

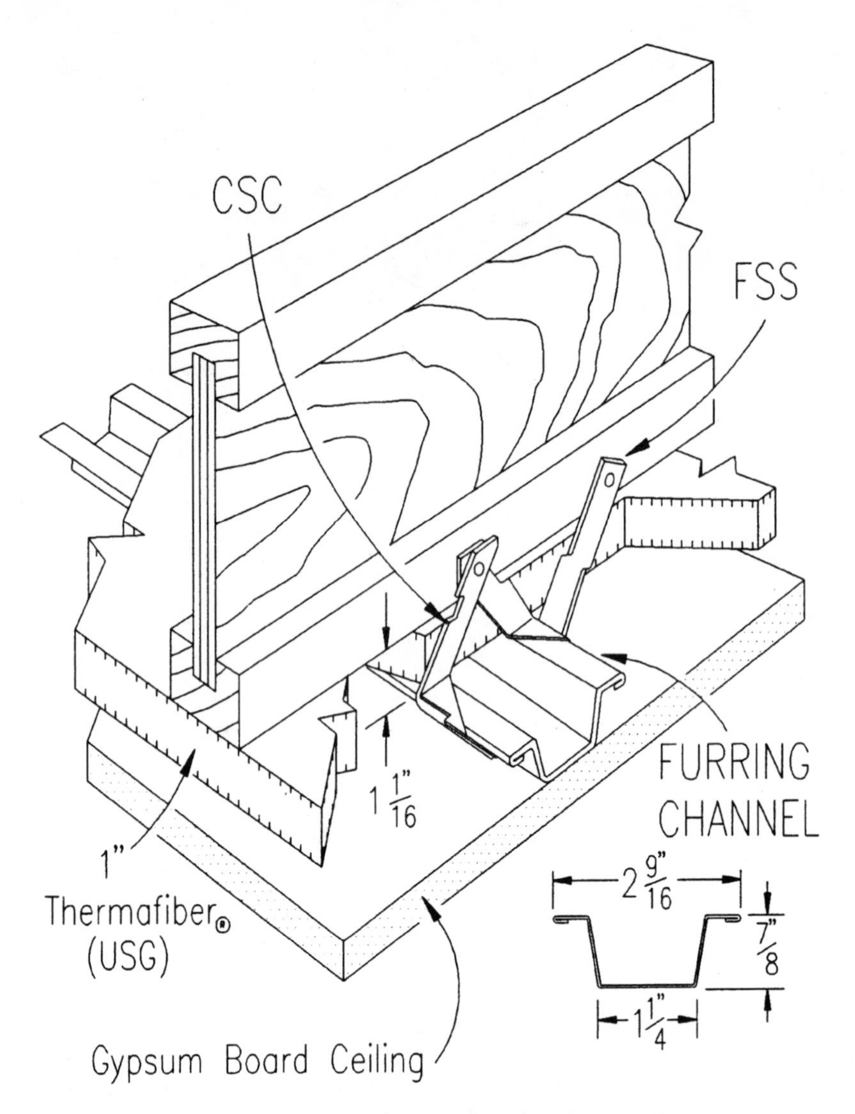

7-25 *CSC and FSS connectors for insulated ceiling applications.*
Courtesy Simpson Strong-Tie Company, Inc.

ST (strap tie) and PA (purlin anchor) series

This series of galvanized metal straps can be used in a variety of framing situations, including ridge ties, joist connections, and wall intersections. The different variations and their uses are as follows: LSTA (light strap tie, revision A) and MSTA (medium strap tie, revision A), which are 1¼-inch-wide, general-purpose strap ties in lengths from 9 up to 36 inches, designed for use with I-joists having a flange narrower than 3½ inches; LSTI (light strap tie I-joist), for use with gun nailing; MSTI (medium strap tie I-joist), to be used for wood members that are 3½ inches wide or wider, and featuring a 3-inch nailing pattern to minimize wood splitting; and PAI (purlin anchor I-joist, see Fig. 7-26) and MPAI (medium purlin anchor I-joist), used specifically for attaching wood I-joists to poured concrete or concrete block walls.

CB (column base)

The same type of column base connectors that have been used for years with solid sawn lumber are also available in sizes for use with composite woods (Fig. 7-27). The standard CB versions are ideal for most applications, with two through-bolt holes for solid attachment to the supported member, and an attached metal plate to prevent contact between the bottom of the post and the concrete slab. Two other versions within this series are the LCB (light column base), which is a lighter, lower-cost post

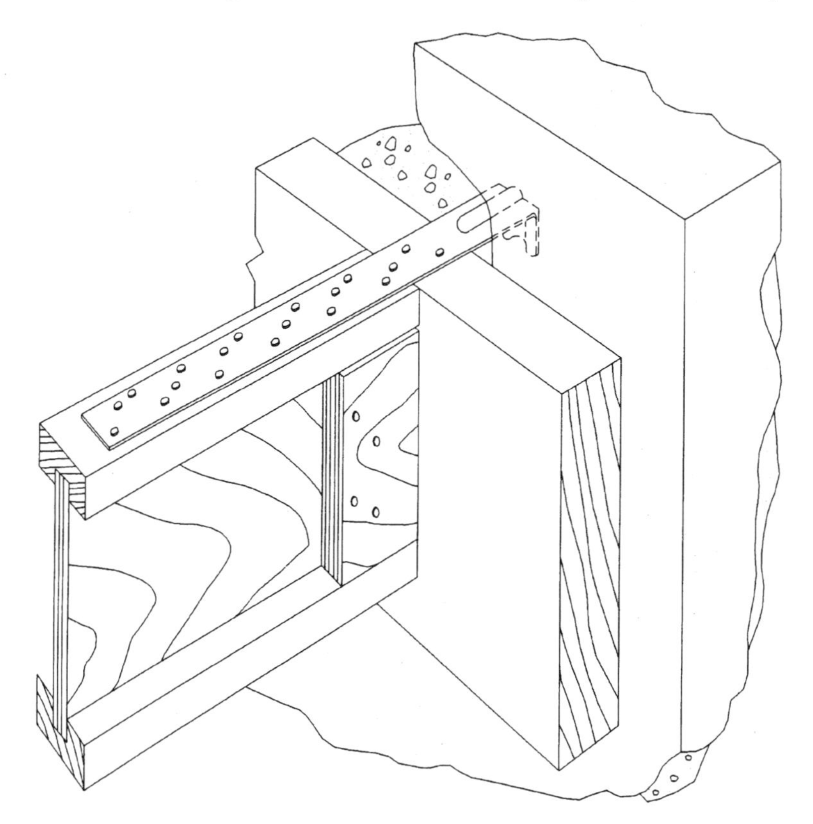

7-26 *Another style of hanger for connecting an I-joist to masonry construction.* Courtesy Simpson Strong-Tie Company, Inc.

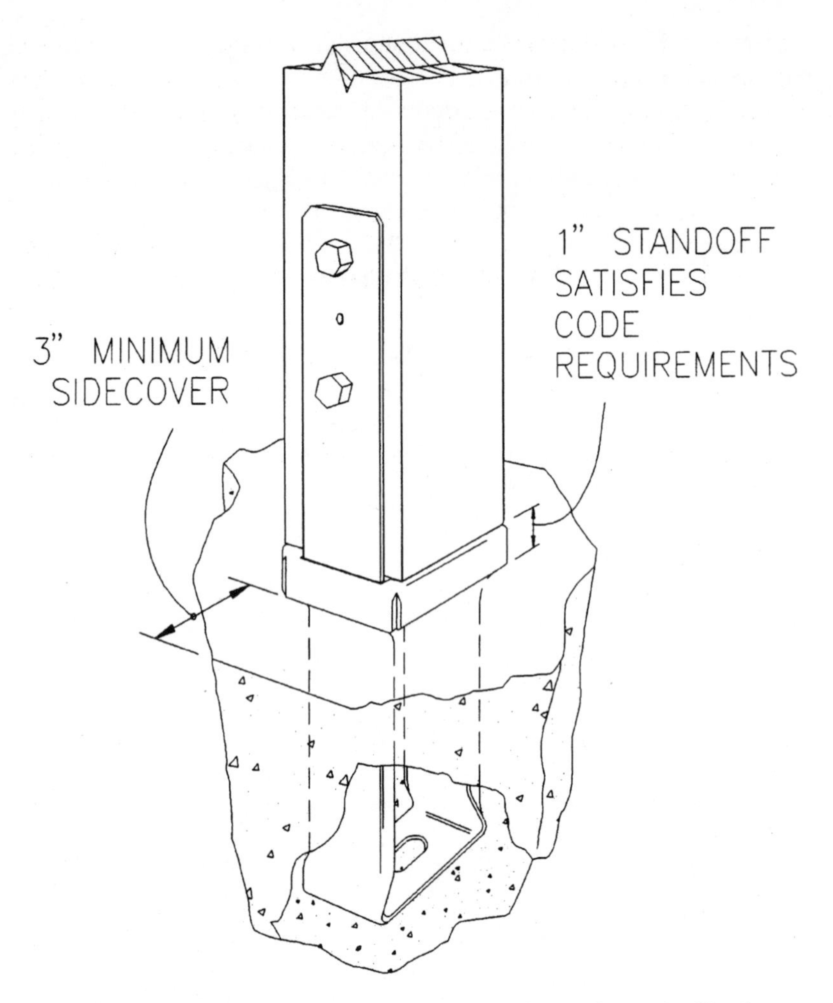

7-27 *A column base connector with a raised metal stand-off to keep the wood above the concrete.* Courtesy Simpson Strong-Tie Company, Inc.

base for use in carports, breezeways, and similar lighter-load applications; and the CBS (column base standoff), which has an attached galvanized plate that raises the post 1 inch above the slab to lessen the risk of contact between the bottom of the post and any surface water, a useful feature in basements and for permanent post or column installations in areas that are exposed to the weather or water splash.

CC (column caps)

As with the column base series that have been on job sites for many years, the equally familiar column cap (Fig. 7-28) is also available in sizes for composite beams and columns. The standard CC consists of a U-shaped seat that is sized for the supported member (the beam or header), and two straps, below the seat and aligned on an axis perpendicular to the seat, that are sized for the supporting post or column. Nail holes are provided for temporary installation and alignment, but bolts are required for permanent installation.

There are several variations of the CC hanger, designed to accommodate a variety of construction applications. The ECC (end column

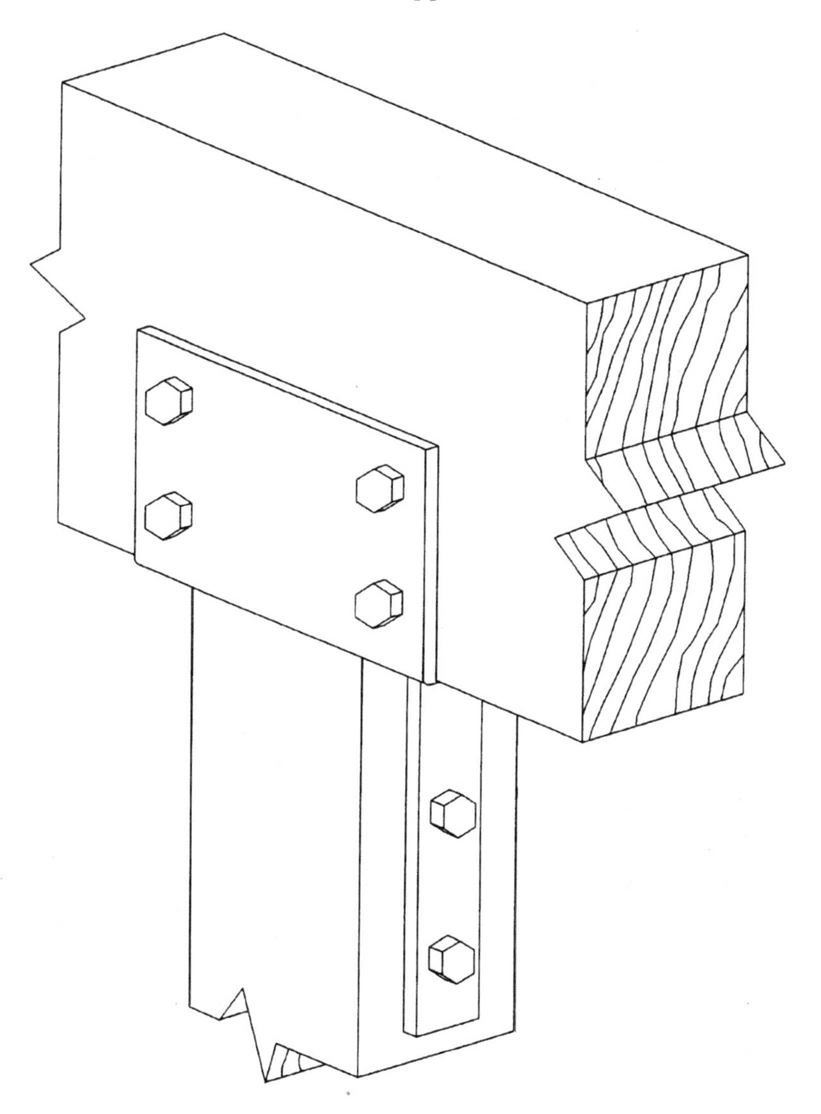

7-28 *A column cap connector, model CC.* Courtesy Simpson Strong-Tie Company, Inc.

cap) is a particularly useful option for situations where the supporting post is at the end of the beam. Here again, the seat is sized for the supported beam and the straps are sized for the supporting column, and the straps can also be ordered rotated 90° to simplify certain installations. The CCO (column cap only) is just the seat for the supported beam, allowing it to be field-welded to pipe or other types of metal columns or supports. CCOB (column cap only, back welded) hangers allow you to have two column caps welded together back to back, for use in cross-beam connection applications. Three other model variations are available factory-welded in many different size combinations for special applications: the CCC (column cap cross) hanger (Fig. 7-29),

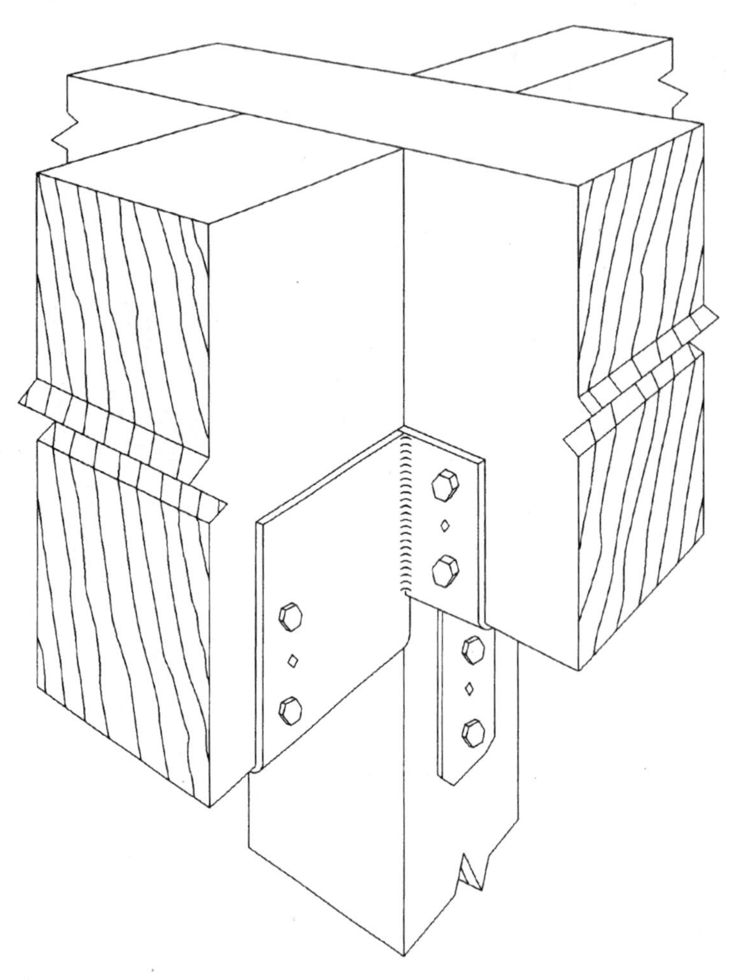

7-29 *Specialized column caps for connecting multiple members in various configurations.* Courtesy Simpson Strong-Tie Company, Inc.

for use in situations where two beams intersect with one another at a cross connection; the CCT (column cap T, see Fig. 7-30), for use where one beam intersects the middle of another beam at a right angle; and the CCL (column cap L, see Fig. 7-31), for beams that intersect one another at the end, in an L shape.

Plated truss connectors

Since the use of manufactured or plated trusses has become such a large part of our industry in the past two decades or so, a number of very strong and versatile metal hangers, clips, anchors, and other connection components have made their way onto the market. This section lists a selection of those connectors from the Simpson Strong-Tie company, from their catalog C-PT94-1.

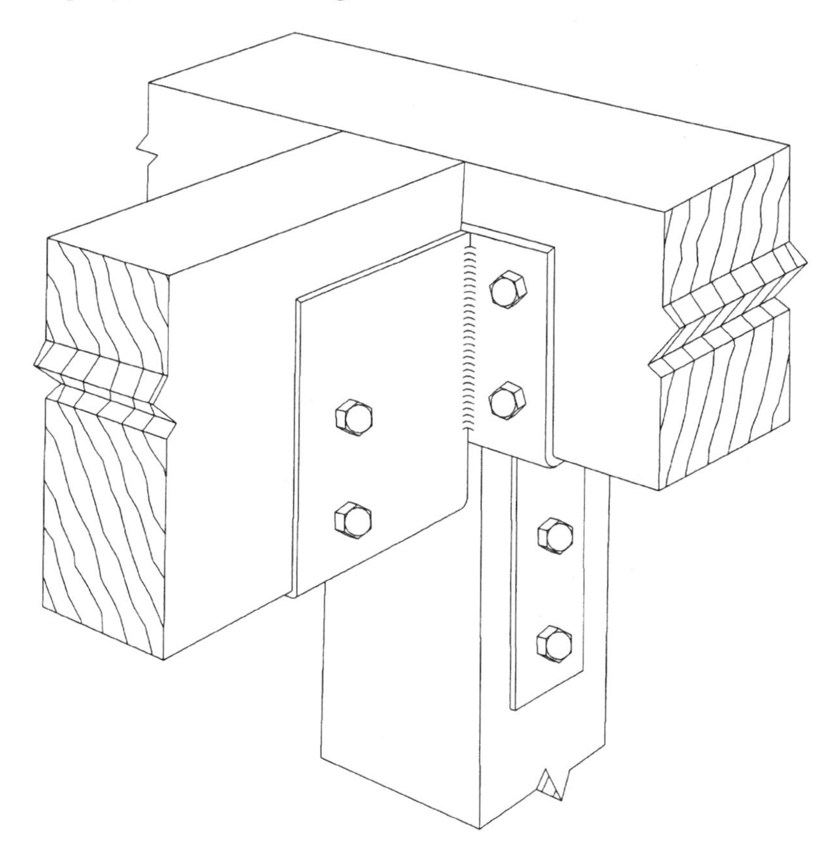

7-30 *Specialized column caps for connecting multiple members in various configurations.* Courtesy Simpson Strong-Tie Company, Inc.

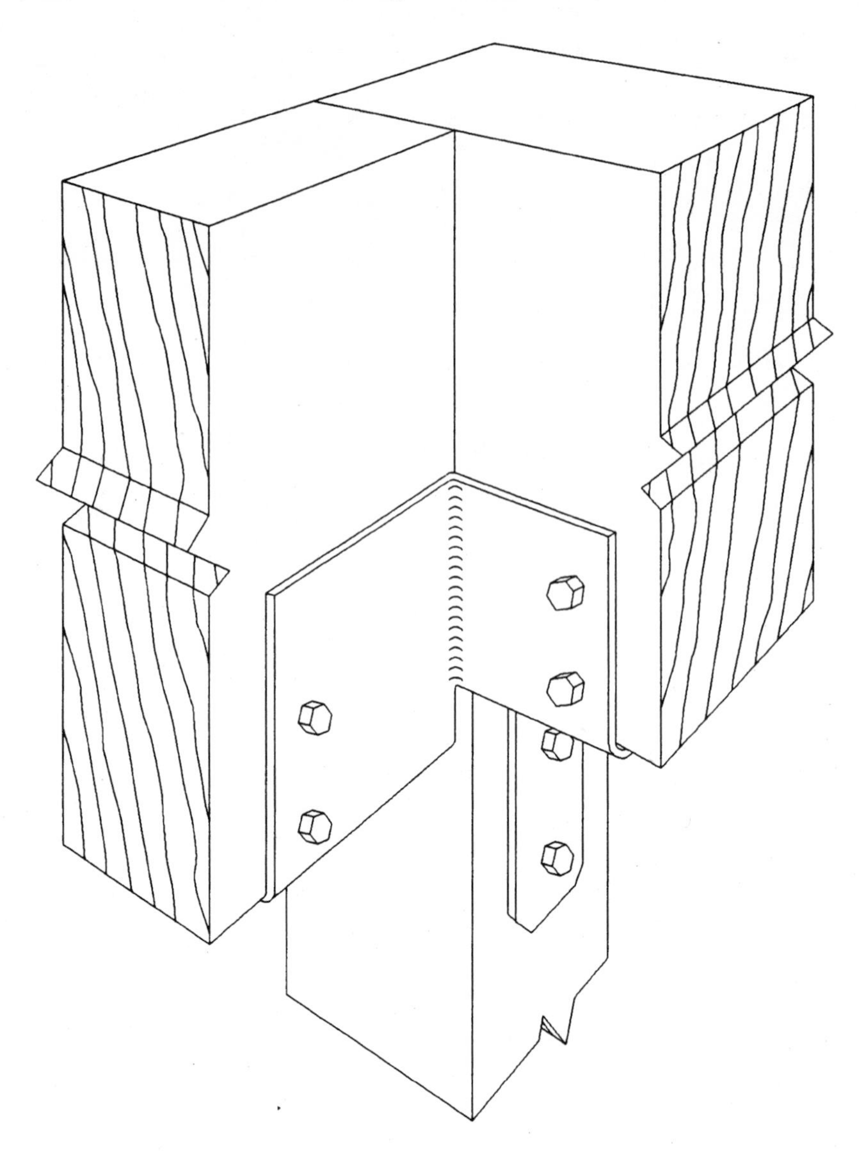

7-31 *Specialized column caps for connecting multiple members in various configurations.* Courtesy Simpson Strong-Tie Company, Inc.

THG (truss girder hanger). This is the basic hanger for attaching a truss to a supporting member (Fig. 7-32). THG hangers are available in a variety of different configurations, and can be used in multiple-truss applications for combining up to four trusses in one hanger. The THGAR and THGAL series (skewed truss girder hanger, right or left)

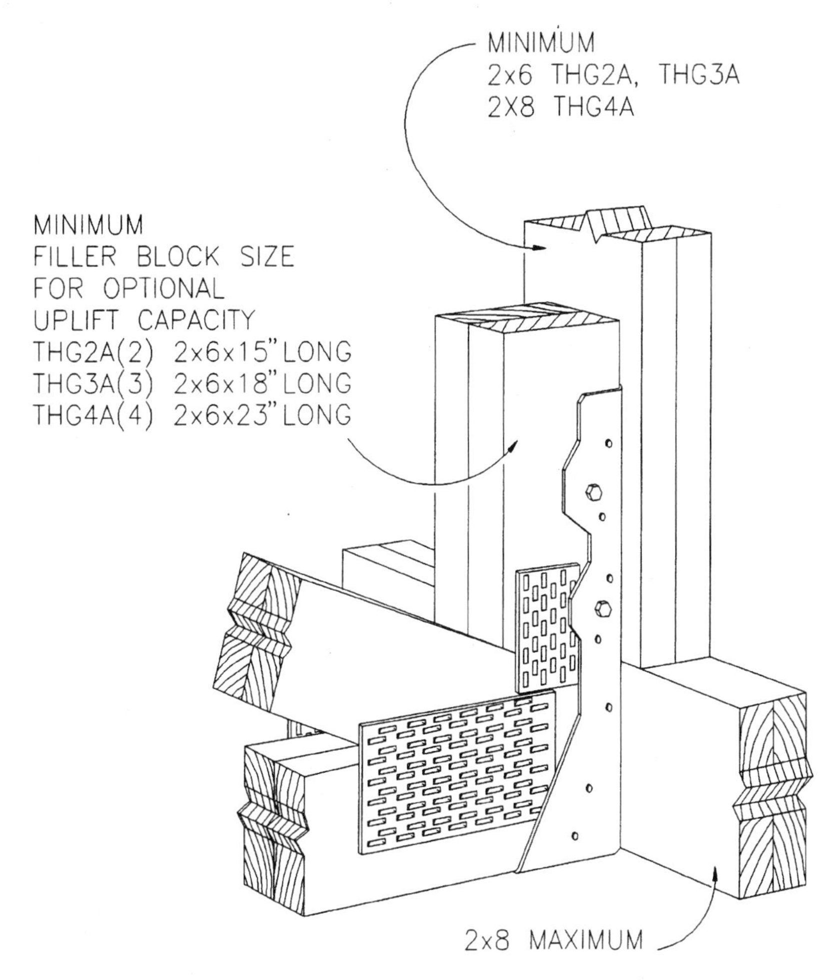

7-32 *THG connectors for truss-to-truss connections.* Courtesy Simpson Strong-Tie Company, Inc.

allows for connections between trusses and supporting members that are skewed at angles other than 90°, or in sloped applications.

THMA (multiple-truss hanger, revision A). This specialized truss hanger (Fig. 7-33) is designed to accommodate the connection of three single trusses to a supporting member, each at a 45° angle to the next. The THMA-2 accommodates one single truss and two truss pairs (truss pairs used in these applications must be fastened to each other as well as to the truss hanger). Both styles of hanger are constructed of 12-gauge steel.

THJR-2 and THJL-2 (multiple-truss hanger, right or left). This is similar to the THMA hanger, but is designed to accommodate the 45°

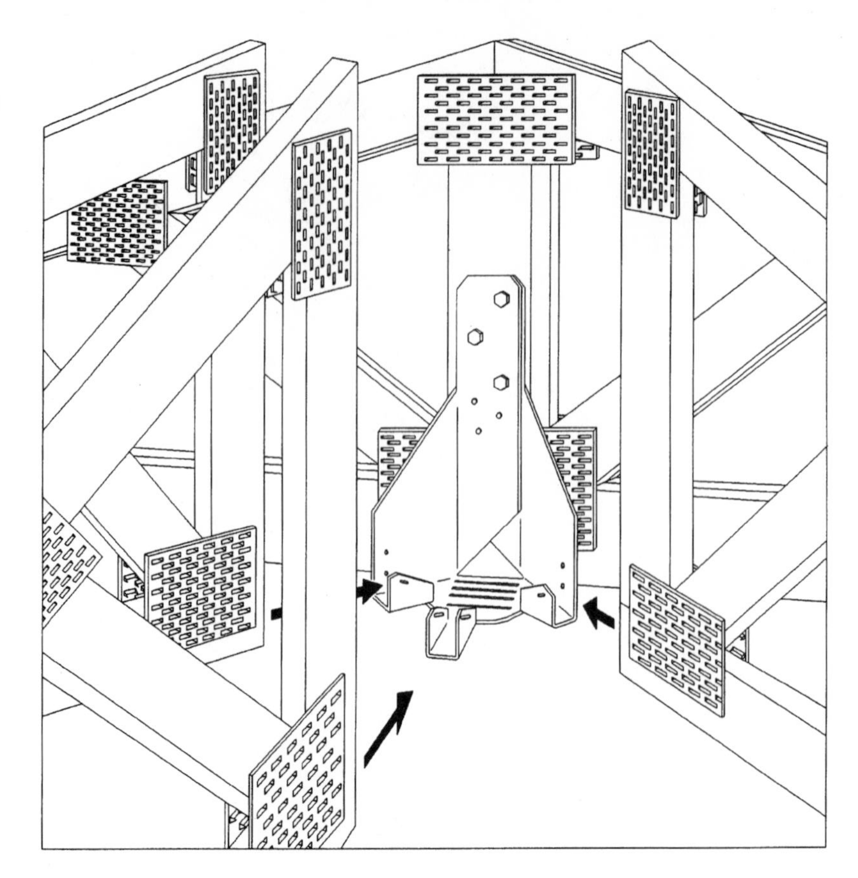

7-33 *A THMA multiple-member hanger.* Courtesy Simpson Strong-Tie Company, Inc.

connection of one single truss and one truss pair (Fig. 7-34). For right-skewed trusses, specify the R version; for left-skewed trusses, specify L.

THJA26 (truss hip/jack girder hanger, ambidextrous). This is a lighter (14-gauge metal) but more versatile hip-and-jack truss connector (Fig. 7-35). It has the advantages of lower cost and the ability to accommodate multiple framing situations with a single connector. The design also allows the hanger to be installed after the trusses are in place.

LTHJR and LTHJL (light truss hip/jack hanger, right or left). The LTHJ (Fig. 7-36) is a lighter-weight, 18-gauge hip/jack connector for lighter truss loads, accommodating the connection of two single trusses. Different versions and options are offered, including the standard R and L models (skewed 45° right or left), and special-order versions with either left or right skews of between 45° and 67½° or hip slopes of up to 45°.

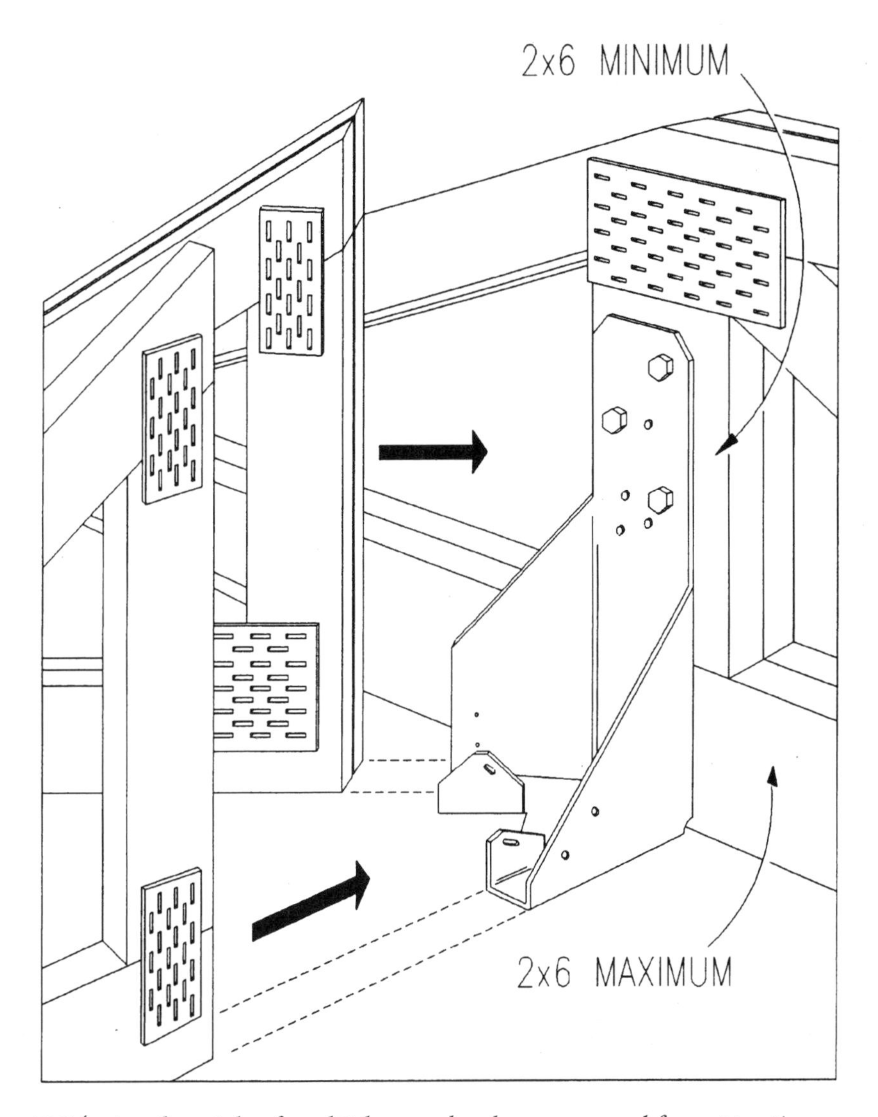

7-34 *Another style of multiple-member hangers, used for supporting two members.* Courtesy Simpson Strong-Tie Company, Inc.

TBE (truss-bearing enhancer). The truss-bearing enhancer (Fig. 7-37) is designed to be used in pairs to help transfer truss load to the wall plates. They are especially useful in situations where the bearing surface is limited, and in areas where an increase in uplift capacity is required. TBE connectors must be installed in pairs.

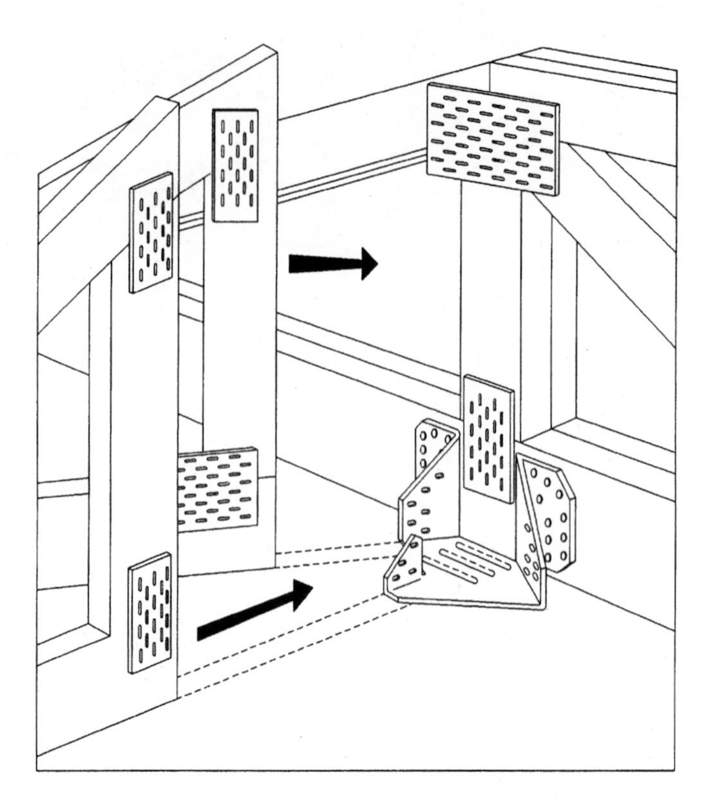

7-35 *The THJA hanger can accept left- or right-mounting applications for two intersecting members.* Courtesy Simpson Strong-Tie Company, Inc.

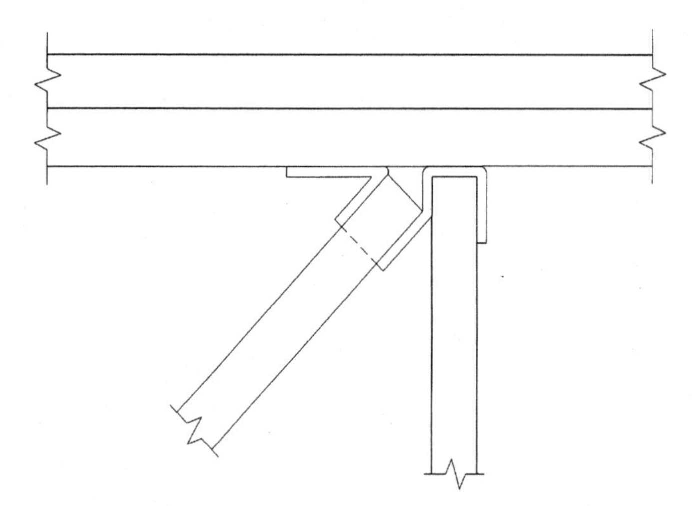

7-36 *A plan view of a two-member connector, the LTHJL (L is for left-hand skew).* Courtesy Simpson Strong-Tie Company, Inc.

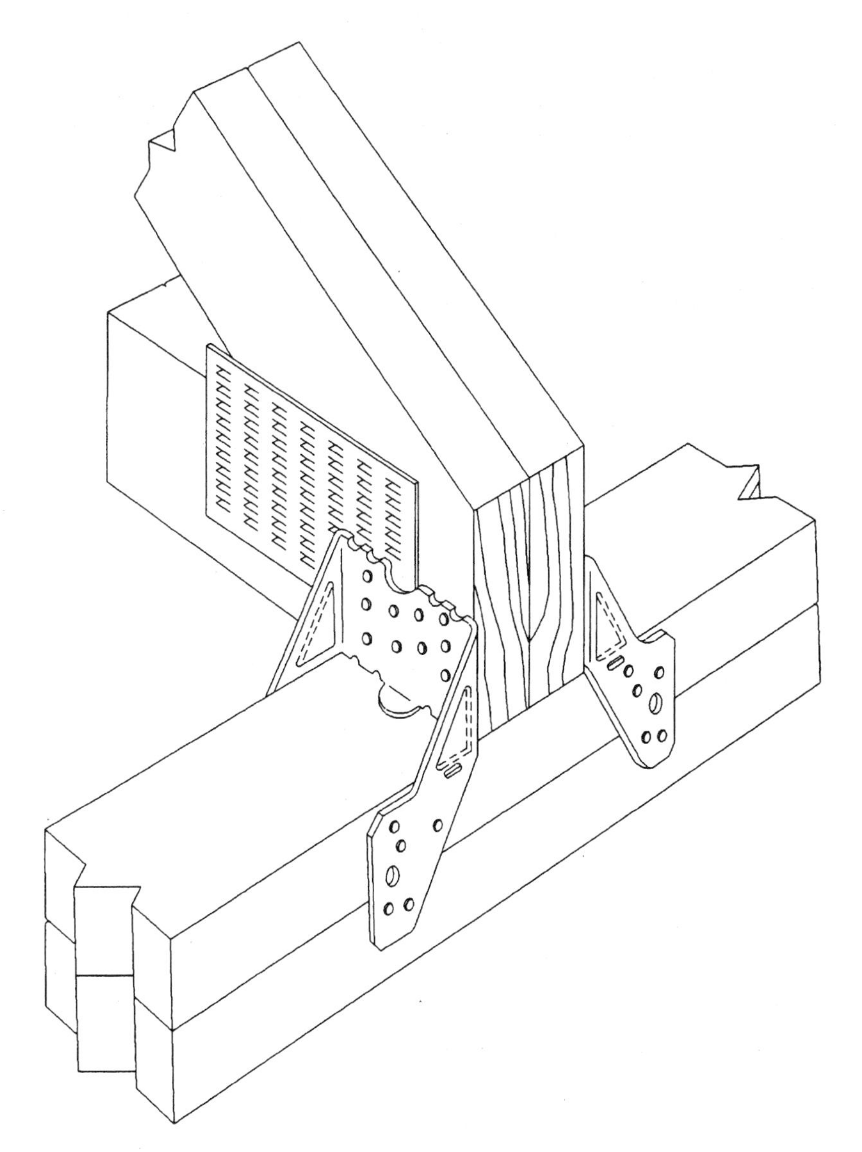

7-37 *Truss-bearing enhancer (TBE), for connecting plated trusses to the wall plates in limited bearing applications. TBEs must be used in pairs.* Courtesy Simpson Strong-Tie Company, Inc.

TC (truss connectors). Truss connectors (Fig. 7-38) are great for making the truss-to-plate connection in areas where additional uplift capacity is required. They also work well for scissor truss connections, since they allow as much as 1¼ inches of horizontal movement in the truss during the installation of roofing materials, while still keeping the wall aligned. Two versions are available: the TC24, for use on 2 × 4 plates, and the TC26, for use on 2 × 6 plates.

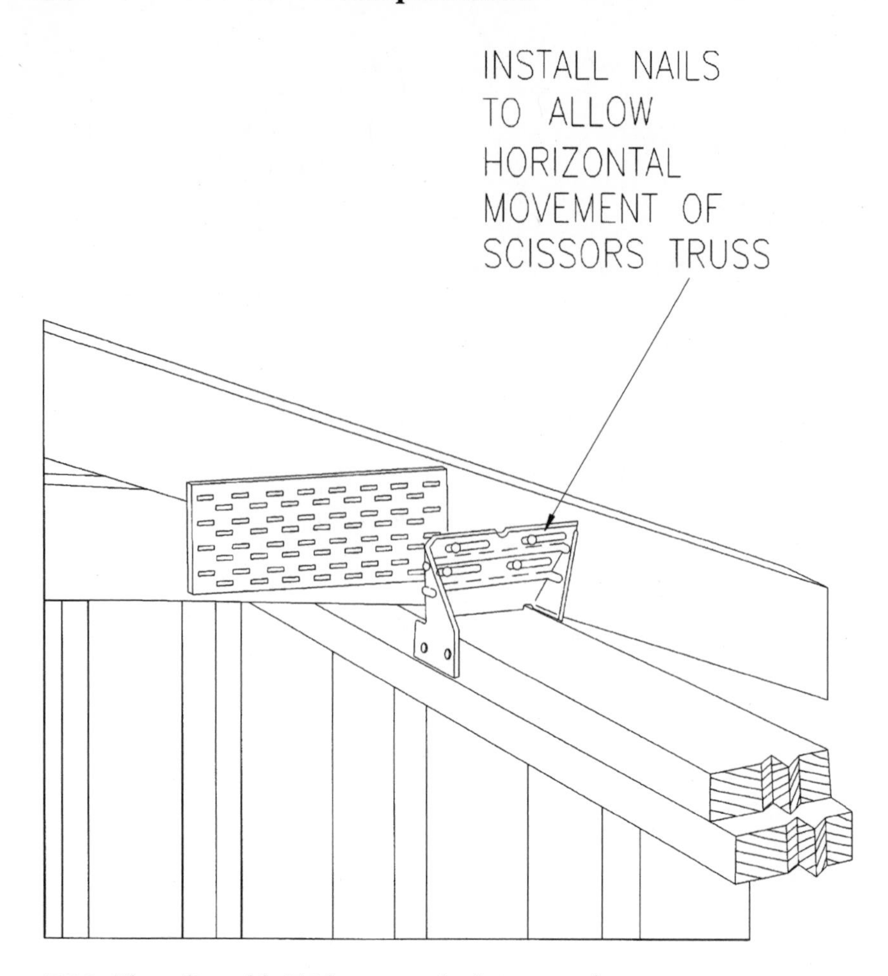

7-38 *The adjustable TC hanger, which is great for scissor trusses.* Courtesy Simpson Strong-Tie Company, Inc.

Other compatible hangers and connectors

In addition to the hangers and connectors specifically designed or modified for use with composite wood and I-beam materials, a number of dimensional lumber connectors are also suitable for use with composite wood. The Simpson Strong-Tie catalog C-96 again offers the largest selection of metal hangers and connectors currently available. What follows is a description of the some of the solid-wood connectors from this catalog that will also work with composite lumber and plated trusses.

ETA (embedded truss anchor). Embedded truss anchors (Fig. 7-39) are used in applications where a plated truss needs to be attached to a poured concrete or poured concrete block wall. The lower portion of the anchor is designed to be embedded in the concrete during the pour, while the upper section is left exposed for later connection to the truss. Variations of the ETA include the HETA (heavy embedded truss anchor), which is constructed of heavier-gauge material for use with heavier loads, and the HETAL (heavy embedded truss anchor, lateral), which has an additional plate near the bottom that, when installed in concrete, increases the anchor's resistance to lateral (side-to-side) movement. The TSS (truss seat snap-in) option works in conjunction with the ETA anchor to create a moisture barrier between the masonry and the bottom chord of the truss.

ACE (post cap plate). These are light-gauge, L-shaped plates used to reinforce the connection between a post and a beam (Fig. 7-40).

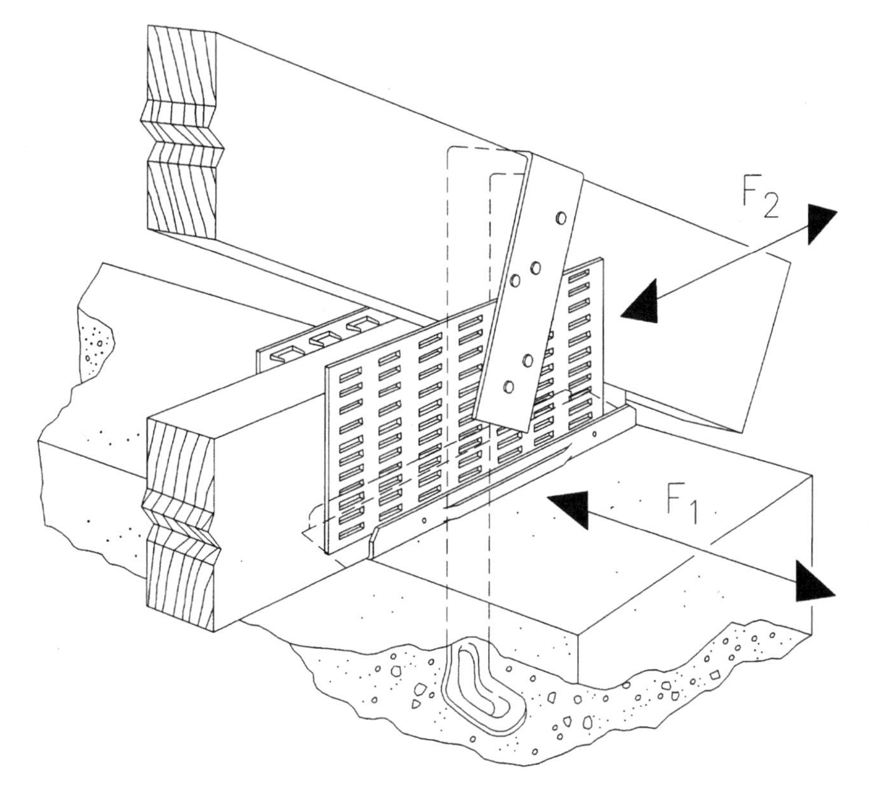

7-39 *A different type of connector for use with trusses in masonry construction. Note the seat that the truss sits in to prevent contact with the masonry.* Courtesy Simpson Strong-Tie Company, Inc.

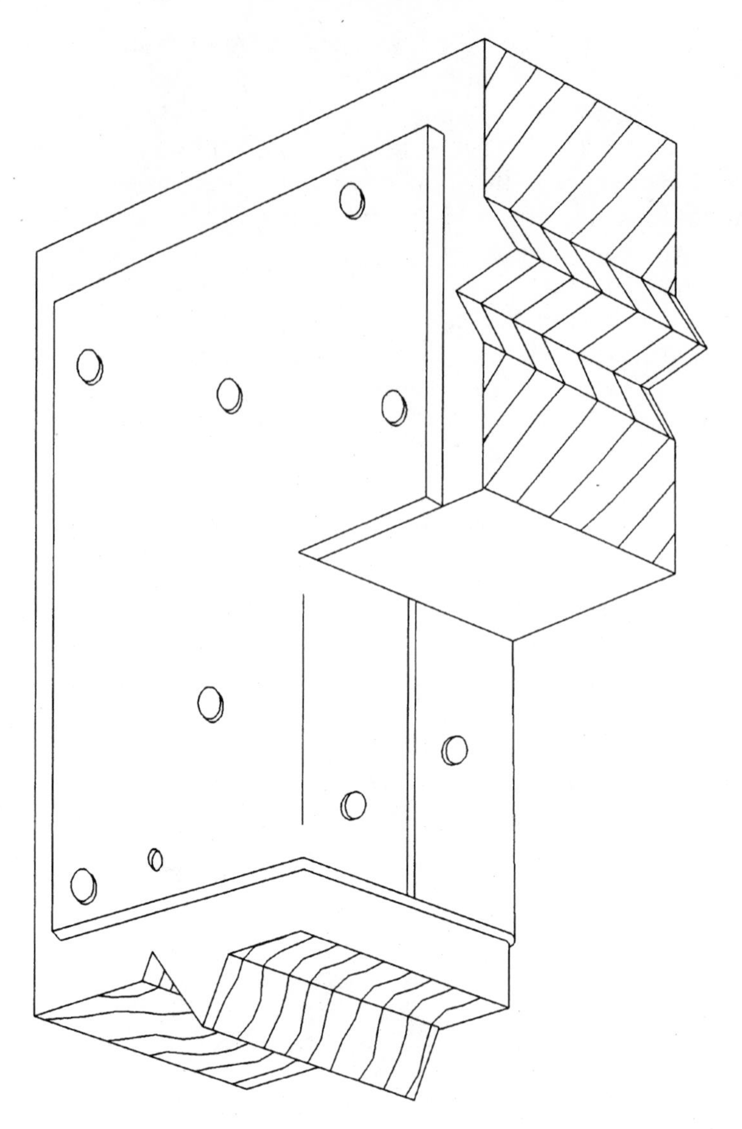

7-40 *A simple ACE hanger for the end of a beam.*
Courtesy Simpson Strong-Tie Company, Inc.

They are available in left and right versions, and should be applied in pairs to achieve the full rated strength.

L (L bracket). This is a handy metal L bracket (Fig. 7-41), available in four different lengths to work with a variety of framing members and construction applications. L brackets are used for reinforcing joints and providing additional support. A variation of the L is the LS

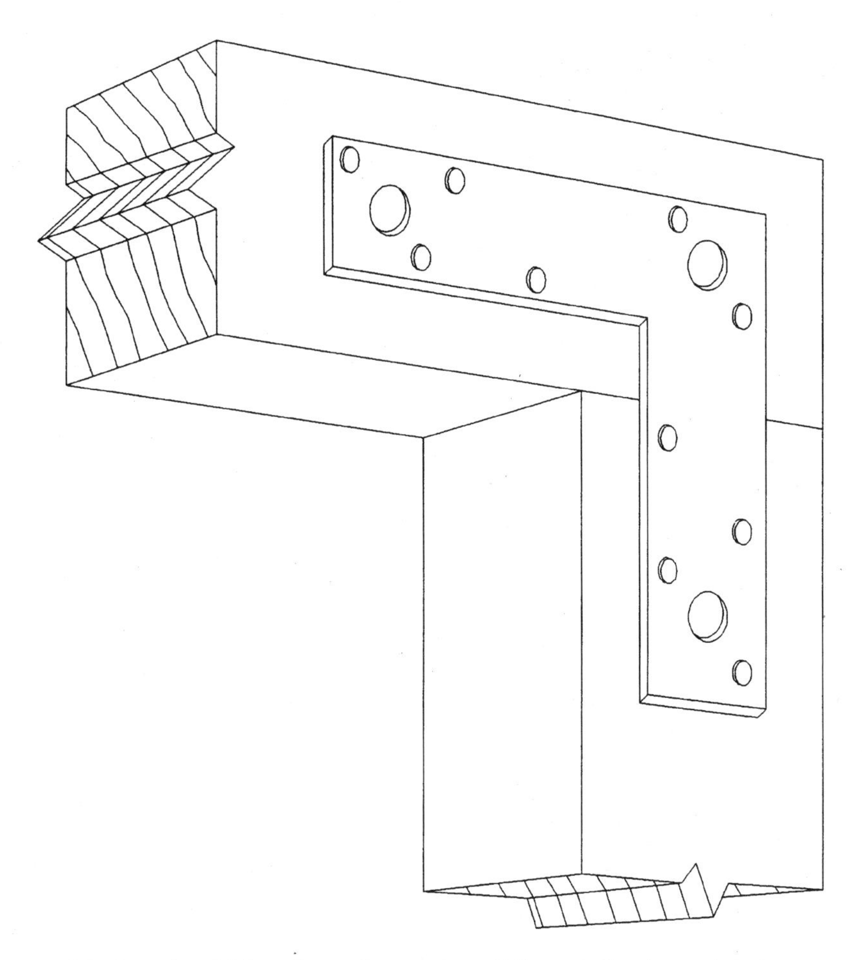

7-41 *Standard L brackets also work well for reinforcing a joint.*
Courtesy Simpson Strong-Tie Company, Inc.

(L bracket, skewable), which is also available in four lengths, with the additional feature of being able to be site-bent to accommodate framing member connections of other than 90°.

A (angle bracket). Angle brackets (Fig. 7-42) are L-shaped metal brackets available in eight different sizes, used as a mechanical reinforcement of various right-angle joints.

H (hurricane ties). This is another handy product (Fig. 7-43) for making and reinforcing a variety of structural member connections, especially at the intersection of rafter to wall plate. H ties greatly increase uplift resistance, and might be required by code in certain applications.

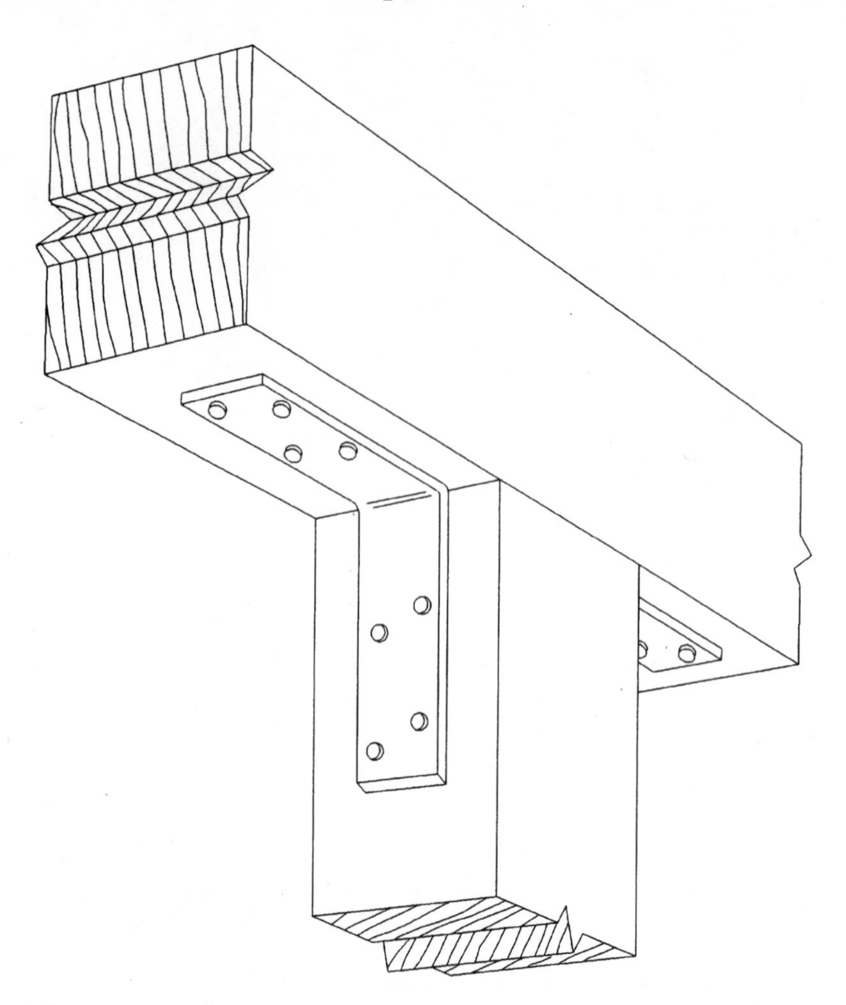

7-42 *Angle brackets for reinforcing a beam-to-post connection.*
Courtesy Simpson Strong-Tie Company, Inc.

Options

The standard hanger or connector is designed around the basic concept of keeping two structural members secured perpendicular to one another in both the vertical and horizontal planes. A perfect example of this is the 90° connection where a floor joist meets a rim joist. When viewed from above or below, the hanger keeps the joist and the rim connected at a right angle. When viewed from the side, the hanger aligns the joist to create a flat horizontal plane in relation to the run of the rim.

There are any number of other connection applications, however, where members meet at a joint that is other than perpendicular, or

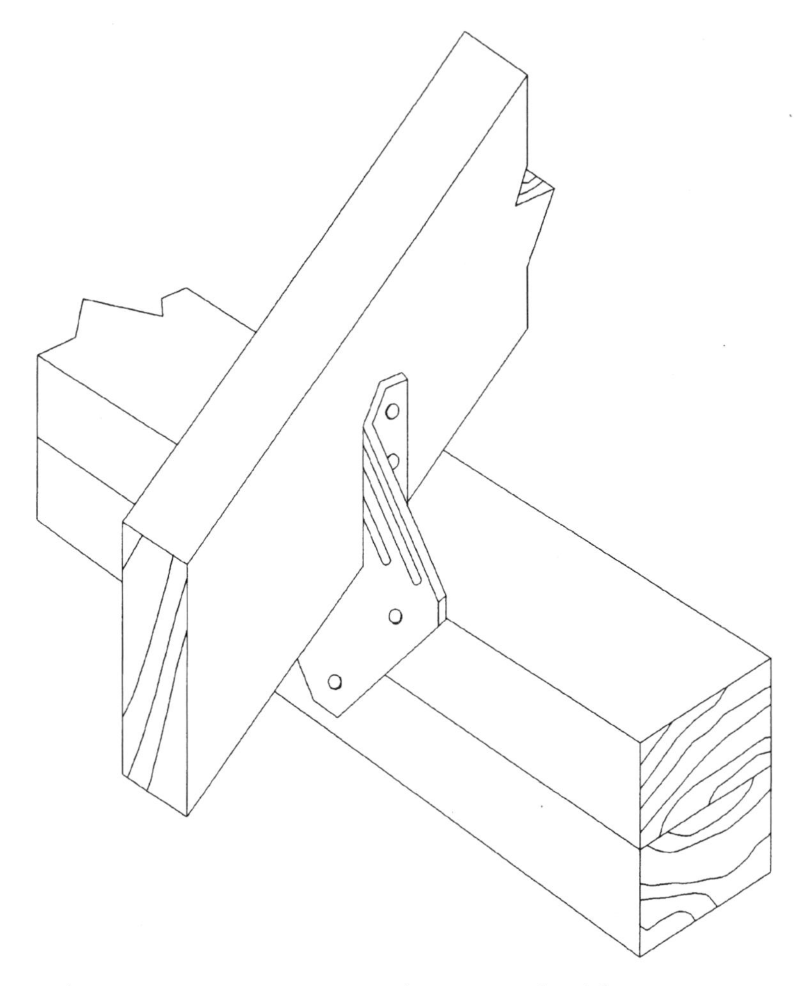

7-43 *A standard hurricane tie for improved uplift resistance.*
Courtesy Simpson Strong-Tie Company, Inc.

where other modifications, such as the location of the nail holes in relation to the hanger flanges, might be necessary or desirable. For that reason, there are several available options in ordering hangers and connectors. Most are stock items, but not all lumberyards carry them, so you might need to order them. Other optional modifications can be done for you at the factory by special order.

It's also important to remember that these hangers are engineered to very strict, tested standards to meet the load and stress rating listed in the catalog. For that reason, you should never modify hangers on site by bending, cutting, welding, or other means, and never use a hanger in an application it's not rated for.

Skewed hangers

One of the most common options is the skewed hanger (Fig. 7-44). This is a hanger that's angled off the perpendicular, as viewed from above or below. A good example of this is a hanger used for connecting a hip rafter to a common rafter or other support at a 45° angle. Depending on the model, many hangers are available in standard 45° skew angles, or by special order in skews of up to 67½°.

Remember that a hanger will be skewed in one direction or the other in relation to the supporting member it's being attached to. For that reason, you need to specify if the hanger skews to the left (designated by an L following the hanger model number) or to the right (designated by an R).

Skewing a hanger will often lessen its load-bearing and stress-resistance characteristics in comparison to those it would have if left at a 90° angle. In the case of standard skewed hangers, you can find modified load tables in the specification catalog; just be sure you are referring to the proper table for the hanger model and angle of skew that relates to the hanger you're using. For custom hangers, request revised load figures from the factory.

Sloped hangers

A sloped hanger (Fig. 7-44) connects one member to another at an angle other than horizontal, as viewed from the side. A common use for a sloped hanger is to connect a rafter and ridge beam, where the rafters slopes down from the ridge. Many styles of hangers and connectors are available, sloped as much as 45°. Here again, refer to that hanger's load tables for specific load information.

Concealed flange

The standard configuration for most hangers is to have the nailing flanges bent outward from the sides of the hanger, leaving the nail heads exposed after the supported member is installed. As an option on some hangers, you can order the nailing flanges bent in, so the supported member conceals the nails. This is useful in a number of applications where the hanger is the same width as the supporting member, leaving no surface to attach the outside nailing flanges to, or in applications where appearance is a factor. Concealed flanges (Fig. 7-45) do not alter the load rating for most hangers, but consult the catalog for specifics.

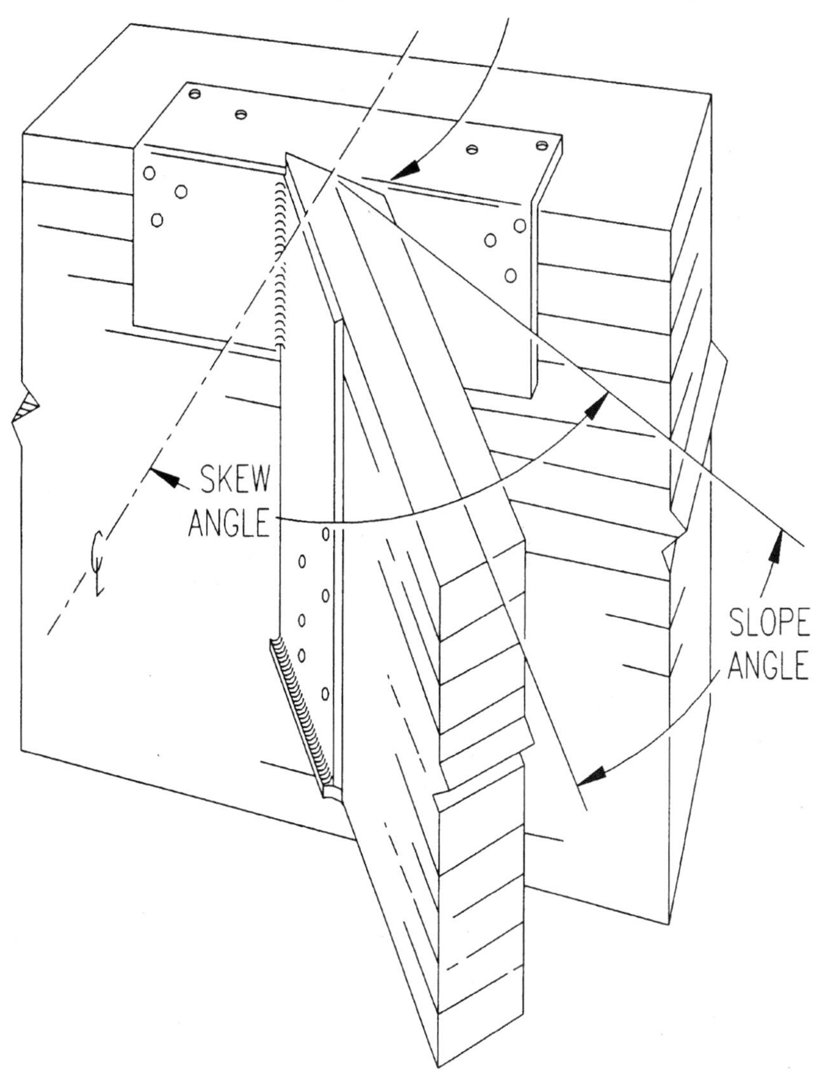

7-44 *An example of slope and skew in a hanger.* Courtesy Simpson Strong-Tie Company, Inc.

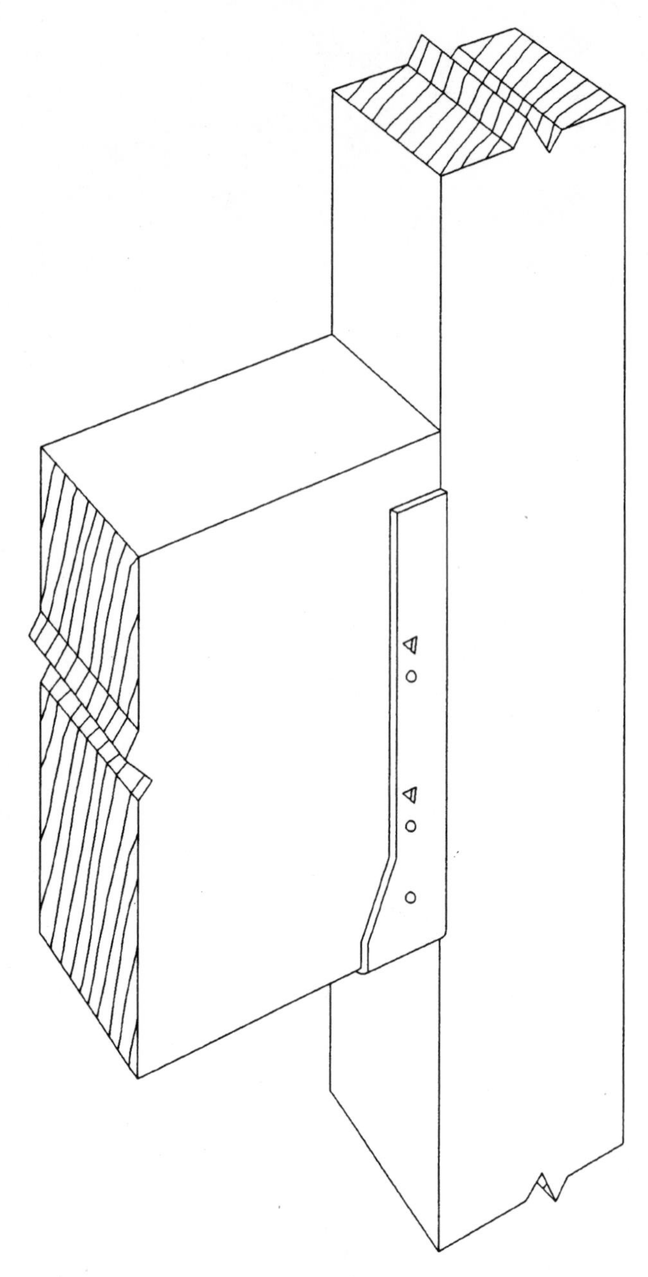

7-45 *Concealed flange hangers are the right choice for applications where both members are the same width.* Courtesy Simpson Strong-Tie Company, Inc.

Saddle hangers

Saddle hangers (Fig. 7-46) are intended to hang over a supporting member and extend down each side of it, like saddle bags on a horse. Most standard top-flange hangers are available as pairs that are welded together at the factory to create a saddle hanger. You'll need to know the model of the hanger you want to use, and the overall width of the member you're hanging them over. For this specification, typically add D (double) to the model number and provide the supporting member width.

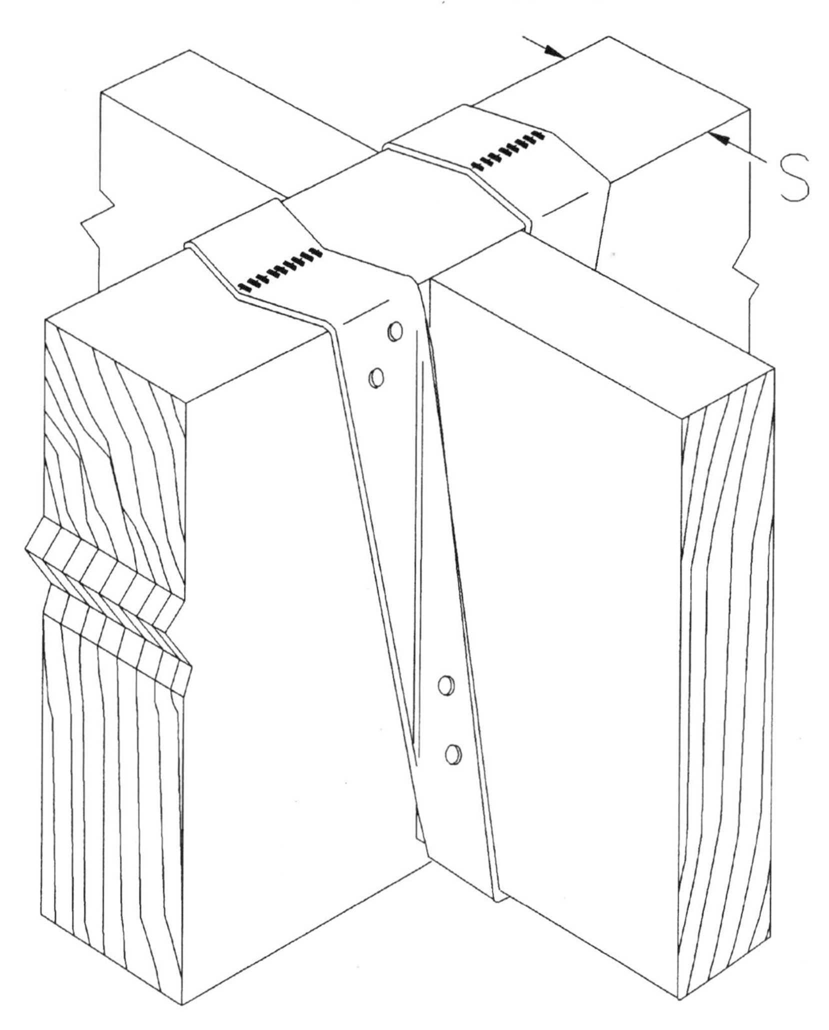

7-46 *Saddle hanger.* Courtesy Simpson Strong-Tie Company, Inc.

Sloped top-flange hangers

You can order certain types of hangers with the top flanges at an angle to the sides and seat of the hanger. These can be used in situations where the supporting member is sloped, but the member you're connecting to it needs to remain plumb (vertical). You can also order sloped top hangers with a seat that's sloped or skewed in relation to the top flange, or both. Some types of hangers can be skewed as much as 84°, with a slope of up to 45°. To order a sloped top-flange hanger (Fig. 7-47), you need to know the angle of the supporting member in relation to the vertical seat and whether you want the top to slope

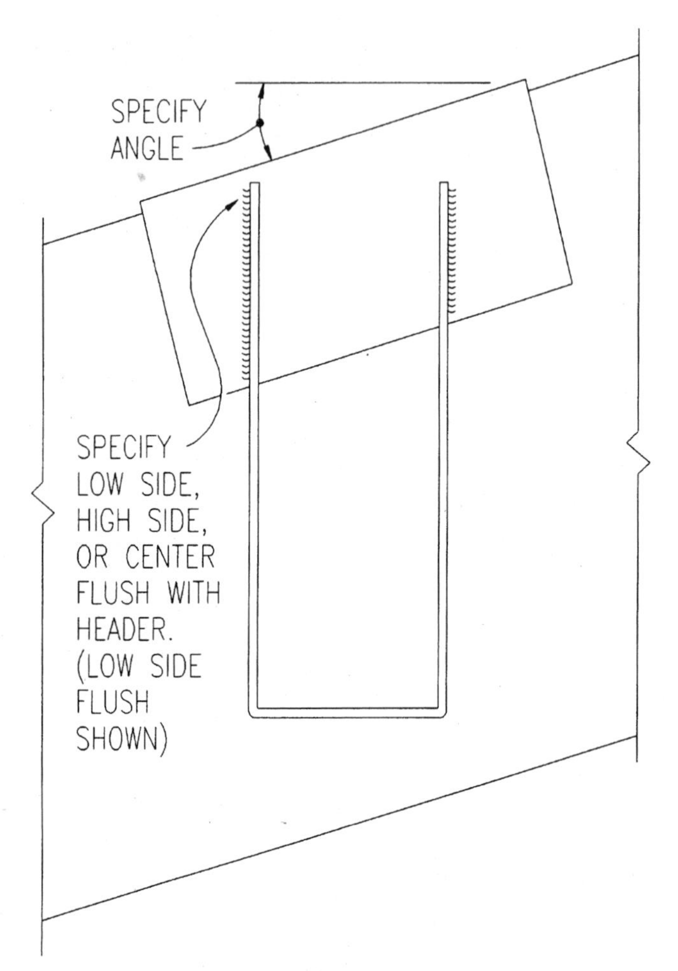

7-47 *A special-order welded hanger that allows the top flange to follow the angle of the roof, while keeping the saddle vertical.* Courtesy Simpson Strong-Tie Company, Inc.

down left or right. You also need to consult the load tables to see how sloping the top flange will alter the hanger's load capacity.

Ridge hanger

A variation of the sloped top-flange hanger is the ridge hanger (Fig. 7-48), which has a top flange that slopes in both directions to accommodate the installation of a supported member directly at the ridge. You can order the hanger so the saddle is centered to the ridge and the top flange slopes equally in each direction, or with a top flange that has two different slopes. You need to specify the exact angle(s) of slope, up to a maximum of 35°.

Offset top flange

You can offset the top flange of certain hangers to the right or left, to allow the hanger to be placed at the end of the header or other supporting

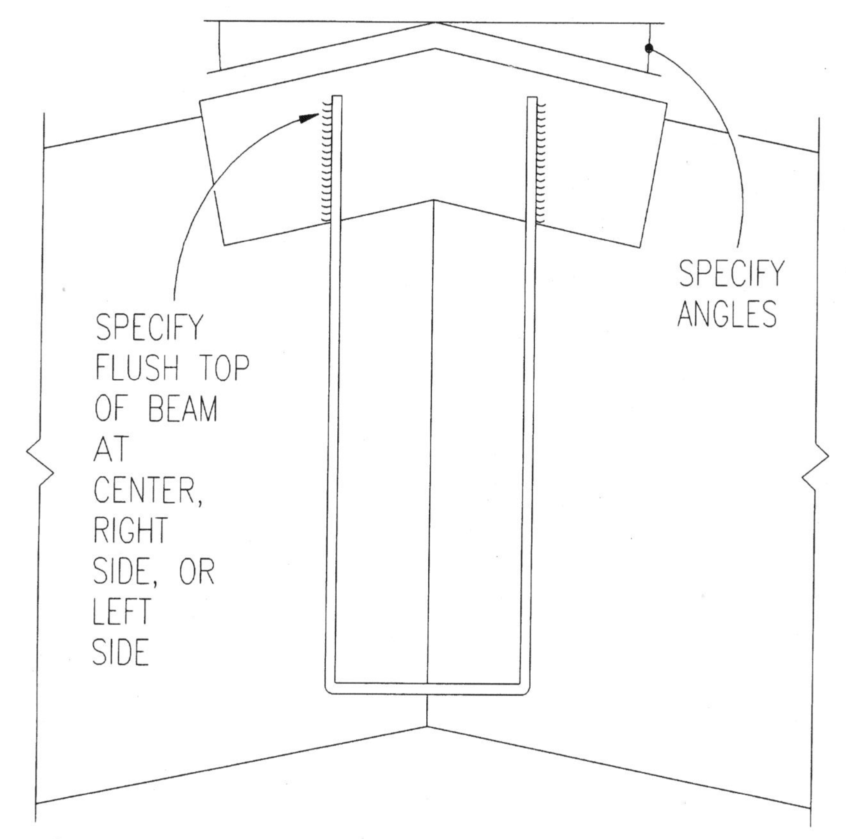

7-48 *For use at the ridge, the top flange can also be ordered with a double slope.* Courtesy Simpson Strong-Tie Company, Inc.

beam without interference from the overhanging portion of the flange. You must, however, specify offset top hangers (Fig. 7-49) as offset left or offset right, with a corresponding reduction in the load-carrying capacity; consult a current Simpson Strong-Tie catalog for specific reductions. Offset top-flange hangers are also available with sloped or skewed seats.

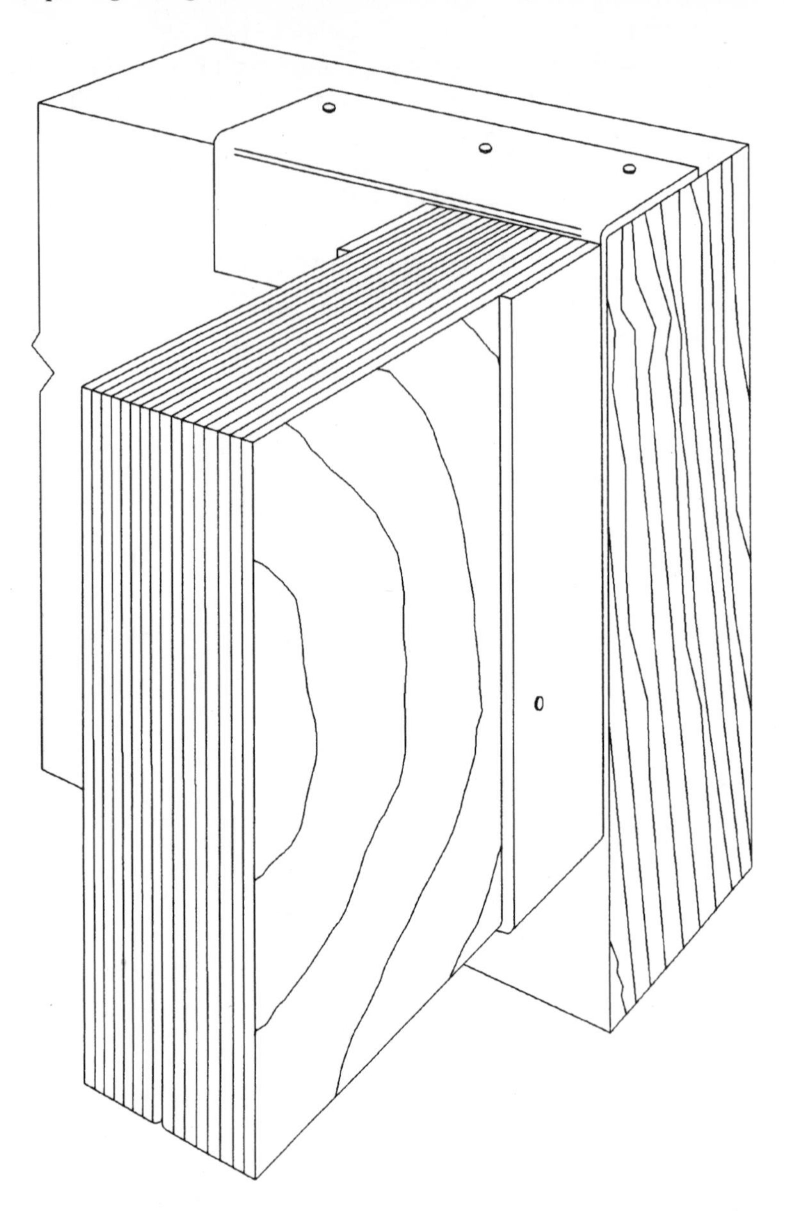

7-49 *Another variation, for use at the end of a beam.*
Courtesy Simpson Strong-Tie Company, Inc.

Open or closed top flange

Top-flange hangers are manufactured with the top flange at a 90° angle to the side flanges. In situations where the supporting member has a beveled top, you can order certain hangers with the top flange either opened past or closed down below 90°, to a maximum of 30° in either direction (a minimum of 60° up to a maximum of 120°). You need to specify if the hanger flange is open (more than 90°) or closed (less than 90°), and the exact angle desired. Once again, the load rating is reduced, which you can recalculate by referring to to the manufacturer's instructions.

Installation

It's important to remember that metal hangers and connectors, of any type and from any manufacturer, play a crucial role in the ultimate strength and stability of the building you're constructing. Some builders view hangers as simply a convenient "pocket" to slip a board into, speeding up the framing process and reducing the number of nails that have to be driven. But hangers are actually very carefully engineered and tested pieces of hardware, each with a specific load and uplift rating and each designed for use with wood members of specific sizes and types.

In order to achieve the full structural rating of any hanger, *you must install it in the manner for which it was engineered, using the proper size, type, and quantity of fasteners* (Fig. 7-50). Improperly installing a hanger or connector will result in a serious reduction of its shear strength and load rating, and can also result in the complete—and possibility catastrophic—failure of the joint. When installing metal hangers and connectors, keep the following warnings in mind:

- Never guess or make assumptions about what a hanger or connector is designed or intended to be used for.
- Fill all the nail and/or bolt holes in each hanger, including the nailing flanges, top flanges, and nailing tabs, if it comes so equipped. Remember, however, that some of these holes could be optional, depending on the load rating or uplift capacity you're trying to achieve, so consult the manufacturer's nailing schedule for specific instructions.
- Never use a hanger for a structural member or application other than the specific one for which it was intended.
- Never cut, bend, weld, drill, or otherwise alter a hanger or connector.

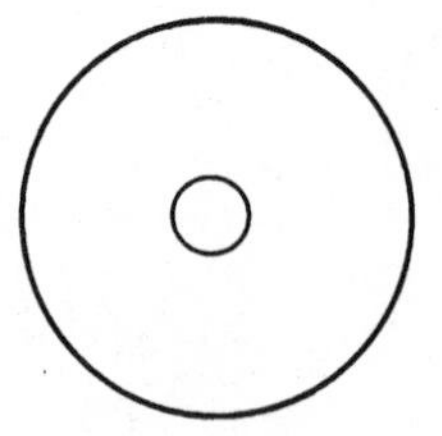

Round Holes
All holes must be filled except for the THAI adjustable height hanger. Refer to load tables for THAI nail quantities.

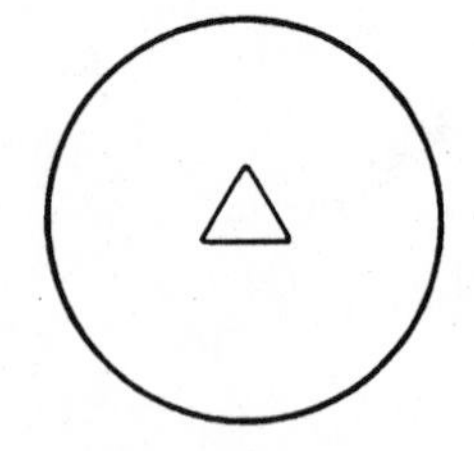

Triangle Holes
Provided on some products in addition to round holes. Round and triangle holes must be filled to achieve the published maximum load value.

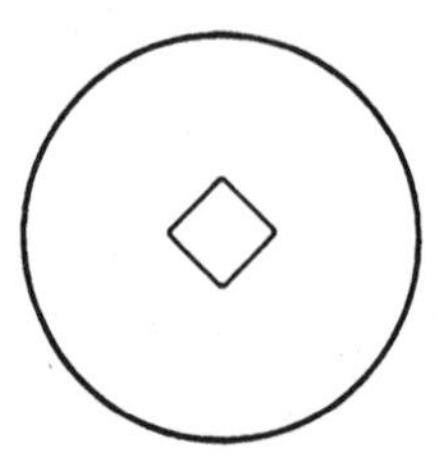

Diamond Holes
Optional holes to temporarily secure connectors to the member during installation.

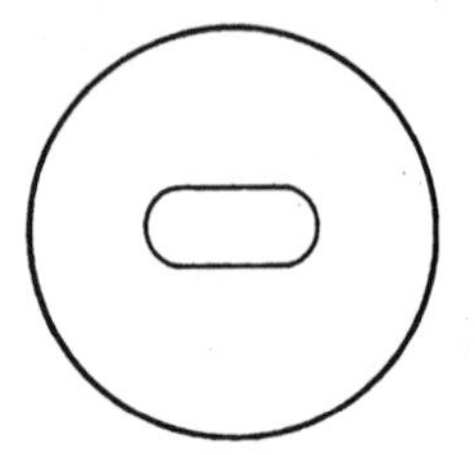

Obround Holes
Used to provide easier nailing access in tight locations.

7-50 *Examples of the four types of holes found in Simpson Strong-Tie hangers.* Courtesy Simpson Strong-Tie Company, Inc.

- If you have any doubts or questions about how or where a specific type of hanger should be used, or, conversely, what the proper type of hanger is for a specific framing situation, consult the manufacturer's catalog or contact the manufacturer directly.

Fasteners

Having the proper type of fastener is equally as important as using the proper hanger. Hanger manufacturers will specify in their catalogs the correct size, type, and quantity of fasteners for use with a particular hanger, and you need to comply with those specifications exactly in order to achieve the full rated load capacity of the hanger or connector. Substituting one type of nail for another might be allowed in some instances, and the manufacturer will typically provide a substitution chart or instructions that addresses this issue.

Most hangers and connectors are intended to be installed with hand-driven (hammered) nails. Air-gun-driven nails might deflect against the metal components of the hanger, lessening the nail's penetration into the structural members and weakening the joint, as well as posing a serious potential injury risk to the installer. Gun-driven nails might, however, still be an acceptable alternative in some situations and, here again, the manufacturer should specifically list whether or not that's an installation option.

In addition to hand- or gun-driven nails, screws offer a third fastener option for some hangers and connectors. Simpson, for example, has created an 8 × 1¼-inch screw with a low, Phillips drive head that works well in certain applications. You should be able to find these screws available wherever you purchase your hangers.

Epoxy-Tie adhesive system

Epoxy-Tie (Fig. 7-51) is a patented Simpson Strong-Tie product designed for high-strength structural anchoring. You can use it for fastening bolts into concrete and masonry walls; for doweling rebar; for repairing cracks in brick, ceramic, unreinforced masonry, and concrete; and for filling in nonstructural cracks up to ¾ inch wide.

The Epoxy-Tie system consists of a two-part epoxy adhesive that's blended in equal parts as needed, then applied. It is cleverly designed for fast, easy, and accurate measuring and application of the epoxy product, with a dual container system that holds the epoxy resin and hardener in separation, and a special disposable mixing nozzle that automatically blends the two together as they're dispensed, ensuring thorough and mess-free mixing.

The ET1.7 kit contains 1.7 ounces of epoxy adhesive for small jobs, and includes a plastic plunger that fits into the pair of tubes for dispensing. For larger jobs, there's the ET22 (22 fluid ounce) or the ET56 (56 fluid ounce) container, each of which fits into special dispensing applicators that somewhat resemble a traditional caulking gun. All three of these products are designed for concrete temperatures of 40° and above. There is also a cold-weather formula, ET22C, for use with concrete temperatures of 25 to 45°.

To use the Epoxy-Tie adhesive in concrete (Fig. 7-52), simply drill a hole in the concrete using a masonry bit, then clean out the hole using a small wire brush and a blast of oil-free compressed air. Dispense the epoxy into the hole, watching to see that the black and

Epoxy-Tie™ Adhesive

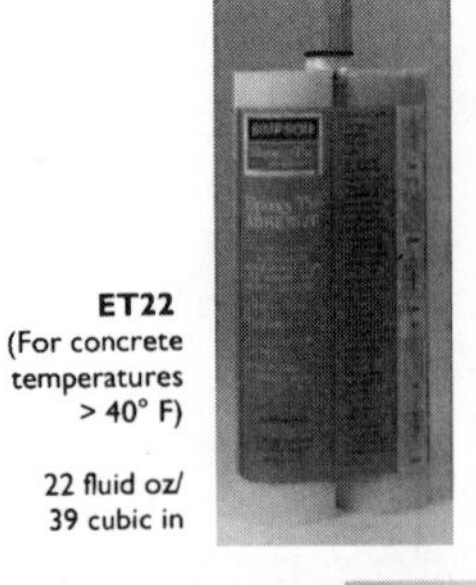

ET22
(For concrete temperatures > 40° F)

22 fluid oz/
39 cubic in

ET22C
(For concrete temperatures 25° F to 45° F)

22 fl oz/
39 cubic in

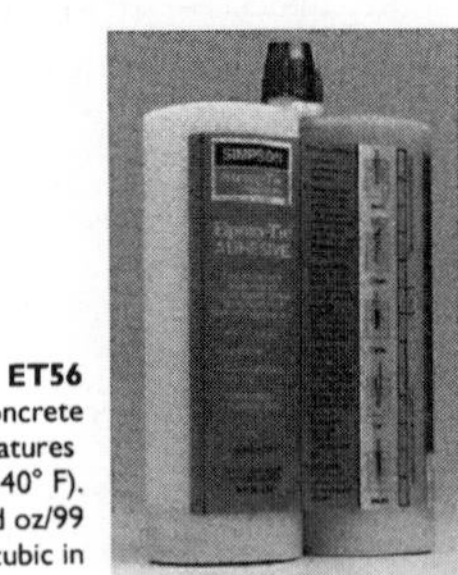

ET56
(For concrete temperatures > 40° F).
56 fluid oz/99 cubic in

EDT22 Dispensing Tool
For use with ET22 and ET22C

Limited one year warranty against defects in material and workmanship

ET1.7KT Kit
Includes an ET1.7 cartridge, a twin plunger and two nozzles

EDT22P Pneumatic Dispensing Tool
For use with ET22 and ET22C

Limited one year warranty against defects in material and workmanship

EDT56P Pneumatic Dispensing Tool
For use with ET56

Limited one year warranty against defects in material and workmanship

EMN22 Mixing Nozzle
For ET22 and ET22C
(EMN56 for ET56 similar)

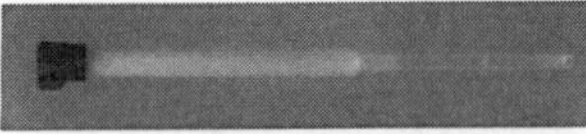

ETB Brushes
For cleaning holes

ETS Screens
For use with brick

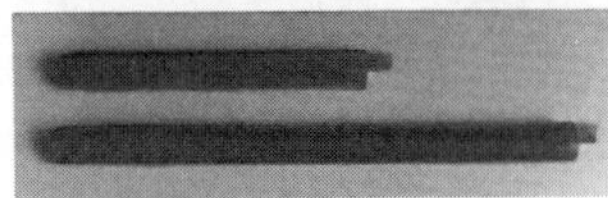

EMN1.7-R Mixing Nozzle
2 replacement nozzles for use with the ET1.7.

7-51 *The Epoxy-Tie adhesive system for masonry. The three styles of guns (center) accept the disposable epoxy bottles for ease of use.*
Courtesy Simpson Strong-Tie Company, Inc.

Epoxy-Tie™ Adhesive

Preparation

» Insert cartridge into dispensing tool.

» Remove plugs from cartridge and purge some epoxy off to the side to set the cartridge wip at the same level. Attach a clean mixing nozzle which is free of gelled or hardened material.

» Tighten retaining nut on nozzle. DO NOT OVER-TIGHTEN.

» **Dispense bead of Epoxy-Tie off to the side to check for proper mixture (a uniform gray color) before using.**

» Caution: Epoxy will start to harden in the mixing nozzle after 7-8 minutes. Epoxy will harden faster as the air temperature increases. Replace nozzle to avoid blowouts.

» Note: When installing the Epoxy-Tie, holes must be free of standing water, frost or ice prior to injection of the epoxy.

» If using a pneumatic dispensing tool, air pressure must be regulated at 80-100 psi.

» Store a partially-used cartridge up to one year at a temperature above 45°F.

Installation into Concrete

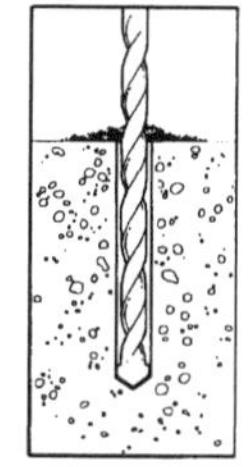

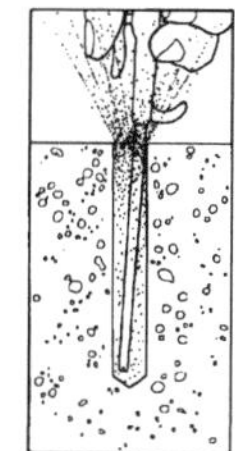

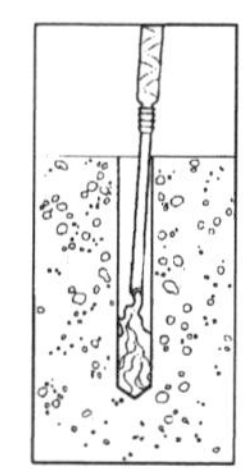

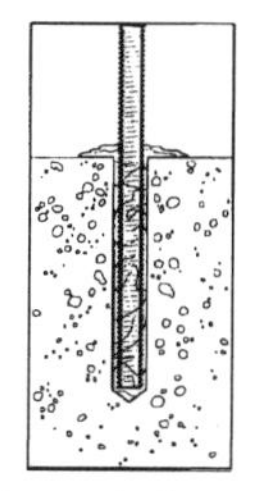

Drill—Drill hole to the specified diameter and depth.

Clean—Remove dust from hole with oil-free compressed air. Clean with nylon brush and blow out remaining dust. Dust left in hole will reduce the epoxy's holding capacity.

Fill—Dispense bead of ET off to the side to check for proper mixture (a uniform gray color) before using. Fill hole halfway, starting from bottom of hole to avoid air pockets. Withdraw nozzle as hole fills up.

Insert—Anchors must be clean and oil free. Insert anchor, turning slowly until the anchor hits the bottom of the hole. Do not disturb during set time.

Installation into Brick or Masonry

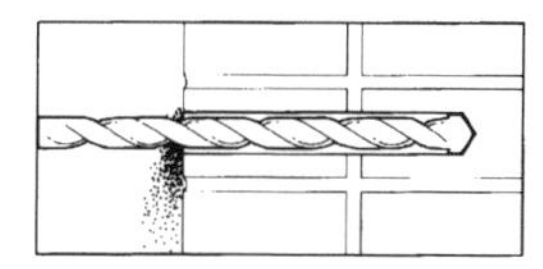

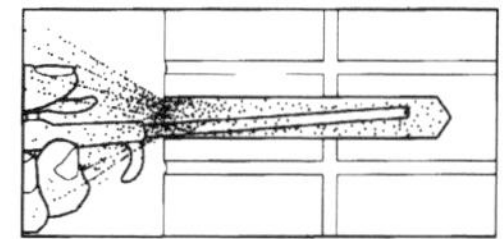

Drill—Drill hole to specified diameter and depth.

Clean—Remove dust from hole with oil-free compressed air. Clean with nylon brush and blow out remaining dust. Dust left in hole will reduce the epoxy's holding capacity.

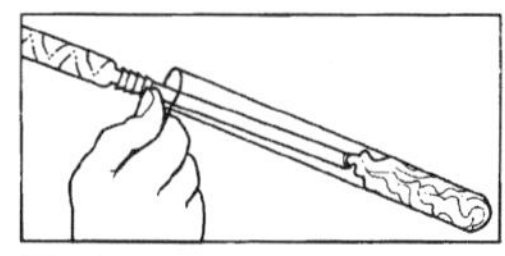

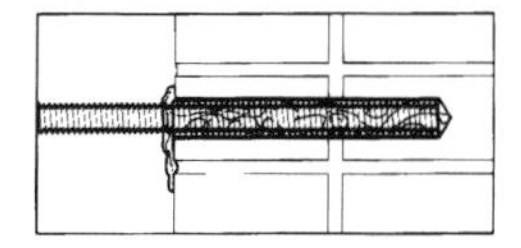

Fill—Dispense bead of ET off to the side to check for proper mixture, shown by a uniform gray color before using. Fill screen with ET. Always start by filling from the bottom of the screen to avoid air pockets. Insert screen into hole.

Insert—Anchors must be clean and oil free. Insert anchor, turning slowly until anchor hits the bottom of the screen. Do not disturb during set time.

7-52 *An example of how the Epoxy-Tie system is used in both concrete and brick.* Courtesy Simpson Strong-Tie Company, Inc.

white components of the mix blend together into a uniform gray color (this ensures a complete and proper mix). Fill the hole about halfway, then install the anchor, twisting it slowly until you feel it hit the bottom of the hole. Allow the epoxy mix to set up according to the manufacturer's directions before placing any load on the anchor—anywhere from 4 to 24 hours, depending on the type of formula and the temperature of the concrete.

The process is a little different for installing in brick or masonry. After drilling and cleaning the hole, you must dispense the epoxy mixture into a special screen tube, and insert the tube into the hole. Twist the anchor into the epoxy-filled tube until it hits bottom.

In addition to the Epoxy-Tie adhesive and a variety of anchors, hole-cleaning brushes and additional screen tubes are available from the manufacturer.

8

Computerized design services for engineered lumber

It seems that there are very few areas in our lives today that are not touched by computers, and the construction industry is certainly no exception. Computerized estimating programs have been a common sight in construction companies for years, as have specialized programs for construction-related accounting, payroll, and other such chores.

Then of course there's CAD (computer-aided design). The CAD programs of just a few short years ago were the sole province of big architecture firms, and were well out of reach of the average builder, from the standpoint of both the expense and the incredible learning curve involved. But that's changed as well; whether you're working with an IBM-compatible or a Macintosh, with DOS or Windows 95, there are now a multitude of very sophisticated CAD programs available for home computers as well as small computerized contracting offices.

As has happened in so many other areas of the construction industry where high-tech is meeting the hammer and nail, the computer might well offer one of the most exciting innovations for today's builders and remodelers. The computer technology now available at some local lumberyards is putting the power of some very sophisticated engineering right into the hands of the average contractor.

Computer design and engineering is certainly nothing new, as any architect or engineer can attest. Builders have reaped the benefits of it for years, even though you might not have been directly

8-1 *The growing use of engineered lumber framing systems has given rise to sophisticated CAD programs for sizing and design.*

involved, or indeed even been aware of it. Roof truss design is all done by computer, for example, and recently more and more cabinet layout and design is being done by computer.

With the advent of more engineered wood products and systems (Fig. 8-1), such as parallel strand lumber (PSL), laminated strand lumber (LSL), and I-joists, builders can also access the design capabilities of manufacturers such as Georgia-Pacific, Boise Cascade, and Trus Joist MacMillan. But until very recently, that's meant sending plans away to a central computer design area and getting back an engineered lumber design that wasn't always what you wanted.

TJ-Xpert software

Trus Joist MacMillan has proven to be a leading innovator in computer design, as well as I-joist technology and the manufacture of engineered lumber products. They have taken computer design and engineering technology a step further and placed it directly into the lumberyards themselves. In the last couple of years, they have begun providing many of their distributors with TJ-Xpert, a computer software program specific to their products. It's a great marketing and sales tool for Trus Joist, certainly, but it's also a real benefit to both the retailer and the builder.

"Proprietary software like our TJ-Xpert lets architects, engineers, and building product retailers make the most of engineered lumber's environmental, structural, and economic benefits," says Bill Bolduc, Trus Joist MacMillan's manager of engineering development. Bolduc says TJ-Xpert's computer-aided design capabilities help dealers satisfy an increasing demand for quality and service, and can help both establish and maintain relationships with customers long into the future.

Trus Joist MacMillan's proprietary computer design and engineering software provides complete analysis of floor and roof systems. It calculates structural loads, distributes the loads to the individual members (Fig. 8-2) while sizing them, and develops a complete material list and framing plot. According to Bolduc, Trus Joist MacMillan product dealers rely on the value this software adds to their services. "Most engineered lumber manufacturers have software that will solve complicated design and engineering problems one of two ways," he says. "The dealer asks 'If I put this beam here, will the building fall to the ground?' And a majority of programs will answer either 'yes' or 'no.'"

TJ-Xpert is, according to the manufacturer, the only software in the industry that takes that kind of interaction two or three steps further. "Our users can produce design reports that show critical shear moment and deflection—all suitable for submission to reviewing engineers and building officials," Bill Bolduc says.

8-2 *Computerized sizing programs can calculate the proper type of engineered lumber for any application.* Courtesy Trus Joist MacMillan

TJ's software creates a totally integrated framing design, including I-joists, beams, blocking, and hangers. Incorporated within the software is the entire line of Truss Joist products, including TJIs in different depths and with different top and bottom flange widths to accommodate a variety of loads; Micro-Lam and Parallam beams (their trade names for LVL and PSL engineered lumber, respectively); as well as a 25-page catalog of Simpson hangers, straps, ties, and other hardware. The program is constantly updated with new versions—the one currently in use is the third upgrade in about the last year—and it meets or exceeds all current ICBO structural codes.

Parr Lumber, a large West Coast retailer with stores in Oregon and Washington, as well as overseas distribution, is one of the retailers currently authorized by Trus Joist MacMillan to own and use its software. (Availability of the software is carefully monitored, with one-time loading capability and a variety of security codes required for use.) Parr lumber is a perfect example of how a quality product, excellent manufacturer and dealer support, thorough training, and the power of a sophisticated computer program can all come together to help you, the contractor, build a better house for less money. And it's all happening right down the street.

Each of Parr's retail lumberyards employs at least one TJ-Xpert designer—typically someone with a strong building or engineering background, as well as specialized training in the TJ software package—who works directly with builders to design TJ systems for new homes and remodels. Local design centers such as these are becoming more and more accessible in all parts of the country.

"What we're able to provide," says Denny Day, a TJ designer at the Parr Lumber center in Bend, Oregon, "is instant engineering. The whole TJ system is simple to use in virtually any framing situation, and the builder has access to me and to the computer throughout the whole framing process."

The computerized design process

The process begins when a builder brings a set of plans into the retailer. The designer begins by inputting dimensions for the house floor plan into the computer. This process alone takes about two hours, and is similar to entering the key coordinates for a CAD drawing.

Once the house layout is in the computer (Fig. 8-3), the designer begins by selecting from a variety of options. The first is to allow the computer to "float," or create its own layout. The system default in

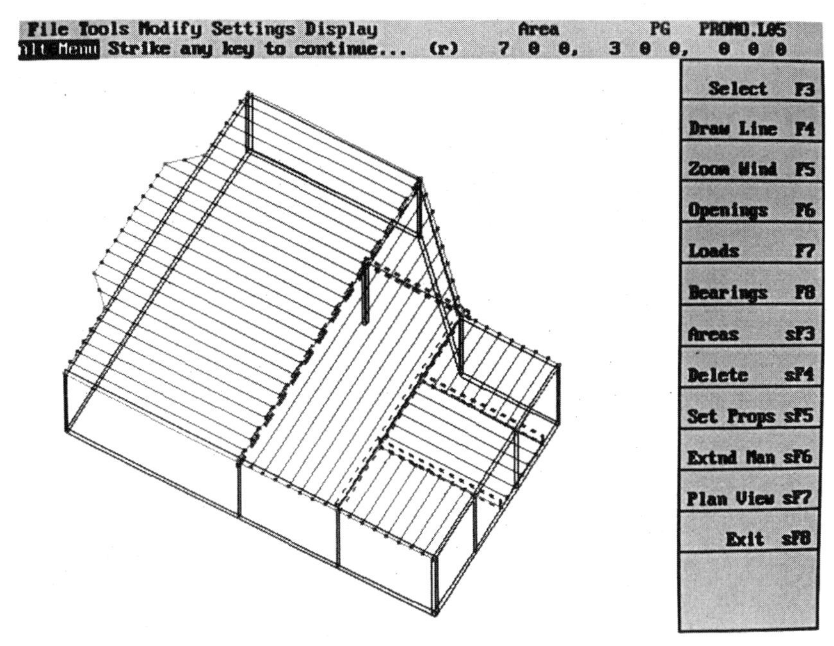

8-3 *An example of a computerized layout for engineered lumber in a floor framing system.* Courtesy Trus Joist MacMillan

this mode is to create the most cost-effective layout that meets current building codes.

"This standard layout works fine for many homes," said Day, "but occasionally there are some drawbacks. The computer's layout might not be the best one for the builder to actually work with on site. The computer might initially design a layout that mixes different spacings depending on the span: 16 inches on center in one area, with 24-inch spacing right next to it. That's where the experience of the designer and input of the builder come into play. I can work directly with the builder and discuss the layout with him, and we usually agree to make the spacing uniform to simplify the framing process."

Working closely with the builder, Day continues to refine the design, a process that takes a total of about three to four hours for the typical house. As with most CAD systems, he can portray the drawing in three dimensions on the screen, and rotate the viewing angle and perspective as desired. The system, however, isn't yet able to print these 3-D views.

The finished design is first saved into the computer system for future retrieval and then printed out on a special TJI design sheet, which is another nice feature of the new system. Each structural member on the drawing is identified with a specific code letter or number, and the code is keyed to a legend on the drawing (Fig. 8-4). This enables the builder to instantly identify I-beams, LVL or PSL beams, conventional lumber framing and other types of beams, hangers and strapping, and even the location of blocking.

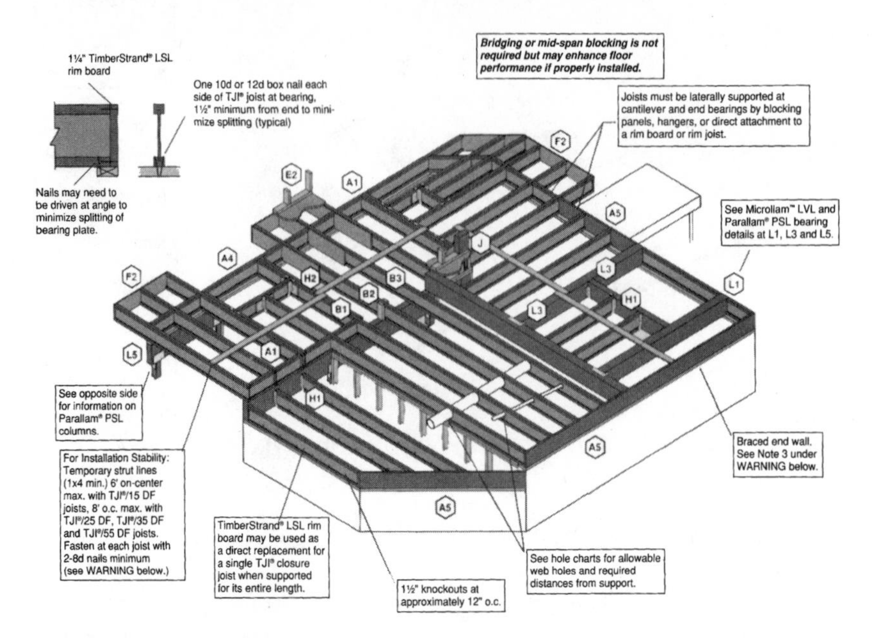

8-4 *This sample layout of a typical floor framing system shows the various components in the system. Letters and numbers relate to printed examples of framing applications, and the computer indicates which details are used at which locations.* Courtesy Trus Joist MacMillan

The drawing is also keyed to a set of detailed, standard TJI layout and framing instructions on the reverse side of the drawing (Fig. 8-5). These show how each section should be assembled, what hangers to use where, safety precautions specific to the use of TJIs during erection, and the nailing schedule for each section.

Two copies of these drawings, with all the appropriate stamps, warnings, disclaimers, and engineering data, are provided for the builder's use in obtaining building permits. For tracts where the same layout will be used for several houses, the computer can print the layout on vellum instead of paper—still with all the keyed legends and instructional drawings—so the builder can easily make multiple copies.

"Anyone can use the system," says Day, "even if you've never worked with TJIs before. It's like building a big model."

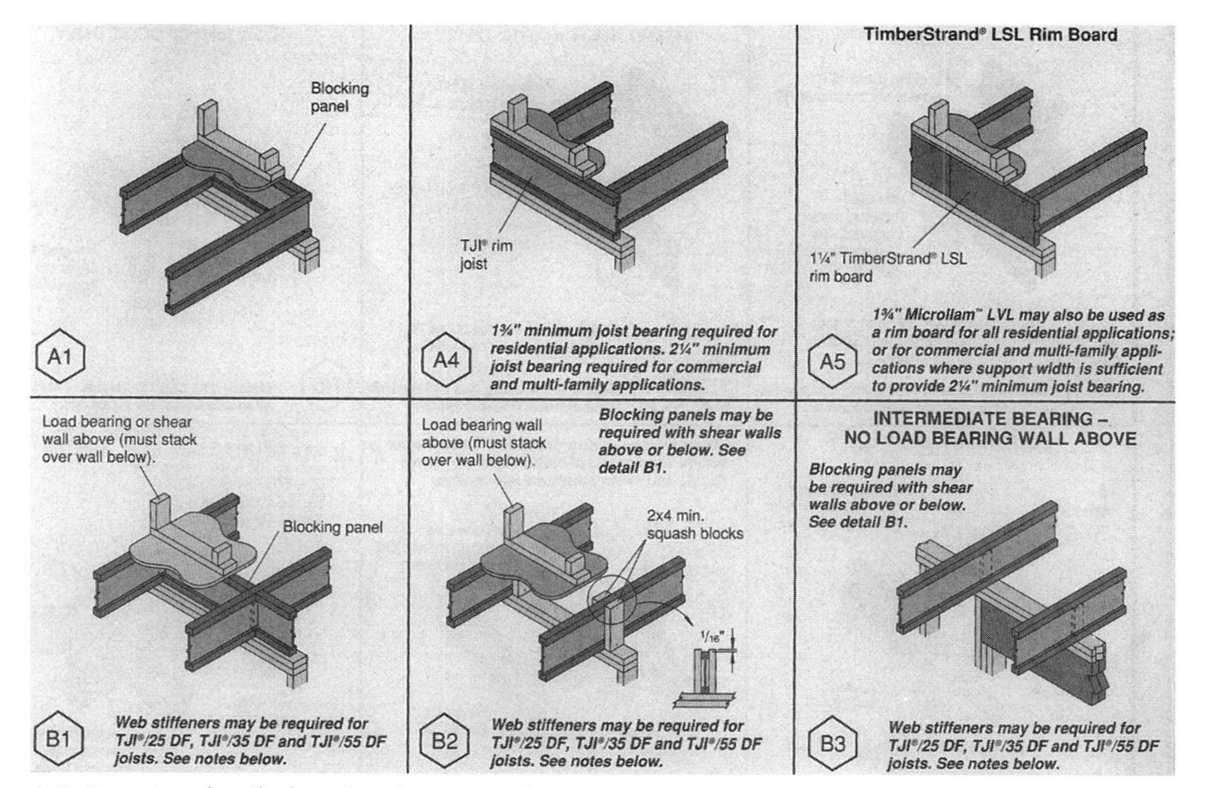

8-5 *Framing details, keyed to the system drawing.* Courtesy Trus Joist MacMillan

Custom designs

Changing the spacing of joists is just one of dozens of options open to the designer and builder (Fig. 8-6). Current code, for example, calls for a maximum horizontal member deflection of L/360. Using this formula, a joist spanning 12 feet (144 inches) can have a maximum deflection at its center point of 0.4 inches ($^{144}/_{360}$). The TJ software improves on that deflection rating by using a recommended default of L/480. In the previous example, the joist deflection would be reduced to 0.3 inches ($^{144}/_{480}$), resulting in a stronger, stiffer floor framing system that's sure to please the new owner of the house.

The designer can override these defaults to drop back to the code requirement minimum, or increase them further for even less deflection. Built-in pricing for the various TJI products also allows the designer to instantly project the cost differential for the builder between code minimum and any upgrades in framing member size or spacing.

The "instant engineering" features of this software also offer the builder a number of other time- and labor-saving advantages, says Day. "Suppose the builder is right in the middle of laying out the second-floor framing, and the plumber points out that one of the joists sits where the toilet will be. The builder can call me here at the yard, and I'll pull his design up on the computer. Within minutes, I can tell

8-6 *Powerful computerized design and sizing programs make quick work of sizing I-joists for various floor and roof framing applications.*

him if it's okay to move the joist and increase the spacing to clear the plumbing. If the spacing can't be increased, I can have someone deliver him an additional joist so he can slip one in on each side of the pipes. Either way, he has an accurately framed building with a minimum of down-time."

An important consideration in situations such as this is that the software is constantly updated to reflect the latest building-code design standards. This means that the designer can immediately print out an amended joist layout, complete with all the appropriate documentation, and get it quickly into the builder's hands. That way, when the building inspector arrives on the construction site and points out that the builder has made changes to the original joist spacing on the plans, the builder has the documentation to satisfy the inspector and ensure passing inspection.

Another feature of the software is the ability to quickly make point-load calculations. Frustrating changes that occur during construction—moving a bathtub location, for example, or changing a lightweight fiberglass shower to a mud-set marble enclosure—can be easily accommodated. Here again, the builder only has to call the designer and specify the details of the change, and the design can be instantly altered to reflect the exact live and dead load requirements over the specific area involved. A new drawing is printed, the builder makes the alteration with the assurance that the building is still structurally sound, and job-site records have the necessary documentation to satisfy the building inspector.

Pricing and material lists

Using TJ-Xpert keys dealers and builders into products manufactured by three specific companies: Trus Joist MacMillan's I-joists and engineered lumber products, Simpson Strong-Tie's hangers and connectors, and a different line of hangers and connectors from Kant-Sag. Assuming you want to use the products from these suppliers (that's what the dealer you're working with sells and is designing around, and they're all industry-leading products, so you probably do), then the dealer's software package can supply you with a couple of other great time-savers, namely material and price lists. Here again, the price is ideal; it's free.

After the design has been created on the computer and the details worked out between the builder and the designer, a few more keystrokes are all that's needed to create the material list. This is a detailed breakdown of exactly which engineered lumber products are specified, including the quantity and length of each. Hangers are calculated for each framing connection on the plan and, here again, that includes the model number and quantity of each. The material list also specifies special-order hangers—skew hangers, sloped hangers, welded brackets—as well as blocking panels, web stiffeners if needed, reinforcing materials, and plywood.

Once the material list is complete, the computer can check it against the inventory lists for the lumberyard itself. Within minutes, you will know which pieces on the list are in stock and which ones, if any, will have to be ordered. The computer knows where the special-order items are warehoused and, either by computer or with a simple phone call, you can quickly find out how long it will take to get the necessary products delivered to your site.

Finally, there's the pricing. The computer contains the current price listings for all the Trus Joist and Simpson products specified, and will print you out a detailed listing. You'll know exactly what you're paying for everything from the largest beam to the smallest hanger, again within moments.

As with any computer system, another advantage here is that all the material is stored on disk at the lumberyard, ready for instant retrieval, including price and material lists. This allows you and the designer to instantly calculate how changes in the building design will affect the price of the engineered lumber materials you've specified.

Suppose, for example, that you've created a design for a new house and given the clients a cost estimate. They would like to know how much more it would cost to upgrade the joists over the garage to allow them to convert the attic space in that area to an office at a future date. Or maybe they're considering switching to cement-tile roofing instead of the laminated composition shingle you had specified initially, and now you need to know how much larger the rafters will need to be and how much it will add to the cost. A quick call to the lumberyard, a few keystrokes on the computer, and you have the new calculations and the new prices at your fingertips.

Other Trus Joist MacMillan software

In addition to the TJ-Xpert program, Trus Joist offers three other packages for use with their products. They are all available to qualified product users and suppliers, and can be ordered directly from the manufacturer (see Appendix E).

TJ-Spec. TJ-Spec software is designed for professional architects or engineers involved in sizing and specifying Trus Joist products, including TJI joists, Microlam LVL, Parallam PSL, TimberStrand LSL, and Open Web Truss members. The software allows you to size floor and room member sizes, and set up specific deflection criteria. It also performs a number of standard calculations including shear, reaction, and deflection, performs column analysis, and can assess the effect of holes of different sizes, shapes, and locations on the structural integrity of the TJI web. The accuracy of all results is warranted by Trus Joist.

TJ-Beam. This software is very similar to TJ-Spec, although it lacks a couple of the more sophisticated engineering features found in that software package. It will calculate and design all the company's products, with the exception of the Open Web Truss members.

TJ-Dealer. This package is intended specifically for use by dealers of Trus Joist products. It tracks inventory by size, length, and quantity in stock, and lets the dealer track incoming shipments and special orders. It also tracks a variety of product cutting lists.

Software from other manufacturers

In addition to the software packages available from Trus Joist MacMillan, at least three other manufacturers offer software packages. This appears to be another growing trend, as more and more contractors and retailers switch to computer systems. Look for a lot more free or very low cost software packages to be offered by a variety of manufacturers in the near future.

Boise Cascade

Boise Cascade offers BC Calc, which you can use to calculate joist and beam sizes for anything in the complete Boise Cascade line.

BC Calc allows you to input a given span and load, and the software selects the appropriate product to meet those specific conditions. You can also do the reverse, inputting the product and getting back the various span and load applications for which the product is appropriate. A third option is to let the computer select the shallowest-depth joist that will work for a given span and load, which allows you to do a little cost comparison between minimum sizes and the next sizes up. BC Calc is available free of charge to anyone who requests it from the company (see Appendix E for contact information). There is a DOS-based version, as well as one that works with Windows.

Also available from Boise Cascade is BC Framer, which is essentially a drawing tool and calculation software. BC Framer is not a stand-alone program, but instead works in conjunction with the highly popular AutoCAD program.

BC Framer allows you to place Boise Cascade I-joists and other products directly onto your AutoCAD-created drawing, and lets you see the layout in both two and three dimensions. You can put everything Boise Cascade sells into the program, and the software will prepare a complete material list for you after all of the calculations and sections are completed. BC Framer also contains the BC Calc program, so there's no need to order that one separately.

According to a company spokesman, BC Framer is available for a slight fee to all distributors of Boise Cascade products, to some retailers, and to some architects and designers. If you're interested, contact the company for specific details on how to obtain the program (see Appendix E).

Louisiana-Pacific

Louisiana-Pacific offers a program called Wood-E, again free to anyone who requests it. Wood-E is another sizing program similar to the others already described, and works as part of Louisiana-Pacific's SolId Start engineered lumber framing system. A nice feature of Wood-E is a stand-alone program that contains drawings of engineered lumber installations and standard applications, which you can print out and place on your own drawings for use as construction details. For more information on Wood-E, contact Louisiana-Pacific directly (see Appendix E).

Simpson Strong-Tie

If you draw a lot of plans and specifications for your jobs, or if your primary occupation is as a designer, architect, or engineer, Simpson Strong-Tie offers a very nice stand-alone software package that has drawings of all of their hanger and connector products.

Two software libraries are available, providing installation drawings from the company's Wood Construction Connectors, Plated Truss Connectors, and Composite Wood Products Connectors catalogs. Both libraries provide detail drawings in two-dimensional perspective for you to include in your own plan sheets. Windows-compatible and DXF-compatible versions are available.

Simpson also currently offers Connector Update, a newsletter of information for the construction industry. The newsletter is free, and offers a variety of technical and installation tips, as well as information of general interest. Contact Simpson Strong-Tie directly for complete details on ordering either the software or the newsletter (see Appendix E for the address and phone number).

No mention of computers would be complete these days without also mentioning the Internet, that vast and rapidly growing network of information available to almost anyone with a computer, modem, and telephone.

The Internet is, as of this writing, still somewhat of an untamed and unexplored wilderness. It can be confusing and daunting, but it offers some absolutely wonderful rewards for anyone with a little time, a little patience, and a bit of an adventurous spirit. It's growing at an absolutely astounding rate—doubling in size every four months was one recent estimate—and promises to do so for the foreseeable future. It will eventually slow down and settle into a manageable growth pattern, but it certainly doesn't appear to be just another fad that will fade away in a short time.

For builders, the Internet offers a wealth of opportunities to gather information and stay abreast of current products, techniques, and trends. You have access to information from manufacturers, retailers, the government, educational institutions—you name it, you can probably access it. You can download pictures to show clients or suppliers; you can access charts, graphs, statistical information, and load tables; and you'll see the latest information from testing labs and product research centers. You can "talk" electronically with product reps and other contractors, often getting the information you need in a matter of minutes instead of days or weeks.

If you own a fairly up-to-date computer and modem, and have either a low-cost Internet service provider in your area or a toll-free number to one of the commercial services, it's certainly something worth checking into. I've provided a few random Internet addresses in Appendix E.

9

Designing and specifying structural engineered lumber

Using I-joists, laminated beams and lumber, and all the other new and innovative engineered lumber products (Fig. 9-1) is both very new and very familiar to anyone who's been around a construction site for a while. The tools are the same as you've always used—circular saw, hammer, string line, nail gun, and all the rest—and there's still the smell of sawdust in the air. The construction sequence is the same as well, so there's no great learning curve to be concerned with. In short, don't worry that these products are something you'll need special training for—something that only "the other guy" knows how to specify, order, and install.

On the other hand, there are definitely differences in how you need to deal with engineered lumber. There are specific things to be concerned about when ordering these materials, and when storing them on the job site. There are also some important distinctions between framing with solid sawn lumber and framing with engineered lumber. In this chapter and the next, you'll learn some of the basic techniques for ordering, storing, and framing with this new generation of lumber products. Remember that these are just generic instructions; the requirements for your specific project might be

9-1 *Framing with engineered lumber is similar to conventional solid sawn lumber in many ways, but there are a few new techniques to learn.*

different. *Always refer to the manufacturer's detailed guidelines for the design and installation of any specific engineered lumber product.*

What makes a strong floor?

One of the biggest customer complaints and one of the most frequent reasons for costly call-backs on a new home is the floor. People hate a floor that squeaks or feels springy underfoot, and builders are constantly fielding complaints about them. So what goes into the construction of a quiet, solid floor? There's no perfect solution for every situation, and floor performance is the end result of lots of factors that all have to work together. You must, of course, build a floor that meets all the requirements of the building codes and satisfies the demands of the loads being placed on it. But that alone might not be enough for everyone. A code-minimum floor could be acceptable to one homeowner, but seem woefully underbuilt to another. Remember that the "feel" of a floor is often highly subjective.

In general, though, most people agree that a floor that feels good underfoot, one that speaks of sound construction practices and a builder who is quality-conscious, has very little deflection and movement, and makes absolutely no noise as you move across it. But cost cannot be overlooked either; more joists and more supports equals more money, so a good floor can be something of a balancing act as well. When weighing cost versus performance in a floor system, it's important to remember to look at all the floor framing options and alternatives. Decreasing the on-center spacing instead of the size of the joist (or vice versa) might be a better, more cost-effective solution.

With all that in mind, here are some floor system tips worth remembering for your next house:

- Deeper joists reduce deflection. With engineered lumber I-joists, the width of the web and the thickness of the top and bottom flanges dictate the overall size of the joist, and the higher (deeper) the joist is, the less it will bend or deflect when a load is placed on it. Deeper joists are, of course, also more expensive, so you have to balance cost against performance.
- Reducing the center-to-center spacing of the joists will also decrease deflection. The closer the joists are to one another, the more they share the overall load placed on the floor. As you increase the load-sharing, you create a floor that acts more and more like a single member, rather than a combination of individual pieces. As load-sharing is increased, deflection is reduced.
- A thicker subfloor reduces deflection. Here's the third element in the floor performance equation. If you increase the thickness of the material you use for the subfloor, you again improve the load-sharing capabilities of the overall floor system, and reduce the deflection.
- Decrease support beam spacing to reduce deflection. This is option number four for decreased floor deflection, and means adding extra girders or beams to better support the joists. This might be an easy solution in some houses or some areas of a house, but impossible in other homes or in certain floor areas.
- Adhesives improve floor performance dramatically, since they stiffen the floor framing and create a floor system that interacts and moves together. Make it a common practice to apply a bead of construction adhesive to the joists immediately prior to installing your subfloor panels, and always use the manufacturer's recommended size, quantity, and spacing of fasteners.
- Blocking will stiffen a floor. Installing full-height blocking between the joists greatly lessens lateral movement, and again causes the joists to react more as a unit than as individual members, which helps disperse and equalize the load. Installing metal cross-bridging or wood strips on the underside of the I-joist bottom flanges will also decrease deflection.

- Dead loads dampen floor movement. Installing cross walls, partition walls, and even cabinets—any random dead load—will help to deaden a floor system and reduce vibration. This is not to say that you need to break up your homes with a lot of cross walls, but rather points out that large open spaces are more prone to vibration and floor movement. You might want to consider a somewhat oversized floor system in those areas.
- Ceilings dampen vibration. A thick, well-installed ceiling on one floor level will also help deaden vibration in the floor above it. This is especially worth remembering for a large, open floor over a basement, where structural floor supports and ceiling finish might be minimal.
- Materials and workmanship play a key role in how well a floor—and indeed the entire house—will look and perform. Take the time to apply adhesives correctly and nail off all areas to the manufacturer's requirements before moving on to the next area. Use good-quality subfloor panels, and strongly consider using engineered lumber in the floor system instead of solid sawn material.

Using engineered lumber

Throughout the book, I've discussed a variety of reasons for using engineered lumber in framing your house. Before looking at the specific handling and installation techniques for this material, I'll summarize the main advantages of engineered lumber framing:

Uniform sizing of structural members. Have you ever tried to install a unit of 2 × 10 floor joists, and had them range in depth from 9⅛ up to 9½ inches? Those ups and downs play havoc with good framing and blocking practices, and can seriously disrupt the appearance of finished floors and ceilings. Engineered lumber is precise in the consistency of its dimensions.

Dimensional stability and consistent moisture content. Engineered lumber pieces are factory-produced to consistent, uniform, tested levels of moisture content, unlike much of the "pond-dried" solid sawn lumber we've had to contend with in recent years. They will not crack, warp, twist, cup, or bow because of inconsistent drying or the effects of natural wood fiber differences in the rings. Engineered lumber members will also not shrink and move as the result of drying, and fasteners hold better and more consistently.

Engineering out defects. In a piece of solid sawn lumber, a knot or other defect in the wood member creates a weak spot, and all weak spots are potential sources of structural failure. With engineered lumber, defects in the wood are removed early on in the manufacturing process; in the case of laminated veneer lumber, any single defect is supported by the solid wood laminations on either side of it. Either way, the structural member has more uniform strength, and the possibility of a structural failure is greatly diminished.

Longer lengths and lighter weight. Engineered lumber members are manufactured in lengths of up to 66 feet, something that's a virtual impossibility with natural solid sawn lumber. And even at greater lengths, the material is lighter per square foot than solid lumber.

Consistent performance. With engineered lumber pieces, you know what to expect. Each piece is engineered and manufactured the same, and it performs the same, piece after piece.

Full engineering support. Most manufacturers of engineered lumber provide software to consumers or dealers of their products, so, whether it's one piece of lumber or an entire structural system, you have engineering support nearby that's fast and accurate. You'll never have to wonder if a piece of lumber of a particular size is adequate to span a particular distance, or where blocking should be placed, or exactly how to support a given load; you'll have an entire picture of your project and its assembly procedures on paper right in front of you.

Reduced call-backs. Consistent performance means a structure you can depend on, with greatly reduced squeaks, sags, and vibrations. And that equates to fewer call-backs and repairs.

Lower installed costs. Factoring in the reduced labor costs that result from easier, more consistent installation as well as a reduction in costly call-backs, studies have shown that a house framed with engineered lumber products is actually less expensive to build than one framed with conventional solid sawn lumber.

Reduced impact on natural resources. The engineered lumber products I've discussed in this book are all designed around the use of small, rapidly growing trees, and many of these products use wood species that have previously been considered unsuitable for structural lumber, including aspen, yellow poplar, and cottonwood. Using these renewable resources greatly helps to conserve the dwindling number of old-growth trees left in our forests.

Selecting and specifying engineered lumber

As with conventional solid sawn lumber, engineered lumber members, including joists, beams, rafters, headers, and other structural members, *must be sized to support specific loads over specific distances, all under specific conditions.* Every manufacturer of engineered lumber products has performed exhaustive engineering studies on their materials, and has developed comprehensive design booklets for use by builders and designers. These design booklets, as well as the computer software that's also offered by many manufacturers, give you all the information you need to select specific structural members for use under a variety of conditions and applications. But before you can use this information, you need to know what those conditions and applications are, at least as they relate to your specific project.

The following sections list the types of information you need to know in order to accurately size different types of structural members for a residential or commercial structure. To obtain specific load information, consult a building official in the area where the building is to be erected, or refer to the building codes manual that has been accepted for use in the city, county, or state where you're building. You can also consult a qualified structural engineer, architect, or designer.

Remember that the loads used to calculate the structural integrity of a building change from area to area due to differences in snow loads, soil conditions, wind conditions, temperatures, and a variety of other factors—what works in Florida won't work in Minnesota, for example—so it's important to obtain load information from the local jurisdiction where you're building. Also, you don't necessarily need to know all this information for every application you're designing structural members for.

Amount of live load (LL). Live load, commonly abbreviated LL, is the weight of the occupants of the building and all movable, nonpermanent loads, such as furniture or items in storage. When designing for a residential structure, a live load of 40 pounds per square foot is relatively common. For commercial structures, the design criteria for live loads are typically much higher, taking into consideration things such as increased occupant load; increased weight from displays, merchandise, or office furniture and equipment; and especially the increase in load arising from the placement and storage of file cabinets and bulk paper. For industrial applications, where the weight of large pieces of machinery or equipment is often concentrated in a specific area—known as a *point load*—it is often necessary to known the size and weight of the specific pieces of equipment, as well as their exact location in the building. Certain high-weight equipment that is placed permanently in the building and never shifted is considered to be a dead load rather than a live load.

Amount of dead load (DL). A building's dead load, commonly abbreviated DL, is the load that's exerted on that building by the structural members themselves, or by any other permanent, nonmovable load. This includes the wall and roof framing, roofing and siding, floor coverings, bathroom fixtures, heating and air conditioning equipment, fireplaces, appliances, cabinets, and, as mentioned in the last section, specific pieces of commercial or industrial equipment that, once installed, are not moved.

Deflection (L/x). Deflection is the total distance in inches that a structural member will sag off the horizontal at a point midway between its supports. Deflection is represented as L/x, where L is the joist length and x is the deflection factor. The current building code in most areas is L/360, which is calculated as:

length of member between supports (in inches) ÷ 360 = maximum deflection (in inches)

Most charts and software programs used for sizing engineered lumber members have at least two possible deflection options: the code minimum of L/360 and L/480, which results in less deflection and a stiffer floor (see the section later in the chapter, *Live-load deflection*).

Span. The distance the structural member spans between supports. Span is usually figured as horizontal distance, even though the member can be installed on an angle, as with rafters. Refer to the instructions that accompany the specific design table you're using.

Spacing. The horizontal center-to-center spacing between the structural members (joists or rafters).

Intended use of the member. Are you designing the piece of lumber as a floor joist, rafter, header, etc?

Opening width. When calculating the size of a header, you need to know the exact size of the opening that the header will span.

Building width. Certain header design tables also require that you factor in the overall width of the building. This is necessary to allow for the amount of additional dead load being placed on the walls of a building by the roof trusses or roof framing—a load that increases proportionally as the overall width of the building increases.

Snow load. When designing a structure, it's important to consider the projected amount of snow it might need to support. These requirements will obviously vary from region to region. Most of Texas probably has no snow-load requirements, for example, while parts of Oregon might have moderate snow loads and parts of Montana might have severe load requirements; all are based on historical snowfall data for the region.

Wind load. Wind loading, or the action of the wind on specific parts of a structure, is another important consideration in some parts of the country, particularly coastal areas. Most of the manufacturer's design tables do not specifically list wind load as a factor in designing and specifying their structural members, so you might need to refer to the building code book for your area to determine what factor to use in increasing the design load, or else contact the manufacturer's design department directly for specific advice.

Roof pitch. As a structural member moves from horizontal (e.g., a floor joist) to inclined (e.g., rafters), the loads imposed on it change, as does its ability to support those loads. For this reason, if you are calculating the size of a rafter, you'll also need to know the pitch of the roof.

Roofing materials. Most span tables used to design rafters are based on standard-weight roofing materials such as wood shingles or shakes, composition shingles, and standard-gauge metal. Heavier roofing materials such as clay or cement tiles might require a different load table or load category, or special calculations by the manufacturer.

Once you have all the necessary information, you need to select which manufacturer will supply the members you're using. You can then use that manufacturer's sizing software or the charts listed in their design manuals to determine exactly what size member you need to fit a specific application.

Live-load deflection

As mentioned earlier in this chapter, one of the hallmarks of a well-built home is the structural integrity of the floor. Build a house with a floor that doesn't squeak, bounce, or vibrate, and you've gone a long, long way toward having a client who is satisfied with his or her home.

The list near the beginning of this chapter explains all the different elements that go into a solid, quiet floor structure. Of those elements, the one that is the single most important is live-load deflection, or the amount a floor bends or sags between supports under the weight of the home's users. The manufacturers of engineered lumber products, particularly I-joists, also recognize the importance of live-load deflection, and have made things a little easier on contractors and building designers by devising design charts that offer you at least two and sometimes three different options.

The first option is L/360, which is the standard set by the building codes in most areas for the maximum live-load deflection allowable. Remember that this deflection is the most allowed by the building codes, making joists that meet this standard as small as the code allows for a particular installation; they're consequently the weakest allowable joists, those with the most allowable sag.

To better understand the calculation, let's apply it to a joist with a 20-foot span. Since the calculations are based on inches, the first thing to do is convert the span into inches: 20 feet × 12 inches per foot = 240 inches. Using the deflection formula L/360, you now need to divide 240 (the L length of the joist in inches) by 360 (the deflection factor), to arrive at 0.67. Using this formula, the maximum deflection that the building codes will allow a 20-foot-long joist to have, then, is 0.67 inches at the center of the span.

In order to create a stiffer floor system, you'd need to use a stronger floor joist, or one with less tendency to sag. With this in mind, the manufacturers also provide a different deflection rating in their catalogs and their software, which is L/480. Using the previous example of a 20-foot joist span, applying this new formula to it, you'd arrive at the following: 240 ÷ 480 = 0.5. Using this joist would result in a floor with a maximum deflection at the center span of 0.5 inches, as opposed to 0.67 inches, an obvious improvement.

In the Boise Cascade Design Guide tables, they offer you a third option for an even stiffer floor, what they refer to as a "four star" floor system: a "live-load deflection at L/960 to give a floor that is much stiffer for the more discriminating purchaser." That standard can, of course, be applied to joists from any manufacturer, but including the table makes selecting upgraded joists from this manufacturer a little easier.

Again applying the previous example, you'd now have 240 ÷ 960 = 0.25 inches, reducing the live-load deflection to a mere ¼ inch at the center of a 20-foot joist, which would make for an extremely stiff and solid floor.

Each upgrade in joist size comes, naturally, with a jump in price, so these tables are not just for your information; they also help sell engineered lumber. How you balance the equation between cost and floor performance is up to you, and, as with most building material upgrades, it will probably ultimately depend on the cost of the house and the budget of the purchaser. For most starter homes, the code-minimum joist, with an L/360 live-load deflection, will result in perfectly satisfactory performance. The L/480 standard is a good upgrade for homes of all types and price ranges, and spending the sometimes considerable extra amount to get up to the L/960 standard is, as Boise Cascade puts it, probably only for "the more discriminating purchaser."

Sizing engineered lumber I-joists for floor framing

To accurately size wooden I-joists for use in a floor system (Fig. 9-2), you need to have five pieces of information: the live load, dead load, unsupported span for the joist, intended center-to-center spacing between the joists, and desired amount of live-load deflection. For the examples in the next several sections, I will assume you're building a house with the following specifications:

Live load. 40 pounds per square foot (psf)
Dead load. 10 psf
Span. 20 feet
Center-to-center spacing. 16 inches
Live-load deflection. L/360

Sizing TJI joists manufactured by Trus Joist MacMillan

If you would like to use TJI I-joists from Trus Joist MacMillan, refer to their software or their *Specifier's Guide, Residential Applications,* available either from any lumberyards that carry TJIs or directly from the manufacturer.

In that booklet, you should first find the section entitled TJI Joist Residential Floor Span Charts (Fig. 9-3). Refer next to the charts headed 40 PSF Live Load and 10 PSF Dead Load, then look specifically at the chart labeled L/360 Live Load Deflection (Code Minimum).

Now that you have the correct chart to work from, look at the area on the chart labeled O.C. Spacing, and find the column for the joist spacing you'd like to use, in this case 16" O.C. Read down that column until you find the first entry that meets or exceeds the 20-foot span you're looking for, in this case 20'-5". Now read across to the left side of the chart, and you'll see that it will take an 11⅞-inch-deep joist, series TJI/15 DF, to span the necessary distance on the centers for the deflection you specified.

While you're still on that chart for a L/360 live-load deflection, you can also quite easily explore a couple of options. Let's say you want to consider dropping your spacing down to 12 inches on center. By moving to the 12" O.C. column and reading down to the first span that meets or exceeds your 20 feet, you'll come to 20'-11". Reading across to the left, you'll see that you can decrease the joist to a less expensive series (11⅞-inch joist, TJI/Pro series). You could also increase the spacing to 24 inches on center and go up two joist series

9-2 *I-joists are light and easy to handle on the job and, properly sized, they'll support heavy loads over long spans.* Courtesy Trus Joist MacMillan

on the chart (11 7⁄8-inch joist, TJI/35 DF series, maximum span 20 feet, 2 inches).

Now you just need to call your lumber dealer and obtain the current pricing for all three series of joists. By multiplying the price per joist by the number of joists needed at each of the three spacings (and

TJI® Joist Residential Floor Span Charts

40 PSF LIVE LOAD, 10 PSF DEAD LOAD (12 PSF DEAD LOAD AT TJI®/55 DF JOISTS)

(Example: Single layer glue-nailed wood sheathing and direct applied ceiling)

L/360 LIVE LOAD DEFLECTION (Code Minimum)

JOIST DEPTH	JOIST SERIES	O.C. SPACING			
		12" o.c.	16" o.c.	19.2" o.c.	24" o.c.
9½"	TJI®/Pro™	17'-11"	15'-8"	14'-4"	12'-9"
	TJI®/15 DF	18'-9"	17'-2"	16'-3"	15'-0"
	TJI®/25 DF	19'-7"	17'-11"	16'-11"	15'-9"
11⅞"	TJI®/Pro™	20'-11"	18'-1"	16'-6"	14'-9"
	TJI®/15 DF	22'-4"	20'-5"	[illegible]	15'-0"
	TJI®/25 DF	23'-4"	21'-4"	[illegible]	18'-9" (6)
	TJI®/35 DF	25'-3"	[illegible]	21'-8"	20'-2" (6)
	TJI®/55 DF	28'-8"	[illegible]	24'-7"	22'-10"
14"	TJI®/25 DF	26'-6"	[illegible]	22'-10" (6)	18'-11" (6)
	TJI®/35 DF	28'-8"	26'-1"	24'-7" (6)	21'-3" (6)
	TJI®/55 DF	32'-6"	29'-7"	27'-11"	25'-11" (5)(6)
16"	TJI®/25 DF	29'-5"	26'-10" (6)	23'-9" (6)	18'-11" (6)
	TJI®/35 DF	31'-9"	28'-11" (6)	26'-7" (6)	21'-3" (6)
	TJI®/55 DF	36'-0"	32'-9"	30'-10" (5)	26'-9" (5)(6)

L/480 LIVE LOAD DEFLECTION

JOIST DEPTH	JOIST SERIES	O.C. SPACING			
		12" o.c.	16" o.c.	19.2" o.c.	24" o.c.
9½"	TJI®/Pro™	16'-2"	14'-9"	13'-11"	12'-9"
	TJI®/15 DF	17'-0"	15'-6"	14'-8"	13'-7"
	TJI®/25 DF	17'-9"	16'-2"	15'-3"	14'-2"
11⅞"	TJI®/Pro™	19'-3"	17'-7"	16'-6"	14'-9"
	TJI®/15 DF	20'-3"	18'-5"	17'-5"	15'-0"
	TJI®/25 DF	21'-1"	19'-3"	18'-2"	16'-11" (6)
	TJI®/35 DF	22'-10"	20'-9"	19'-7"	18'-2"
	TJI®/55 DF	25'-11"	23'-7"	22'-2"	20'-7"
14"	TJI®/25 DF	24'-0"	21'-10"	20'-7" (6)	18'-11" (6)
	TJI®/35 DF	25'-11"	23'-7"	22'-2"	20'-8" (6)
	TJI®/55 DF	29'-5"	26'-9"	25'-2"	23'-4" (5)
16"	TJI®/25 DF	26'-7"	24'-3"	22'-10" (6)	18'-11" (6)
	TJI®/35 DF	28'-8"	26'-1"	24'-7" (6)	21'-3" (6)
	TJI®/55 DF	32'-6"	29'-7"	27'-10"	25'-10" (5)(6)

40 PSF LIVE LOAD, 22 PSF DEAD LOAD (24 PSF DEAD LOAD AT TJI /55 DF JOISTS)

(Example: Single layer glue-nailed wood sheathing with 1½" lightweight concrete and direct applied ceiling)

L/360 LIVE LOAD DEFLECTION (Code Minimum)

JOIST DEPTH	JOIST SERIES	O.C. SPACING			
		12" o.c.	16" o.c.	19.2" o.c.	24" o.c.
9½"	TJI®/Pro™	16'-3"	14'-1"	12'-10"	11'-6"
	TJI®/15 DF	18'-7"	16'-8"	15'-2"	12'-1"
	TJI®/25 DF	19'-5"	17'-8"	16'-3"	12'-11"
11⅞"	TJI®/Pro™	18'-10"	16'-3"	[illegible]	12'-0"
	TJI®/15 DF	22'-1"	18'-2"	[illegible]	12'-1"
	TJI®/25 DF	23'-1"	20'-10" (6)	[illegible]	15'-3" (6)
	TJI®/35 DF	24'-11"	[illegible]	21'-5" (6)	17'-2" (6)
	TJI®/55 DF	28'-0"	[illegible]	24'-0" (5)	21'-9" (5)(6)
14"	TJI®/25 DF	26'-3" (6)	[illegible]	19'-1" (6)	15'-3" (6)
	TJI®/35 DF	[illegible]	25'-9" (6)	21'-5" (6)	17'-2" (6)
	TJI®/55 DF	31'-9"	28'-11" (5)	27'-2" (5)(6)	21'-9" (5)(6)
16"	TJI®/25 DF	28'-10" (6)	22'-11" (6)	19'-1" (6)	15'-3" (6)
	TJI®/35 DF	31'-4" (6)	25'-9" (6)	21'-5" (6)	17'-2" (6)
	TJI®/55 DF	35'-2"	32'-0" (5)(6)	27'-2" (5)(6)	21'-9" (5)(6)

L/480 LIVE LOAD DEFLECTION

JOIST DEPTH	JOIST SERIES	O.C. SPACING			
		12" o.c.	16" o.c.	19.2" o.c.	24" o.c.
9½"	TJI®/Pro™	16'-2"	14'-1"	12'-10"	11'-6"
	TJI®/15 DF	17'-0"	15'-6"	14'-8"	12'-1"
	TJI®/25 DF	17'-9"	16'-2"	15'-3"	12'-11"
11⅞"	TJI®/Pro™	18'-10"	16'-3"	14'-10"	12'-0"
	TJI®/15 DF	20'-3"	18'-2"	15'-2"	12'-1"
	TJI®/25 DF	21'-1"	19'-3"	18'-2" (6)	15'-3" (6)
	TJI®/35 DF	22'-10"	20'-9"	19'-7" (6)	17'-2" (6)
	TJI®/55 DF	25'-11"	23'-7"	22'-2"	20'-7" (5)
14"	TJI®/25 DF	24'-0"	21'-10" (6)	19'-1" (6)	15'-3" (6)
	TJI®/35 DF	25'-11"	23'-7" (6)	21'-5" (6)	17'-2" (6)
	TJI®/55 DF	29'-5"	26'-9"	25'-2" (5)	21'-9" (5)(6)
16"	TJI®/25 DF	26'-7" (6)	22'-11" (6)	19'-1" (6)	15'-3" (6)
	TJI®/35 DF	28'-8"	25'-9" (6)	21'-5" (6)	17'-2" (6)
	TJI®/55 DF	32'-6"	29'-7" (5)	27'-2" (5)(6)	21'-9" (5)(6)

Although the L/480 Live Load Deflection charts will usually provide better floor performance than the L/360 Live Load Deflection charts, the resulting performance still may not be adequate for your project. See page 3 for *A WORD ABOUT FLOOR PERFORMANCE*, or contact your Trus Joist MacMillan representative for assistance.

GENERAL NOTES:

1. Span charts assume composite action with single layer of the appropriate span-rated, glue-nailed wood sheathing for deflection only. **Spans shall be reduced 5" where sheathing panels are nailed only.**
2. Spans shown are clear distances between supports and reflect the most restrictive of simple or multiple span applications, based on uniformly loaded joists and include allowable increases for repetitive member use.
3. For loading conditions not shown, refer to allowable uniform load tables on page 14.

WEB STIFFENER REQUIREMENTS

End Bearings: Web stiffeners are not required at end bearings of TJI® floor joists listed in this guide **except** in hangers when the following conditions exist:

4. **All Joists:** Web stiffeners are required in hangers when the sides of the hanger do not laterally support the TJI® joist top flange.
5. **TJI®/55 DF Joists Only:** Web stiffeners are required in hangers when the TJI®/55 DF joist span is greater than the spans shown in the following chart:

JOIST SERIES	40 PSF LIVE LOAD, 12 PSF DEAD LOAD				40 PSF LIVE LOAD, 24 PSF DEAD LOAD			
	12" o.c.	16" o.c.	19.2" o.c.	24" o.c.	12" o.c.	16" o.c.	19.2" o.c.	24"o.c.
TJI®/55 DF	Not Required	Not Required	28'-8"	22'-11"	Not Required	28'-0"	23'-3"	18'-7"

6. **Intermediate Bearings:** At intermediate supports where the joists are continuous span, web stiffeners are required **only** if the intermediate bearing width is less than 5¼" **and** the span on either side of the intermediate bearing is greater than the spans shown in the following chart:

JOIST SERIES	40 PSF LIVE LOAD, 10 PSF DEAD LOAD*				40 PSF LIVE LOAD, 22 PSF DEAD LOAD**			
	12" o.c.	16" o.c.	19.2" o.c.	24" o.c.	12"o.c.	16" o.c.	19.2" o.c.	24" o.c.
TJI®/Pro™	WEB STIFFENERS NOT REQUIRED				WEB STIFFENERS NOT REQUIRED			
TJI®/15 DF	WEB STIFFENERS NOT REQUIRED				WEB STIFFENERS NOT REQUIRED			
TJI®/25 DF	Not Required	24'-3"	20'-2"	16'-1"	26'-1"	19'-6"	16'-3"	12'-11"
TJI®/35 DF	Not Required	27'-8"	23'-1"	18'-5"	29'-9"	22'-4"	18'-7"	14'-10"
TJI®/55 DF	Not Required	Not Required	Not Required	25'-8"	Not Required	31'-4"	26'-1"	20'-10"

*12 PSF Dead Load at TJI®/55 DF joists. **24 PSF Dead Load at TJI®/55 DF joists

7. Long term deflection under dead load which includes the effect of creep, common to all wood members, has not been considered for any of the above applications. ▬ shaded spans reflect initial dead load deflection exceeding 0.33", which may be unacceptable. For additional information contact your Trus Joist MacMillan representative.

9-3 *Floor span charts for sizing TJI I-joists from Trus Joist MacMillan.* Courtesy Trus Joist MacMillan

factoring in the increased or decreased labor to install them), you can perform some simple arithmetic and determine which is the most cost-effective route to take for the floor you're framing.

Now let's consider another option. Suppose it's possible to add an additional girder under the floor at the mid-span point of the joists, which would cut the span from 20 feet to 10 feet. Still in the same chart for live load, dead load, and live-load deflection, you can move up each column (12, 16, and 24" O.C.) until you arrive at the span that meets or exceeds the new span of 10 feet, which in this example would be all the way back up to the top of the chart. A quick look tells you that you could increase your spacing to 24 inches on center, drop all the way down to a 9½-inch-deep joist, series TJI/Pro, and still be able to span up to 12 feet, 9 inches. By factoring in the reduced price of these joists and reducing the installation labor of going to 24-inch spacing, then weighing that against the additional labor and material costs of adding the girder, you can again easily determine which is the most cost-effective way to frame the floor. Now perhaps you'd like to consider framing a floor that's stiffer than the minimum standard specified by the building codes. Still on the same page in the design booklet and still under the same heading for a 40-psf live load and a 10-psf dead load, move to the right-hand table, labeled L/480 Live Load Deflection. Follow the same procedure as before, and look at the column labeled 16" O.C. You'll see that you need to jump up two joist series to span the same distance at the same on-center spacing as you needed for an L/360 floor system: an 11⅞-inch joist, TJI/35 DF series, which gives you a maximum span of 20 feet, 9 inches.

Sizing LPI joists manufactured by Louisiana-Pacific

To use the same example with another manufacturer, let's look at the LPI I-joists manufactured by Louisiana-Pacific. Refer to their booklet *LPI (TM) Joists, Technical Data for Residential Floor and Roof Applications.* First, find the correct section in the book, the one that deals with floor joists. In L-P's case, they offer a couple of different charts for reference, and a little different way of making joist comparisons.

Look at the page labeled Maximum Simple (Single) Floor Spans (Fig. 9-4), then refer to the upper chart: 40-PSF Live Load, 10-PSF Dead Load. As with the Trus Joist MacMillan charts, look under 16-inch o.c. spacing, then read down the left-hand column, headed L/360. The first span you'll find that meets or exceeds your requirement is 20 feet, 3 inches. Read to the left and you'll see that you need an 11⅞-inch LPI 26-A series joist for that span. The right-hand column under the same 16-inch spacing is for the L/480 live-load deflection, and you'll see that you need to move down two more lines to find a joist that meets that stiffer deflection (11⅞-inch joist, LPI 36-A, maximum span of 20 feet, 9 inches).

L-P's booklet also offers another floor-joist chart to simplify comparisons, labeled LPI Floor Joist Quick Reference Chart (Fig. 9-5). You'll see that this chart is laid out a little differently. Read down the far left column under Joist Span (FT), and read down to 20. Now read across to the right, to the intersection with the column headed 16" under the heading O.C. Joist Spacing. You'll see that, for a floor with a deflection rating of L/360, you would need either an 11⅞-inch series 26-A joist or an 11⅞-inch series 36-A joist for a deflection rating of L/480.

On this chart, note that each of the span/spacing intersections have two boxes under each live-load deflection instead of one. This is simply another way of comparing joists of different series numbers for different floor stiffnesses. The upper box in each column lists the joist that is the minimum in the product line that will work for the given span. The lower box shows an optional joist that will give you a slightly stiffer floor, although no deflection-rating formula is actually given.

By using either one of these charts, you can make the same types of adjustments and consider similar "what ifs" as you did with the TJI charts. Their Quick Reference Chart makes it especially easy to compare different spacings and different spans; then it's just a matter of making the necessary price comparisons.

Sizing BCI joists manufactured by Boise Cascade

Another manufacturer that deals in wood I-joists is Boise Cascade. To specify their I-joist product, called a BCI joist, you'll want to refer to their booklet *Boise Cascade Engineered Wood Products Design Guide for Architects, Engineers & Designers*. Go to the section entitled Floor Framing and you'll see another multilevel chart, again laid out a little differently (Fig. 9-6).

First you need to read the footnotes to see that you're in the correct chart, based on 40-psf live load and 10-psf dead load. The footnotes also explain that the one-star Code Approved section of the chart uses a deflection of L/360, while the next two sections, Three Star and Four Star, use live-load deflections of L/480 and L/960, respectively.

For the example, you'll need to look at the top section of the chart, Code Approved. Look in the far left-hand column under O.C. Spacing and read down to 16. Now follow that line to the right until

MAXIMUM SIMPLE (SINGLE) FLOOR SPANS

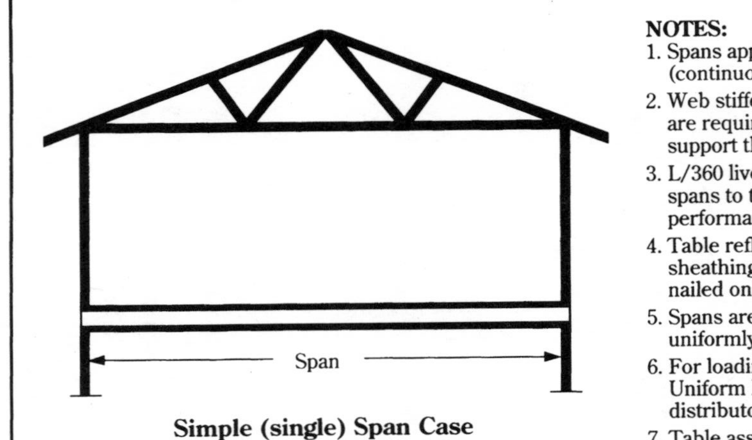

Simple (single) Span Case

NOTES:

1. Spans apply to simple (single) span applications only. For multiple (continuous) span cases, see page 6.
2. Web stiffeners are NOT required for the spans listed below. Web stiffeners are required if used with hangers if the sides of the hanger do not laterally support the LPI Joist top flange.
3. L/360 live load deflection may be used per the code. For stiffer floors limit spans to those shown in the L/480 column. See commentary about floor performance on page 3.
4. Table reflects composite action with a single layer of OSB (or equal) sheathing, **nailed and glued** to top flange of LPI Joist. When sheathing is nailed only, the spans must be reduced by 6".
5. Spans are based on clear distance between bearings with joists loaded uniformly. See page 23 for LPI Joist minimum bearing length requirements.
6. For loading or spacing conditions not shown above, use the Maximum Uniform Floor Load tables on page 7 or contact your Louisiana-Pacific distributor.
7. Table assumes that repetitive member criteria is applicable.

40 PSF LIVE LOAD; 10 PSF DEAD LOAD (15 PSF DEAD LOAD FOR LPI 56A)

		O.C. JOIST SPACING							
		12"		16"		19.2"		24"	
JOIST DEPTH	JOIST SERIES	L/360	L/480	L/360	L/480	L/360	L/480	L/360	L/480
9-1/2"	LPI 26A	17'-0"	16'-10"	17'-0"	15'-4"	16'-0"	14'-5"	14'-10"	13'-4"
	LPI 30A	17'-0"	17'-0"	17'-0"	16'-2"	16'-10"	15'-3"	15'-8"	14'-1"
11-7/8"	LPI 26A	21'-3"	20'-1"	20'-3"	18'-4"	19'-1"	17'-3"	17'-9"	16'-0"
	LPI 30A	21'-3"	21'-1"	21'-3"	19'-3"	20'-0"	18'-1"	18'-7"	16'-9"
	LPI 36A	23'-9"	22'-10"	23'-0"	20'-9"	21'-7"	19'-6"	20'-0"	18'-1"
	LPI 56A	29'-0"	26'-2"	26'-4"	23'-9"	24'-10"	22'-5"	23'-0"	20'-9"
14"	LPI 30A	25'-1"	23'-11"	24'-1"	21'-9"	22'-8"	20'-5"	19'-10"	19'-0"
	LPI 36A	28'-0"	25'-10"	25'-11"	23'-5"	24'-5"	22'-0"	22'-8"	20'-6"
	LPI 56A	32'-10"	29'-8"	29'-10"	26'-11"	28'-1"	25'-4"	26'-0"	23'-6"
16"	LPI 30A	28'-7"	26'-4"	26'-7"	24'-0"	24'-10"	22'-7"	19'-10"	19'-10"
	LPI 36A	31'-6"	28'-6"	28'-8"	25'-11"	27'-0"	24'-4"	23'-10"	22'-7"
	LPI 56A	36'-3"	32'-9"	32'-11"	29'-9"	31'-0"	28'-0"	28'-9"	25'-11"

9-4 *Floor span charts for sizing LPI I-joists from Louisiana-Pacific.* Courtesy Louisiana-Pacific

40 PSF LIVE LOAD; 25 PSF DEAD LOAD

		O.C. JOIST SPACING							
JOIST DEPTH	JOIST SERIES	12"		16"		19.2"		24"	
		L/360	L/480	L/360	L/480	L/360	L/480	L/360	L/480
9-1/2"	LPI 26[A]	17'-0"	17'-0"	16'-2"	15'-4"	15'-2"	14'-5"	13'-7"	13'-5"
	LPI 30[A]	17'-0"	17'-0"	16'-11"	16'-2"	16'-2"	15'-2"	14'-5"	14'-1"
11-7/8"	LPI 26[A]	21'-3"	20'-1"	19'-4"	18'-4"	17'-7"	17'-3"	15'-3"	15'-3"
	LPI 30[A]	21'-3"	21'-1"	20'-4"	19'-3"	19'-0"	18'-1"	15'-3"	15'-3"
	LPI 36[A]	23'-9"	22'-10"	21'-11"	20'-9"	20'-7"	19'-6"	18'-4"	18'-1"
	LPI 56[A]	27'-11"	26'-3"	25'-4"	23'-10"	23'-9"	22'-5"	22'-1"	20'-9"
14"	LPI 30[A]	25'-1"	23'-11"	22'-11"	21'-9"	19'-1"	19'-1"	15'-3"	15'-3"
	LPI 36[A]	27'-4"	25'-9"	25'-0"	23'-5"	22'-11"	22'-1"	18'-4"	18'-4"
	LPI 56[A]	31'-8"	29'-8"	28'-6"	26'-11"	26'-11"	25'-4"	25'-0"	23'-6"
16"	LPI 30[A]	28'-0"	26'-4"	22'-11"	22'-11"	19'-1"	19'-1"	15'-3"	15'-3"
	LPI 36[A]	30'-2"	28'-6"	27'-7"	25'-10"	22'-11"	22'-11"	18'-4"	18'-4"
	LPI 56[A]	34'-10"	32'-9"	31'-7"	29'-9"	29'-8"	27'-11"	26'-0"	25'-11"

9-4 *(Continued)*

you encounter the first span distance that meets or exceeds your span requirement of 20 feet. In this case, you'll come to 21'-8", and you need to read up that column to find that the span and spacing required a 14-inch series 40 joist.

With this particular chart layout, you can also continue along the line to see what else, if anything, in the BCI product line will also satisfy your requirements. Moving further to the right along the 16-inch spacing line, you'll encounter 20'-9" in the next section, which lists an 11⅞-inch series 45 joist. One more section to the right gives you a third alternative for the 20-foot span: an 11⅞-inch series 60 joist, which will handle a span of up to 22 feet, 5 inches.

This chart is another easy one to read if you're making comparisons. Move up to 12-inch spacing and read across to see what joists in the three different series offered by this manufacturer would also work for your 20-foot span, or read along the 24-inch line to see what will work with that increased on-center spacing. You can also read the next two sections down to see what's required for stiffer floors, then make the calls and get prices for the different sizes and series numbers.

Sizing engineered lumber I-joists for roof framing

Using any one of the three booklets described in the previous sections, you can also easily size I-joists for use as rafters, which is becoming an increasingly popular application for them. As with the sizing for floor framing you need to know several specific pieces of information: the live and dead loads, intended spacing, pitch of the roof, and unsupported span.

There are a few things worth noticing about the roof framing charts as opposed to the ones used earlier for floor framing. For one thing, you don't have a choice in deflection; the manufacturers make the obvious assumption that, since no one lives on the roof, the amount of flex in the framing members is of much less concern than it is for floor framing. The deflection in these tables is set at L/240 for live load and L/180 for total load, which you can readily see is much higher than for floor joists. Using the earlier example of a 20-foot floor joist, the allowable deflection in a rafter under total load would be 1.33 inches (240 ÷ 180), which is quite a difference from the 0.50 to 0.67 inches used for the various floor calculations.

LPI FLOOR JOIST QUICK REFERENCE CHART

40 PSF LIVE LOAD; 10 PSF DEAD LOAD (15 PSF DEAD LOAD FOR LPI 56^{A})

JOIST SPAN (FT)	O.C. JOIST SPACING							
	12"		16"		19.2"		24"	
	L/360	L/480	L/360	L/480	L/360	L/480	L/360	L/480
10	9-1/2"-26^{A}	9-1/2"-26^{A}	9-1/2"-26^{A}	9-1/2"-26^{A}	9-1/2"-26^{A}	9-1/2"-26^{A}	9-1/2"-26^{A}	9-1/2"-26^{A}
	9-1/2"-30^{A}	9-1/2"-30^{A}	9-1/2"-30^{A}	9-1/2"-30^{A}	9-1/2"-30^{A}	9-1/2"-30^{A}	9-1/2"-30^{A}	9-1/2"-30^{A}
11	9-1/2"-26^{A}	9-1/2"-26^{A}	9-1/2"-26^{A}	9-1/2"-26^{A}	9-1/2"-26^{A}	9-1/2"-26^{A}	9-1/2"-26^{A}	9-1/2"-26^{A}
	9-1/2"-30^{A}	9-1/2"-30^{A}	9-1/2"-30^{A}	9-1/2"-30^{A}	9-1/2"-30^{A}	9-1/2"-30^{A}	9-1/2"-30^{A}	9-1/2"-30^{A}
12	9-1/2"-26^{A}	9-1/2"-26^{A}	9-1/2"-26^{A}	9-1/2"-26^{A}	9-1/2"-26^{A}	9-1/2"-26^{A}	9-1/2"-26^{A}	9-1/2"-26^{A}
	9-1/2"-30^{A}	9-1/2"-30^{A}	9-1/2"-30^{A}	9-1/2"-30^{A}	9-1/2"-30^{A}	9-1/2"-30^{A}	9-1/2"-30^{A}	9-1/2"-30^{A}
13	9-1/2"-26^{A}	9-1/2"-26^{A}	9-1/2"-26^{A}	9-1/2"-26^{A}	9-1/2"-26^{A}	9-1/2"-26^{A}	9-1/2"-26^{A}	9-1/2"-26^{A}
	9-1/2"-30^{A}	9-1/2"-30^{A}	9-1/2"-30^{A}	9-1/2"-30^{A}	9-1/2"-30^{A}	9-1/2"-30^{A}	9-1/2"-30^{A}	9-1/2"-30^{A}
14	9-1/2"-26^{A}	9-1/2"-26^{A}	9-1/2"-26^{A}	9-1/2"-26^{A}	9-1/2"-26^{A}	9-1/2"-26^{A}	9-1/2"-26^{A}	9-1/2"-30^{A}
	9-1/2"-30^{A}	9-1/2"-30^{A}	9-1/2"-30^{A}	9-1/2"-30^{A}	9-1/2"-30^{A}	9-1/2"-30^{A}	9-1/2"-30^{A}	11-7/8"-26^{A}
15	9-1/2"-26^{A}	9-1/2"-26^{A}	9-1/2"-26^{A}	9-1/2"-26^{A}	9-1/2"-26^{A}	9-1/2"-30^{A}	9-1/2"-30^{A}	11-7/8"-26^{A}
	9-1/2"-30^{A}	9-1/2"-30^{A}	9-1/2"-30^{A}	9-1/2"-30^{A}	9-1/2"-30^{A}	11-7/8"-26^{A}	11-7/8"-30^{A}	11-7/8"-30^{A}
16	9-1/2"-26^{A}	9-1/2"-26^{A}	9-1/2"-26^{A}	9-1/2"-30^{A}	9-1/2"-26^{A}	11-7/8"-26^{A}	11-7/8"-26^{A}	11-7/8"-30^{A}
	9-1/2"-30^{A}	9-1/2"-30^{A}	9-1/2"-30^{A}	11-7/8"-26^{A}	9-1/2"-30^{A}	11-7/8"-30^{A}	11-7/8"-30^{A}	14"-30^{A}
17	9-1/2"-26^{A}	9-1/2"-30^{A}	9-1/2"-26^{A}	11-7/8"-26^{A}	11-7/8"-26^{A}	11-7/8"-26^{A}	14"-30^{A}	14"-30^{A}
	9-1/2"-30^{A}	11-7/8"-26^{A}	9-1/2"-30^{A}	11-7/8"-30^{A}	11-7/8"-30^{A}	11-7/8"-30^{A}	16"-30^{A}	16"-30^{A}
18	11-7/8"-26^{A}	11-7/8"-26^{A}	11-7/8"-26^{A}	11-7/8"-26^{A}	11-7/8"-26^{A}	11-7/8"-30^{A}	11-7/8"-36^{A}	11-7/8"-36^{A}
	11-7/8"-30^{A}	11-7/8"-30^{A}	11-7/8"-30^{A}	11-7/8"-30^{A}	11-7/8"-30^{A}	11-7/8"-36^{A}	14"-36^{A}	14"-36^{A}
19	11-7/8"-26^{A}	11-7/8"-26^{A}	11-7/8"-26^{A}	11-7/8"-30^{A}	11-7/8"-26^{A}	11-7/8"-36^{A}	11-7/8"-36^{A}	14"-36^{A}
	11-7/8"-30^{A}	11-7/8"-30^{A}	11-7/8"-30^{A}	11-7/8"-36^{A}	11-7/8"-30^{A}	14"-30^{A}	14"-36^{A}	16"-36^{A}
20	11-7/8"-26^{A}	11-7/8"-26^{A}	11-7/8"-26^{A}	11-7/8"-36^{A}	11-7/8"-30^{A}	14"-30^{A}	11-7/8"-36^{A}	14"-36^{A}
	11-7/8"-30^{A}	11-7/8"-30^{A}	11-7/8"-30^{A}	14"-30^{A}	11-7/8"-36^{A}	16"-30^{A}	14"-36^{A}	16"-36^{A}
21	11-7/8"-26^{A}	11-7/8"-30^{A}	11-7/8"-30^{A}	14"-30^{A}	11-7/8"-36^{A}	16"-30^{A}	14"-36^{A}	16"-36^{A}
	11-7/8"-30^{A}	14"-30^{A}	14"-30^{A}	16"-30^{A}	14"-30^{A}	14"-36^{A}	16"-36^{A}	14"-56^{A}
22	11-7/8"-36^{A}	11-7/8"-36^{A}	11-7/8"-36^{A}	14"-36^{A}	14"-36^{A}	14"-36^{A}	14"-36^{A}	16"-36^{A}
	14"-30^{A}	14"-30^{A}	14"-30^{A}	16"-30^{A}	11-7/8"-56^{A}	11-7/8"-56^{A}	16"-36^{A}	14"-56^{A}

23	11-7/8"-36[A]	14"-30[A]	11-7/8"-36[A]	14"-36[A]	14"-36[A]	16"-36[A]	11-7/8"-56[A]	14"-56[A]
	14"-30[A]	14"-36[A]	14"-30[A]	16"-30[A]	11-7/8"-56[A]	14"-56[A]	14"-56[A]	16"-56[A]
24	14"-30[A]	14"-36[A]	14"-30[A]	16"-30[A]	14"-36[A]	16"-36[A]	14"-56[A]	16"-56[A]
	14"-36[A]	16"-30[A]	14"-36[A]	16"-36[A]	11-7/8"-56[A]	14"-56[A]	16"-56[A]	–
25	14'-30[A]	16"-30[A]	14"-36[A]	16"-36[A]	16"-36[A]	16"-36[A]	14"-56[A]	16"-56[A]
	14"-36[A]	14"-36[A]	16"-30[A]	14"-56[A]	14"-56[A]	16"-56[A]	16"-56[A]	–
26	14"-36[A]	16"-30[A]	16"-30[A]	14"-56[A]	16"-36[A]	16"-56[A]	14"-56[A]	–
	16"-30[A]	11-7/8"-56[A]	11-7/8"-56[A]	16"-56[A]	16"-56[A]	–	16"-56[A]	–
27	14"-36[A]	16"-36[A]	16"-36[A]	16"-56[A]	16"-36[A]	16"-56[A]	16"-56[A]	–
	16"-30[A]	14"-56[A]	14"-56[A]	–	16"-56[A]	–	–	–
28	14"-36[A]	16"-36[A]	16"-36[A]	16"-56[A]	16"-56[A]	16"-56[A]	–	–
	16"-30[A]	14"-56[A]	14"-56[A]	–	–	–	–	–
29	16"-36[A]	14"-56[A]	14"-56[A]	16"-56[A]	16"-56[A]	–	–	–
	11-7/8"-56[A]	16"-56[A]	16"-56[A]	–	–	–	–	–
30	16"-36[A]	16"-56[A]	16"-56[A]	–	16"-56[A]	–	–	–
	14"-56[A]	–	–	–	–	–	–	–
31	16"-36[A]	16"-56[A]	16"-56[A]	–	16"-56[A]	–	–	–
	14"-56[A]	–	–	–	–	–	–	–
32	14"-56[A]	16"-56[A]	16"-56[A]	–	–	–	–	–
	16"-56[A]	–	–	–	–	–	–	–

NOTES:

1. Web stiffeners are not required for use with the table above. Web stiffeners are required when LPI Joists are used with hangers if the sides of the hangers do not laterally support the LPI Joist top flange.
2. L/360 live load deflection may be used per the code. For stiffer floors choose an LPI Joist from the L/480 column. See commentary about floor performance on page 3.
3. Selecting the LPI Joist in the lower box will result in a stiffer floor and should provide more satisfaction to the end user.
4. Spans apply to both simple span (see detail on page 5) and multiple span (see detail on page 6) applications.
5. Spans shown are clear spans between supports based on uniformly distributed loads only.
6. Table reflects composite action with a single layer of OSB (or equal) nailed and glued to the I-joist top flange.
7. Table assumes that repetitive member criteria is applicable.

9-5 *Another style of sizing chart for LPI I-joists.* Courtesy Louisiana-Pacific

O.C. spacing	40 SERIES – 1½" FLANGE WIDTH			45 SERIES – 1¾" FLANGE WIDTH				60 SERIES – 2⁵⁄₁₆" FLANGE WIDTH				
	9½"	11⅞"	14"	9½"	11⅞"	14"	16"	11⅞"	14"	16"	18"	20"
★ CODE APPROVED ★												
12	18' - 1"	21' - 7"	24' - 6"	19' - 2"	22' - 10"	25' - 11"	28' - 9"	24' - 8"	28' - 0"	30' - 0"	30' - 0"	30' - 0"
16	16' - 6"	19' - 6"	21' - 8"	17' - 5"	20' - 9"	23' - 7"	26' - 2"	22' - 5"	25' - 5"	28' - 2"	30' - 0"	30' - 0"
19.2	15' - 5"	17' - 10"	19' - 9"	16' - 5"	19' - 7"	21' - 10"	24' - 5"	19' - 10"	21' - 10"	24' - 11"	29' - 0"	30' - 0"
24	13' - 9"	15' - 10"	17' - 5"	13' - 9"	15' - 10"	17' - 5"	19' - 11"	15' - 10"	17' - 5"	19' - 11"	24' - 8"	27' - 10"
32	10' - 4"	11' - 10"	13' - 1"	10' - 4"	11' - 10"	13' - 1"	14' - 11"	11' - 10"	13' - 1"	14' - 11"	18' - 5"	20' - 10"
★★★ THREE STAR ★★★												
12	16' - 4"	19' - 5"	22' - 1"	17' - 3"	20' - 7"	23' - 5"	25' - 11"	22' - 3"	25' - 3"	28' - 0"	30' - 0"	30' - 0"
16	14' - 10"	17' - 8"	20' - 1"	15' - 8"	18' - 8"	21' - 3"	23' - 6"	20' - 2"	22' - 11"	25' - 4"	27' - 9"	30' - 0"
19.2	14' - 0"	16' - 8"	18' - 11"	14' - 9"	17' - 7"	19' - 11"	22' - 1"	18' - 11"	21' - 6"	23' - 10"	26' - 1"	28' - 3"
24	12' - 11"	15' - 5"	17' - 5"	13' - 8"	15' - 10"	17' - 5"	19' - 11"	15' - 10"	17' - 5"	19' - 11"	24' - 1"	26' - 1"
32	10' - 4"	11' - 10"	13' - 1"	10' - 4"	11' - 10"	13' - 1"	14' - 11"	11' - 10"	13' - 1"	14' - 11"	18' - 5"	20' - 10"
★★★★ FOUR STAR ★★★★												
12	11' - 6"	15' - 0"	17' - 1"	11' - 6"	15' - 11"	18' - 1"	20' - 1"	17' - 2"	19' - 6"	21' - 7"	23' - 8"	25' - 8"
16	11' - 5"	13' - 7"	15' - 6"	11' - 6"	14' - 4"	16' - 4"	18' - 1"	15' - 5"	17' - 7"	19' - 6"	21' - 4"	23' - 1"
19.2	10' - 0"	12' - 9"	14' - 6"	10' - 0"	13' - 5"	15' - 3"	16' - 11"	14' - 5"	16' - 5"	18' - 2"	19' - 11"	21' - 7"
24	9" - 10"	11' - 9"	13' - 4"	10' - 0"	12' - 4"	14' - 1"	15' - 7"	13' - 3"	15' - 1"	16' - 9"	18' - 4"	19' - 11"
32	8' - 7"	10' - 4"	11' - 10"	8' - 7"	10' - 4"	11' - 10"	13' - 2"	11' - 2"	12' - 9"	14' - 2"	15' - 7"	16' - 11"

Span tables assume that sheathing is glued and nailed to joists.
Spans represent the most restrictive of simple or multiple span applications.
Span tables are based on a residential floor load of 40 PSF live load and 10 PSF dead load, and a clear distance between supports

★ *Code allowed live load deflection at L/360, see page 3 "about floor performance guide" for additional data.*

★★★ *Live Load deflection at L/480.*

★★★★ *Live Load deflection at L/960 to give a floor that is much stiffer for the more discriminating purchaser.*

9-6 *Floor span charts for sizing BCI I-joists from Boise Cascade.* Courtesy Boise Cascade

As with all framing calculations, dead load in terms of roof framing is the weight of the rafters, roof sheathing, and roofing. Notice that the charts have at least two dead-load options, allowing you to adjust your calculations to reflect heavier roofing materials. The live load refers to the weight of people who will be on the roof structure during the construction, or who'll be up there later on for repair and maintenance. More importantly, live load takes snow into consideration, since it's applied and then removed, as opposed to being permanent. The live load requirements for snow vary with the part of the country you're building in, from as little as a 16 psf in nonsnow areas to 50 psf in heavy snow areas. Consult with your building department for the proper live- and dead-load calculations for your area.

If you have specific point loads on your roof, such as the weight of a furnace, air conditioning unit or other piece of air-handling equipment, solar panels, or other heavy, direct loads, these need to be accounted for separately. You need to know the total weight of the unit, including any ladders, stands, or other support equipment or structures, as well as the size of the area over which the weight will be spread. The design charts will not cover applications such as these, so you'll most likely need to consult with the engineered lumber manufacturer, the equipment manufacturer, or a licensed engineer or architect for a support and reinforcement system that's adequate for the load.

Also note that, even though the rafters are installed on a slope, you're instructed to measure the span distance between supports horizontally, not on an angle. These joists also require a minimum slope of ¼ inch in 12 inches of horizontal run to allow for adequate runoff.

For these examples, we'll select a residential application with the following specifications:

Live load (moderate snow area). 25 psf
Dead load. 15 psf
Roof pitch. 6/12 (6 inches of vertical rise per 12 inches of horizontal run)
Spacing. 24 inches center-to-center
Unsupported span. 16 feet

Sizing TJI joists manufactured by Trus Joist MacMillan

First, find the section of the booklet titled TJI Joist Residential Roof Span Chart (Fig. 9-7). Notice that the top of the chart breaks roof pitches into two categories: a low slope being 6/12 or less, and a high

TJI® Joist Residential Roof Span Chart

RESIDENTIAL ROOF SPAN CHART

Low Slope: 6"/12" or less.
High Slope: Over 6"/12" through 12"/12".

O.C. SPACING	DEPTH	SERIES	SLOPE	DESIGN LIVE LOAD (LL) AND DEAD LOAD (DL) IN PSF													
				NON-SNOW (125%)				SNOW LOAD AREA (115%)									
				16LL 15DL	16LL 20DL	20LL 15DL	20LL 20DL	20LL 15DL	20LL 20DL	25LL 15DL	25LL 20DL	30LL 15DL	30LL 20DL	40LL 15DL	40LL 20DL	50LL 15DL	50LL 20DL
19.2"	9½"	TJI®/Pro™	LOW	N.A.	N.A.	18'-3"	17'-5"	17'-11"	16'-8"	16'-10"	15'-9"	15'-10"	15'-0"	14'-5"	13'-9"	13'-3"	12'-9"
			HIGH	16'-11"	15'-11"	16'-4"	15'-5"	16'-4"	15'-5"	15'-8"	14'-10"	15'-2"	14'-2"	13'-10"	13'-1"	12'-10"	12'-3"
		TJI®/15 DF	LOW	N.A.	N.A.	18'-11"	18'-0"	18'-11"	18'-0"	18'-1"	17'-4"	17'-5"	16'-9"	16'-3"	15'-9"	15'-5"	14'-11"
			HIGH	17'-6"	16'-6"	16'-11"	16'-0"	16'-11"	16'-0"	16'-3"	15'-6"	15'-8"	15'-0"	14'-9"	14'-2"	14'-0"	13'-6"
		TJI®/25 DF	LOW	N.A.	N.A.	19'-10"	18'-11"	19'-10"	18'-11"	19'-0"	18'-2"	18'-3"	17'-7"	17'-1"	16'-6"	16'-2"	15'-8"
			HIGH	18'-4"	17'-3"	17'-9"	16'-9"	17'-9"	16'-9"	17'-0"	16'-3"	16'-5"	15'-9"	15'-5"	14'-11"	14'-8"	14'-2"
	11⅞"	TJI®/Pro™	LOW	N.A.	N.A.	21'-7"	20'-1"	20'-8"	19'-3"	19'-5"	18'-3"	18'-4"	17'-4"	16'-8"	15'-10"	15'-4"	14'-9"
			HIGH	20'-5"	19'-2"	19'-8"	18'-8"	19'-7"	18'-1"	18'-5"	17'-2"	17'-6"	16'-5"	16'-0"	15'-2"	14'-10"	14'-2"
		TJI®/15 DF	LOW	N.A.	N.A.	22'-10"	21'-9"	22'-10"	21'-9"	21'-10"	20'-11"	21'-0"	20'-2"	19'-8"	18'-9"	18'-1"	16'-8"
			HIGH	21'-1"	19'-10"	20'-4"	19'-3"	20'-4"	19'-3"	19'-7"	18'-8"	18'-11"	18'-1"	17'-9"	17'-1"	16'-10"	16'-4"
		TJI®/25 DF	LOW	N.A.	N.A.	23'-11"	22'-9"	23'-11"	22'-9"	22'-10"	21'-11"	22'-0"	21'-2"	20'-7"	19'-4"	17'-10"	16'-6"
			HIGH	22'-1"	20'-10"	21'-4"	20'-2"	21'-4"	20'-2"	20'-6"	19'-6"	19'-9"	18'-11"	18'-7"	17'-11"	17'-8"	17'-1"
		TJI®/35 DF	LOW	N.A.	N.A.	25'-11"	24'-9"	25'-11"	24'-9"	24'-10"	23'-9"	23'-11"	23'-0"	22'-4"	21'-7"	20'-5"	18'-11"
			HIGH	24'-0"	22'-7"	23'-2"	21'-11"	23'-2"	21'-11"	22'-3"	21'-2"	21'-6"	20'-7"	20'-3"	19'-6"	19'-2"	18'-7"
		TJI®/55 DF	LOW	N.A.	N.A.	29'-9"	28'-4"	29'-9"	28'-4"	28'-5"	27'-3"	27'-4"	26'-3"	25'-7"	24'-9"	24'-2"	23'-6"
			HIGH	27'-6"	25'-10"	26'-6"	25'-1"	26'-6"	25'-1"	25'-6"	24'-3"	24'-7"	23'-6"	23'-2"	22'-3"	22'-0"	21'-3"
	14"	TJI®/25 DF	LOW	N.A.	N.A.	27'-4"	26'-0"	27'-4"	26'-0"	26'-2"	25'-0"	25'-2"	23'-2"	21'-1"	19'-4"	17'-10"	16'-6"
			HIGH	25'-3"	23'-9"	24'-5"	23'-1"	24'-5"	23'-1"	23'-5"	22'-4"	22'-8"	21'-8"	21'-3"	19'-8"	18'-11"	17'-2"
		TJI®/35 DF	LOW	N.A.	N.A.	29'-7"	28'-2"	29'-7"	28'-2"	28'-4"	27'-2"	27'-3"	26'-2"	24'-1"	22'-1"	20'-6"	18'-11"
			HIGH	27'-5"	25'-10"	26'-5"	25'-0"	26'-5"	25'-0"	25'-5"	24'-2"	24'-6"	23'-6"	23'-1"	22'-2"	21'-11"	20'-6"
		TJI®/55 DF	LOW	N.A.	N.A.	33'-10"	32'-3"	33'-10"	32'-3"	32'-5"	31'-0"	31'-2"	29'-11"	29'-1"	28'-2"	27'-6"	26'-9"
			HIGH	31'-3"	29'-6"	30'-3"	28'-7"	30'-3"	28'-7"	29'-1"	27'-8"	28'-0"	26'-10"	26'-4"	25'-5"	25'-0"	24'-3"
	16"	TJI®/25 DF	LOW	N.A.	N.A.	30'-6"	29'-0"	30'-6"	29'-0"	29'-0"	25'-10"	25'-10"	23'-2"	21'-1"	19'-4"	17'-10"	16'-6"
			HIGH	28'-2"	26'-6"	27'-2"	25'-9"	27'-2"	25'-9"	26'-2"	24'-11"	25'-3"	23'-1"	22'-0"	19'-8"	18'-11"	17'-2"
		TJI®/35 DF	LOW	N.A.	N.A.	32'-11"	31'-4"	32'-11"	31'-4"	31'-6"	29'-6"	29'-6"	26'-6"	24'-1"	22'-1"	20'-5"	18'-11"
			HIGH	30'-5"	28'-8"	29'-5"	27'-10"	29'-5"	27'-10"	28'-3"	26'-11"	27'-3"	26'-1"	25'-8"	23'-9"	22'-2"	20'-6"
		TJI®/55 DF	LOW	N.A.	N.A.	37'-7"	35'-10"	37'-7"	35'-10"	36'-0"	34'-5"	34'-7"	33'-3"	32'-4"	31'-4"	29'-6"	27'-5"
			HIGH	34'-9"	32'-9"	33'-7"	31'-9"	33'-7"	31'-9"	32'-3"	30'-9"	31'-2"	29'-9"	29'-3"	28'-2"	27'-9"	25'-7"

24"	9½"	TJI®/Pro™	LOW	N.A.	N.A.	16'-8"	15'-7"	16'-0"	14'-11"	15'-0"	14'-1"	14'-2"	13'-5"	12'-10"	12'-3"	11'-10"	11'-4"
			HIGH	15'-8"	14'-9"	15'-1"	14'-4"	15'-1"	13'-11"	14'-3"	13'-3"	13'-7"	12'-8"	12'-5"	11'-9"	11'-6"	10'-11"
		TJI®/15 DF	LOW	N.A.	N.A.	17'-6"	16'-8"	17'-6"	16'-8"	16'-9"	16'-1"	16'-1"	15'-6"	15'-1"	14'-6"	13'-3"	12'-4"
			HIGH	16'-2"	15'-3"	15'-8"	14'-10"	15'-8"	14'-10"	15'-0"	14'-4"	14'-6"	13'-10"	13'-8"	13'-2"	12'-11"	12'-6"
		TJI®/25 DF	LOW	N.A.	N.A.	18'-5"	17'-6"	18'-5"	17'-6"	17'-7"	16'-10"	16'-11"	16'-3"	15'-9"	15'-3"	14'-3"	13'-2"
			HIGH	17'-0"	16'-0"	16'-5"	15'-6"	16'-5"	15'-6"	15'-9"	15'-0"	15'-3"	14'-7"	14'-4"	13'-9"	13'-7"	13'-1"
	11⅞"	TJI®/Pro™	LOW	N.A.	N.A.	19'-4"	18'-0"	18'-6"	17'-3"	17'-4"	16'-3"	16'-5"	15'-6"	14'-10"	14'-2"	13'-2"	12'-3"
			HIGH	18'-11"	17'-7"	18'-3"	16'-10"	17'-6"	16'-2"	16'-6"	15'-4"	15'-8"	14'-8"	14'-4"	13'-4"	12'-9"	11'-7"
		TJI®/15 DF	LOW	N.A.	N.A.	21'-1"	20'-1"	21'-1"	20'-1"	20'-2"	19'-3"	19'-3"	17'-3"	15'-8"	14'-5"	13'-3"	12'-4"
			HIGH	19'-6"	18'-5"	18'-10"	17'-10"	18'-10"	17'-10"	18'-1"	17'-3"	17'-6"	16'-9"	16'-5"	15'-6"	14'-5"	13'-4"
		TJI®/25 DF	LOW	N.A.	N.A.	22'-1"	21'-1"	22'-1"	21'-1"	21'-2"	20'-3"	20'-4"	18'-6"	16'-10"	15'-5"	14'-3"	13'-2"
			HIGH	20'-5"	19'-3"	19'-9"	18'-8"	19'-9"	18'-8"	19'-0"	18'-1"	18'-4"	17'-6"	17'-3"	15'-9"	15'-1"	13'-8"
		TJI®/35 DF	LOW	N.A.	N.A.	24'-0"	22'-10"	24'-0"	22'-10"	23'-0"	22'-0"	22'-1"	21'-2"	19'-3"	17'-8"	16'-3"	15'-1"
			HIGH	22'-3"	20'-11"	21'-5"	20'-4"	21'-5"	20'-4"	20'-7"	19'-8"	19'-11"	19'-0"	18'-8"	17'-9"	17'-0"	15'-5"
		TJI®/55 DF	LOW	N.A.	N.A.	27'-6"	26'-2"	27'-6"	26'-2"	26'-4"	25'-2"	25'-4"	24'-4"	23'-8"	22'-10"	22'-4"	21'-8"
			HIGH	25'-5"	24'-0"	24'-7"	23'-3"	24'-7"	23'-3"	23'-7"	22'-6"	22'-9"	21'-9"	21'-5"	20'-7"	20'-4"	19'-8"
	14"	TJI®/25 DF	LOW	N.A.	N.A.	25'-4"	24'-1"	25'-4"	23'-2"	23'-2"	20'-7"	20'-7"	18'-6"	16'-10"	15'-5"	14'-3"	13'-2"
			HIGH	23'-5"	22'-0"	22'-7"	21'-5"	22'-7"	21'-5"	21'-8"	20'-2"	20'-11"	18'-6"	17'-7"	15'-9"	15'-1"	13'-8"
		TJI®/35 DF	LOW	N.A.	N.A.	27'-5"	26'-1"	27'-5"	26'-1"	26'-3"	23'-7"	23'-7"	21'-2"	19'-3"	17'-8"	16'-3"	15'-1"
			HIGH	25'-4"	23'-11"	24'-6"	23'-2"	24'-6"	23'-2"	23'-6"	22'-5"	22'-9"	21'-9"	20'-10"	19'-0"	17'-9"	16'-4"
		TJI®/55 DF	LOW	N.A.	N.A.	31'-4"	29'-10"	31'-4"	29'-10"	30'-0"	28'-8"	28'-10"	27'-9"	26'-11"	25'-7"	23'-7"	21'-11"
			HIGH	29'-0"	27'-4"	28'-0"	26'-6"	28'-0"	26'-6"	26'-11"	25'-7"	26'-0"	24'-10"	24'-5"	23'-5"	22'-6"	20'-5"
	16"	TJI®/25 DF	LOW	N.A.	N.A.	28'-2"	25'-3"	26'-6"	23'-2"	23'-2"	20'-7"	20'-7"	18'-6"	16'-10"	15'-5"	14'-3"	13'-2"
			HIGH	26'-1"	24'-4"	25'-2"	22'-4"	25'-2"	22'-4"	23'-4"	20'-2"	21'-0"	18'-6"	17'-7"	15'-9"	15'-1"	13'-8"
		TJI®/35 DF	LOW	N.A.	N.A.	30'-6"	28'-10"	30'-4"	26'-6"	26'-6"	23'-7"	23'-7"	21'-2"	19'-3"	17'-8"	16'-3"	15'-1"
			HIGH	28'-2"	26'-7"	27'-3"	25'-9"	27'-3"	25'-9"	26'-2"	24'-11"	25'-3"	22'-8"	20'-10"	19'-0"	17'-9"	16'-4"
		TJI®/55 DF	LOW	N.A.	N.A.	34'-10"	33'-2"	34'-10"	33'-0"	33'-4"	31'-10"	32'-0"	30'-7"	27'-11"	25'-7"	23'-7"	21'-11"
			HIGH	32'-2"	30'-4"	31'-1"	29'-5"	31'-1"	29'-5"	29'-10"	28'-5"	28'-10"	27'-6"	26'-2"	23'-5"	22'-6"	20'-5"

GENERAL NOTES:

1. Roof surface must be sloped ¼" in 12" minimum to provide positive drainage.
2. Maximum deflection is limited to L/180 at total load, and L/240 at live load.
3. For loads not shown, refer to allowable uniform load tables on page 15.
4. Charts are based on a support beam or wall at the high end. Applications utilizing ridge boards are not covered by these charts.
5. Spans shown are horizontal clear distances between supports and reflect the most restrictive of simple or multiple span applications, based on uniformly loaded joists and include allowable increases for repetitive member use.

WEB STIFFENER REQUIREMENTS:

6. **TJI®/Pro™, TJI®/15 DF, TJI®/25 DF and TJI®/35 DF joists:** Web stiffeners are required if the sides of the hanger do not laterally support the TJI® joist top flange. Web stiffeners are also required at all sloped hanger and birdsmouth cut locations.
7. **TJI®/55 DF joists:** Web stiffeners are required at all hanger and birdsmouth cut locations.

9-7 *Roof span charts for sizing TJI I-joists from Trus Joist MacMillan.* Courtesy Trus Joist MacMillan

slope being 6/12 through 12/12 (slopes steeper than 12/12 will require special engineering). For the example, you'll use a low slope.

Begin by reading across the top of the chart. Under Snow Load Area, find the column headed 25LL, 15DL to get the proper live-load and dead-load figures. Now look at the first column on the left side of the chart, labeled O.C. Spacing. Follow that down to the section 24".

Back in the 25LL, 15DL column, read down the column until you pass the line that marks the start of the 24" section (the layout of this particular chart is a little harder to read than some of the others). Once past that line, continue down until you find the first span figure for low-slope roofs that meets or exceeds your desired 16-foot rafter span. Note that the third entry down is 16 feet, 9 inches for a low slope. Read back across to the left, and you'll see that you need a 9½-inch TJI/15 DF series joist.

You can again use the chart to explore your options. Decreasing the on-center spacing to 16 inches, for example, means you could use a less expensive 9½-inch TJI/Pro series joist, which would allow you to span up to 18 feet, 5 inches. You would then simply need to determine the price difference for the joists and the additional labor cost to install them, and decide which is the more cost-effective option.

Sizing LPI joists manufactured by Louisiana-Pacific

Using the same booklet as before, turn to the section marked Maximum Roof Spans (Figs. 9-8 and 9-9). As you'll see at the top of the page, the set of charts on the left all pertain to low-slope roofs of 6/12 or less, so these are the ones you'll use. Go to the bottom chart marked 24" O.C., then use these numbers for your desired 24-inch, center-to-center rafter spacing.

The next column to the right of the joist spacing is broken into two sections, Non-Snow and Snow. Under Snow, find your desired live- and dead-load figures of 25/15, then read across to the right until you find a span that meets or exceeds your desired 16 feet. In this table, it's in the very first column: 16'-8". Now simply read up that column to the top, and you'll see that you need a 9½-inch LPI series 26A joist to satisfy this application.

Sizing BCI joists manufactured by Boise Cascade

First you need to find the section of the booklet labeled Roof Tables (Fig. 9-10). The first set of tables is for roof slopes of 3½ over 12 or less, so you need to read further on to find the ones for steeper pitches. After that, you'll find these tables a little more confusing to use, since the manufacturer apparently assumes that you already know what series I-joist you want to use and only need to know the size. As a result of this method of organization, as well as the general layout of the charts themselves, you'll need to do a little more detective work to find what you're looking for than you did with the first two sets of charts.

The first chart you'll come to is 40 Series BCI Joists, so start there. Find 24" O.C. in the left-hand column, then Snow in the second column, and 25/15 (live load/dead load) in column three. Now read across until you find a span of 16 feet or more that's also under one of the center columns labeled 4/12 to 8/12. The first one you'll encounter that meets the proper span criteria is 17'-10", which when you read up the column tells you that you need an 11⅞-inch BCI 40 series joist.

Since this chart calls for a larger joist than the ones specified by Trus Joist MacMillan or Louisiana-Pacific, you'll probably want to explore further. On the next several pages, under the charts labeled 45 Series BCI Joists and 60 Series BCI Joists, you can follow the same procedure. You'll find that you can also use an 11⅞-inch BCI 45 Series I-joist, or an 11⅞-inch BCI 60 Series I-joist. None of the 9½-inch deep BCI joists will apparently work for your 16-foot span requirements, so you'll need to do some careful cost comparisons before framing this roof.

Other structural engineered lumber applications

In all the design booklets offered by various manufacturers—and there are others besides the three discussed in this chapter—you'll find a number of additional tables that cover many other loads and framing applications (refer to the examples in Figs. 9-11 through 9-16). Different tables are available for heavier live- and dead-load applications, and for a variety of multi- and single-support applications. Simply study the charts, read the footnotes, and determine which one most closely meets the framing situation you're dealing with.

In addition to the design booklets for residential applications, you'll also find other booklets for commercial and industrial construction. These typically cover heavier load applications, and deal with a wide variety of multi-level and long-span applications.

MAXIMUM ROOF SPANS

LOW SLOPE (6:12 OR LESS)

I-JOIST SPACING	DURATION OF LOAD	LOAD LIVE/DEAD PSF	JOIST DEPTH & SERIES											
			9-1/2"		11-7/8"				14"			16"		
			LPI 26[A]	LPI 30[A]	LPI 26[A]	LPI 30[A]	LPI 36[A]	LPI 56[A]	LPI 30[A]	LPI 36[A]	LPI 56[A]	LPI 30[A]	LPI 36[A]	LPI 56[A]
12" o.c.	125% NON-SNOW	20/15	22'-3"	23'-6"	26'-9"	28'-1"	30'-8"	35'-8"	32'-0"	34'-8"	40'-3"	35'-4"	38'-3"	44'-7"
		20/20	21'-2"	22'-7"	25'-5"	26'-9"	29'-3"	33'-11"	30'-5"	33'-2"	38'-5"	33'-6"	36'-6"	42'-5"
	115% SNOW	25/15	21'-2"	22'-6"	25'-8"	26'-11"	29'-3"	34'-1"	30'-6"	33'-3"	38'-8"	33'-9"	36'-8"	42'-9"
		25/20	20'-3"	21'-8"	24'-6"	25'-10"	28'-2"	32'-8"	29'-2"	31'-11"	36'-11"	32'-3"	35'-2"	40'-11"
		30/15	20'-4"	21'-8"	24'-8"	25'-10"	28'-2"	32'-9"	29'-5"	31'-11"	37'-2"	32'-6"	35'-5"	41'-0"
		30/20	19'-8"	20'-11"	23'-7"	24'-10"	27'-2"	31'-6"	28'-2"	30'-8"	35'-8"	31'-1"	34'-0"	39'-6"
		40/15	19'-0"	20'-4"	22'-11"	24'-1"	26'-3"	30'-6"	27'-7"	29'-11"	34'-9"	30'-5"	33'-1"	38'-5"
		40/20	18'-5"	19'-9"	22'-3"	23'-6"	25'-6"	29'-7"	26'-7"	29'-0"	33'-8"	29'-4"	32'-0"	37'-2"
		50/15	17'-11"	19'-2"	21'-7"	22'-10"	24'-11"	29'-0"	25'-10"	28'-4"	32'-9"	28'-8"	31'-3"	36'-4"
		50/20	17'-7"	18'-7"	21'-2"	22'-3"	24'-3"	28'-0"	25'-2"	27'-6"	31'-10"	27'-11"	30'-4"	35'-3"
16" o.c.	125% NON-SNOW	20/15	20'-2"	21'-5"	24'-1"	25"-6"	27'-8"	32'-2"	28'-11"	31"-6"	36'-6"	32'-0"	34'-10"	40'-5"
		20/20	19'-1"	20'-4"	23'-0"	24'-3"	26'-4"	30'-9"	27'-5"	29'-11"	34'-9"	30'-4"	33'-0"	38'-4"
	115% SNOW	25/15	19'-3"	20'-4"	23'-2"	24'-5"	26'-7"	30'-11"	27'-8"	30'-0"	35'-0"	30'-6"	33'-3"	38'-8"
		25/20	18'-4"	19'-7"	22'-1"	23'-4"	25'-4"	29'-7"	26'-5"	28'-9"	33'-5"	29'-4"	31'-10"	36'-11"
		30/15	18'-5"	19'-9"	22'-3"	23'-6"	25'-6"	29'-7"	26'-7"	29'-0"	33'-8"	29'-4"	32'-0"	37'-2"
		30/20	17'-9"	18'-11"	21'-4"	22'-6"	24'-6"	28'-5"	25'-8"	27'-10"	32'-3"	28'-3"	30'-9"	35'-7"
		40/15	17'-2"	18'-4"	20'-10"	22'-0"	23'-10"	27'-8"	24'-10"	27'-0"	31'-5"	27'-6"	29'-10"	34'-10"
		40/20	16'-8"	17'-10"	20'-1"	21'-1"	23'-1"	26'-10"	24'-0"	26'-1"	30'-4"	26'-8"	29'-0"	33'-7"
		50/15	16'-4"	17'-5"	19'-7"	20'-9"	22'-5"	26'-2"	23'-5"	25'-7"	29'-7"	25'-4"	28'-2"	32'-9"
		50/20	15'-10"	16'-10"	19'-0"	20'-1"	21'-10"	25'-4"	22'-11"	24'-11"	28'-10"	23'-4"	27'-5"	31'-11"

19.2" o.c.	125% NON-SNOW	20/15	18'-10"	20'-0"	22'-8"	23'-11"	25'-11"	30'-1"	27'-3"	29'-5"	34'-3"	29'-11"	32'-8"	37'-10"
		20/20	18'-0"	19'-1"	21'-7"	22'-10"	24'-10"	28'-8"	25'-9"	28'-1"	32'-7"	28'-8"	31'-2"	36'-1"
	115% SNOW	25/15	17'-11"	19'-2"	21'-7"	22'-10"	24'-11"	29'-0"	26'-1"	28'-4"	32'-8"	28'-8"	31'-3"	36'-4"
		25/20	17'-2"	18'-4"	20'-10"	21'-11"	23'-10"	27'-8"	24'-9"	27'-0"	31'-4"	27'-6"	29'-10"	34'-9"
		30/15	17'-3"	18'-5"	21'-0"	21'-10"	24'-0"	27'-11"	24'-11"	27'-2"	31'-7"	27'-8"	30'-0"	34'-10"
		30/20	16'-8"	17'-9"	20'-0"	21'-3"	22'-11"	26'-8"	23'-11"	26'-0"	30'-2"	26'-5"	28'-10"	33'-5"
		40/15	16'-2"	17'-2"	19'-7"	20'-6"	22'-4"	25'-10"	23'-4"	25'-3"	29'-5"	24'-10"	28'-1"	32'-7"
		40/20	15'-7"	16'-7"	18'-10"	19'-9"	21'-8"	25'-1"	22'-7"	24'-6"	28'-6"	22'-7"	27'-2"	31'-6"
		50/15	15'-2"	16'-2"	18'-5"	19'-6"	21'-1"	24'-6"	21'-1"	24'-0"	27'-10"	21'-1"	26'-5"	30'-9"
		50/20	14'-10"	15'-9"	17'-11"	18'-11"	20'-5"	23'-10"	19'-5"	23'-3"	27'-1"	19'-5"	24'-9"	29'-10"

24" o.c.	125% NON-SNOW	20/15	17'-4"	18'-7"	21'-0"	22'-0"	24'-1"	27'-10"	25'-0"	27'-2"	31'-9"	27'-10"	30'-2"	35'-2"
		20/20	16'-8"	17'-7"	20'-1"	20'-11"	22'-9"	26'-6"	24'-0"	25'-11"	30'-2"	26'-5"	28'-8"	33'-4"
	115% SNOW	25/15	16'-8"	17'-9"	20'-0"	21'-3"	23'-0"	26'-9"	23'-11"	26'-0"	30'-3"	26'-6"	28'-11"	33'-6"
		25/20	16'-0"	17'-0"	19'-2"	20'-4"	22'-1"	25'-8"	22'-10"	24'-11"	28'-11"	23'-9"	27'-8"	32'-0"
		30/15	15'-10"	16'-11"	19'-3"	20'-3"	22'-0"	25'-7"	23'-0"	25'-1"	29'-2"	24'-1"	27'-8"	32'-3"
		30/20	15'-4"	16-'4"	18'-7"	19'-6"	21'-4"	24'-7"	21'-6"	24'-1"	28'-0"	21'-6"	26'-9"	30'-11"
		40/15	14'-11"	15'-11"	18'-1"	18'-11"	20'-7"	23'-11"	19'-10"	23'-6"	27'-2"	19'-10"	25'-3"	30'-2"
		40/20	14'-5"	15'-4"	17'-3"	18'-0"	20'-1"	23'-4"	18'-0"	22'-9"	26'-5"	18'-0"	23'-0"	29'-3"
		50/15	14'-2"	14'-9"	16'-8"	16'-10"	19'-7"	22'-8"	16'-10"	21'-5"	25'-8"	16'-10"	21'-5"	28'-6"
		50/20	13'-6"	13'-7"	15'-6"	15'-6"	19'-0"	22'-1"	15'-6"	19'-9"	24'-11"	15'-6"	19'-9"	26'-11"

NOTES: **(continued on the following page)**

1. Spans apply to both simple and multiple span applications.
2. Web stiffeners are not required for the spans listed in these tables. Web stiffeners are required at bird's mouth cut locations, at sloped hanger locations and for lateral support of the LPI Joist when used with hangers if the sides of the hanger do not laterally support the LPI Joist top flange.
3. Roof joists must have a minimum slope of 1/4:12 for drainage.

9-8 *Roof span charts for sizing LPI I-joists from Louisiana-Pacific.* Courtesy Louisiana-Pacific

MAXIMUM ROOF SPANS

HIGH SLOPE (OVER 6:12 THRU 12:12)

I-JOIST SPACING	DURATION OF LOAD	LOAD LIVE/DEAD PSF	JOIST DEPTH & SERIES											
			9-1/2"		11-7/8"				14"			16"		
			LPI 26^A	LPI 30^A	LPI 26^A	LPI 30^A	LPI 36^A	LPI 56^A	LPI 30^A	LPI 36^A	LPI 56^A	LPI 30^A	LPI 36^A	LPI 56^A
12" o.c.	125% NON-SNOW	20/15	19'-11"	21'-2"	23'-10"	25'-1"	27'-4"	31'-9"	28'-5"	31'-0"	35'-11"	31'-5"	34'-3"	39'-9"
		20/20	18'-8"	20'-0"	22'-8"	23'-9"	25'-10"	30'-1"	27'-0"	29'-4"	34'-2"	29'-11"	32'-6"	37'-8"
	115% SNOW	25/15	19'-1"	20'-3"	23'-0"	24'-2"	26'-5"	30'-7"	27'-4"	29'-10"	34'-7"	30'-3"	32'-11"	38'-2"
		25/20	18'-1"	19'-4"	21'-11"	22'-11"	25'-0"	29'-0"	26'-1"	28'-5"	32'-11"	28'-11"	31'-5"	36'-4"
		30/15	18'-4"	19'-7"	22'-1"	23'-4"	25'-4"	29'-6"	26'-6"	28'-9"	33'-5"	29'-3"	31'-10"	37'-0"
		30/20	17'-7"	18'-9"	21'-3"	22'-5"	24'-4"	28'-1"	25'-5"	27'-6"	32'-0"	27'-11"	30'-5"	35'-3"
		40/15	17'-3"	18'-5"	20'-10"	21'-11"	23'-11"	27'-9"	24'-10"	27'-0"	31'-6"	27'-6"	29'-10"	34'-8"
		40/20	16'-8"	17'-8"	20'-0"	21'-2"	23'-0"	26'-9"	23'-11"	26'-1"	30'-4"	26'-6"	28'-10"	33'-6"
		50/15	16'-5"	17'-5"	19'-9"	20'-10"	22'-7"	26'-4"	23'-6"	25'-8"	29'-10"	26'-1"	28'-5"	32'-11"
		50/20	15'-10"	16'-10"	19'-2"	20'-3"	22'-0"	25'-5"	22'-10"	24'-11"	28'-11"	25'-2"	27'-5"	31'-10"

I-JOIST SPACING	DURATION OF LOAD	LOAD LIVE/DEAD PSF	9-1/2" LPI 26^A	9-1/2" LPI 30^A	11-7/8" LPI 26^A	11-7/8" LPI 30^A	11-7/8" LPI 36^A	11-7/8" LPI 56^A	14" LPI 30^A	14" LPI 36^A	14" LPI 56^A	16" LPI 30^A	16" LPI 36^A	16" LPI 56^A
16" o.c.	125% NON-SNOW	20/15	18'-0"	19'-1"	21'-6"	22'-10"	24'-10"	28'-8"	25'-10"	28'-1"	32'-8"	28'-7"	31'-1"	36'-1"
		20/20	16'-11"	18'-1"	20'-6"	21'-8"	23'-6"	27'-3"	24'-5"	26'-6"	30'-11"	27'-0"	29'-5"	34'-2"
	115% SNOW	25/15	17'-2"	18'-4"	20'-9"	21'-10"	23'-10"	27'-7"	24'-11"	27'-1"	31'-4"	27'-5"	29'-11"	34'-8"
		25/20	16'-5"	17'-6"	19'-9"	20'-10"	22'-7"	26'-5"	23'-7"	25'-8"	29'-10"	26'-1"	28'-6"	33'-0"
		30/15	16'-8"	17'-8"	20'-0"	21'-2"	23'-0"	26'-9"	23'-11"	26'-1"	30'-4"	26'-6"	28'-10"	33'-6"
		30/20	15'-11"	16'-11"	19'-3"	20'-2"	21'-11"	25'-7"	22'-11"	25'-0"	28'-10"	25'-4"	27'-6"	32'-0"
		40/15	15'-8"	16'-7"	18'-11"	19'-11"	21'-7"	25'-1"	22'-7"	24'-5"	28'-6"	24'-11"	27'-1"	31'-5"
		40/20	15'-0"	16'-1"	18'-1"	19'-0"	20'-9"	24'-1"	21'-9"	23'-7"	27'-5"	24'-0"	26'-2"	30'-4"
		50/15	14'-10"	15'-10"	17'-10"	18'-9"	20'-7"	23'-9"	21'-5"	23'-3"	27'-1"	23'-8"	25'-8"	29'-11"
		50/20	14'-4"	15'-2"	17'-4"	18'-3"	19'-10"	23'-1"	20'-8"	22'-6"	26'-1"	22'-9"	24'-10"	28'-11"

19.2" o.c.	125% NON-SNOW	20/15	16'-10"	17'-11"	20'-3"	21'-4"	23'-4"	27'-0"	24'-3"	26'-6"	30'-7"	26'-9"	29'-2"	33'-10"
		20/20	15'-11"	16'-11"	19'-2"	20'-2"	22'-1"	25'-6"	23'-0"	24'-11"	29'-0"	25'-4"	27'-7"	32'-1"
	115% SNOW	25/15	16'-3"	17'-3"	19'-6"	20'-7"	22'-5"	25'-11"	23'-5"	25'-5"	29'-5"	25'-10"	28'-0"	32'-6"
		25/20	15'-5"	16'-4"	18'-7"	19'-7"	21'-3"	24'-9"	22'-3"	24'-2"	28'-1"	24'-6"	26'-7"	31'-1"
		30/15	15'-7"	16'-6"	18'-10"	19'-10"	21'-7"	25'-1"	22'-7"	24'-5"	28'-5"	24'-11"	27'-0"	31'-6"
		30/20	15'-0"	15'-10"	18'-1"	18'-11"	20'-7"	23'-11"	21'-6"	23'-6"	27'-2"	23'-10"	25'-11"	30'-1"
		40/15	14'-9"	15'-7"	17'-9"	18'-7"	20'-2"	23'-6"	21'-1"	23'-0"	26'-9"	23'-5"	25'-5"	29'-6"
		40/20	14'-2"	15'-0"	17'-1"	18'-0"	19'-5"	22'-7"	20'-4"	22'-1"	25'-8"	22'-7"	24'-5"	28'-6"
		50/15	13'-10"	14'-9"	16'-10"	17'-7"	19'-2"	22'-4"	20'-1"	21'-9"	25'-5"	22'-2"	24'-2"	28'-0"
		50/20	13'-6"	14'-4"	16'-2"	17'-0"	18'-8"	21'-7"	19'-5"	21'-2"	24'-6"	21'-6"	23'-4"	27'-2"

24" o.c.	125% NON-SNOW	20/15	15'-7"	16'-7"	18'-10"	19'-9"	21'-6"	24'-11"	22'-6"	24'-6"	28'-5"	24'-9"	27'-1"	31'-3"
		20/20	14'-8"	15'-9"	17'-10"	18'-9"	20'-4"	23'-9"	21'-2"	23'-1"	26'-10"	23'-5"	25'-7"	29'-9"
	115% SNOW	25/15	15'-0"	15'-11"	18'-2"	19'-1"	20'-9"	24'-0"	21'-7"	23'-5"	27'-2"	23'-10"	25'-11"	30'-1"
		25/20	14'-2"	15'-2"	17'-1"	18'-1"	19'-9"	22'-9"	20'-5"	22'-5"	25'-11"	22'-9"	24'-9"	28'-9"
		30/15	14'-6"	15'-4"	17'-6"	18'-4"	20'-0"	23'-2"	20'-11"	22'-7"	26'-4"	23'-0"	25'-1"	29'-1"
		30/20	13'-10"	14'-7"	16'-8"	17'-7"	19'-1"	22'-3"	19'-10"	21'-8"	25'-2"	22'-0"	23'-11"	27'-9"
		40/15	13'-8"	14'-5"	16'-5"	17'-3"	18'-9"	21'-9"	19'-6"	21'-3"	24'-8"	21'-7"	23'-7"	27'-4"
		40/20	13'-0"	13'-11"	15'-10"	16'-7"	18'-1"	21'-0"	18'-11"	20'-6"	23'-9"	20'-10"	22'-7"	26'-4"
		50/15	12'-11"	13'-9"	15'-5"	16'-5"	17'-8"	20'-8"	18'-6"	20'-2"	23'-4"	20'-0"	22'-3"	25'-10"
		50/20	12'-5"	13'-3"	15'-1"	15'-9"	17'-2"	20'-1"	18'-0"	19'-6"	22'-7"	18'-2"	21'-6"	25'-0"

NOTES: (continued from previous page)

4. Maximum deflection limited to L/240 for live load, and L/180 for total load.
5. Spans are based on the horizontal clear distance between supports with joists loaded uniformly.
6. Tables assume that repetitive member criteria is applicable.
7. Tables do NOT include additional stiffness from composite action with nailed sheathing.
8. Roof applications in high wind areas may require special analysis with reduced maximum spans. Additional connections such as straps and uplift anchors may be required.

9-9 *Roof span charts for sizing LPI I-joists from Louisiana-Pacific.* Courtesy Louisiana-Pacific

40 SERIES BCI JOISTS – $1^1/_2$" FLANGE WIDTH

Maximum clear span in feet and inches, based on horizontal spans.

		Live Load / Dead Load	9½" BCI 40			11⅞" BCI 40			14" BCI 40		
		PSF	Less than 4/12	4/12 to 8/12	Greater than 8/12	Less than 4/12	4/12 to 8/12	Greater than 8/12	Less than 4/12	4/12 to 8/12	Greater than 8/12
12" O.C.	Non-Snow (125%)	20/10	23'-4"	22'-1"	19'-4"	27'-10"	26'-3"	23'-0"	31'-7"	29'-10"	26'-1"
		20/15	22'-1"	20'-7"	17'-11"	26'-4"	24'-7"	21'-5"	29'-11"	27'-11"	24'-3"
		20/20	21'-1"	19'-6"	16'-11"	25'-2"	23'-3"	20'-2"	28'-7"	26'-5"	22'-11"
	Snow (115%)	25/10	22'-2"	20'-0"	17'-11"	26'-5"	23'-11"	21'-4"	30'-0"	27'-1"	24'-3"
		25/15	21'-1"	19'-0"	16'-10"	25'-2"	22'-8"	20'-2"	28'-7"	25'-9"	22'-10"
		30/10	21'-2"	19'-2"	17'-5"	25'-3"	22'-10"	20'-9"	28'-8"	26'-0"	23'-7"
		30/15	20'-3"	18'-3"	16'-5"	24'-3"	21'-10"	19'-8"	27'-6"	24'-10"	22'-4"
		40/10	19'-3"	17'-7"	16'-6"	23'-0"	21'-1"	19'-9"	26'-1"	23'-11"	22'-5"
		40/15	18'-11"	17'-1"	15'-9"	22'-8"	20'-6"	18'-10"	25'-5"	23'-3"	21'-4"
		50/10	17'-9"	16'-3"	15'-9"	21'-3"	19'-6"	18'-10"	24'-2"	22'-2"	21'-5"
		50/15	17'-9"	16'-2"	15'-1"	21'-1"	19'-4"	18'-1"	23'-5"	22'-0"	20'-6"
16" O.C.	Non-Snow (125%)	20/10	21'-1"	19'-3"	16'-11"	25'-3"	22'-9"	20'-0"	28'-8"	25'-10"	22'-8"
		20/15	20'-0"	18'-2"	15'-10"	23'-11"	21'-5"	18'-9"	27'-1"	24'-4"	21'-3"
		20/20	19'-0"	17'-3"	15'-0"	22'-9"	20'-5"	17'-9"	25'-10"	23'-2"	20'-2"
	Snow (115%)	25/10	20'-0"	18'-1"	16'-2"	23'-11"	21'-8"	19'-4"	27'-2"	24'-7"	22'-0"
		25/15	19'-1"	17'-2"	15'-3"	22'-10"	20'-7"	18'-3"	25'-9"	23'-4"	20'-8"
		30/10	19'-2"	17'-4"	15'-9"	22'-11"	20'-9"	18'-10"	25'-10"	23'-6"	21'-4"
		30/15	18'-4"	16'-6"	14'-11"	21'-11"	19'-9"	17'-10"	24'-3"	22'-6"	20'-2"
		40/10	17'-4"	15'-11"	14'-11"	20'-9"	19'-1"	17'-10"	23'-1"	21'-8"	20'-3"
		40/15	17'-1"	15'-6"	14'-3"	19'-10"	18'-6"	17'-0"	22'-0"	21'-1"	19'-4"
		50/10	16'-0"	14'-9"	14'-3"	19'-0"	17'-7"	17'-1"	21'-1"	20'-1"	19'-5"
		50/15	15'-9"	14'-7"	13'-8"	18'-3"	17'-6"	16'-4"	20'-3"	19'-11"	18'-7"

19.2" O.C.	Non-Snow (125%)	20/10	19'-10"	18'-1"	15'-11"	23'-8"	21'-4"	18'-9"	26'-11"	24'-3"	21'-4"
		20/15	18'-9"	17'-1"	14'-11"	25'-5"	20'-2"	17'-7"	25'-6"	22'-10"	20'-0"
		20/20	17'-10"	16'-3"	14'-1"	21'-4"	19'-2"	16'-8"	24'-4"	21'-9"	18'-11"
	Snow (115%)	25/10	18'-9"	17'-0"	15'-2"	22'-6"	20'-4"	18'-2"	25'-2"	23'-1"	20'-8"
		25/15	17'-11"	16'-1"	14'-4"	21'-2"	19'-3"	17'-2"	23'-6"	21'-11"	19'-5"
		30/10	17'-11"	16'-3"	14'-9"	21'-3"	19'-5"	17'-8"	23'-7"	22'-1"	20'-1"
		30/15	17'-2"	15'-6"	13'-11"	20'-0"	18'-7"	16'-8"	22'-2"	21'-1"	19'-0"
		40/10	16'-3"	14'-11"	14'-0"	19'-0"	17'-10"	16'-9"	21'-1"	20'-4"	19'-1"
		40/15	15'-8"	14'-6"	13'-4"	18'-1"	17'-4"	16'-0"	20'-1"	19'-8"	18'-2"
		50/10	15'-0"	13'-9"	13'-4"	17'-4"	16'-6"	16'-0"	19'-3"	18'-10"	18'-3"
		50/15	14'-5"	13'-8"	12'-10"	16'-8"	16'-4"	15'-4"	18'-5"	18'-2"	17'-5"
24" O.C.	Non-Snow (125%)	20/10	18'-4"	16'-9"	14'-9"	21'-11"	19'-10"	17'-5"	24'-11"	22'-6"	19'-9"
		20/15	17'-4"	15'-10"	13'-9"	20'-9"	18'-8"	16'-3"	23'-4"	21'-2"	18'-6"
		20/20	16'-6"	15'-0"	13'-1"	19'-8"	17'-9"	15'-5"	21'-10"	20'-2"	17'-6"
	Snow (115%)	25/10	17'-4"	15'-8"	14'-1"	20'-4"	18'-10"	16'-10"	22'-6"	21'-5"	19'-1"
		25/15	16'-4"	14'-11"	13'-3"	18'-11"	17'-10"	15'-10"	21'-0"	20'-4"	18'-0"
		30/10	16'-5"	15'-0"	13'-8"	19'-0"	18'-0"	16'-4"	21'-1"	20'-6"	18'-7"
		30/15	15'-5"	14'-4"	12'-11"	17'-10"	17'-2"	15'-5"	19'-10"	19'-4"	17'-7"
		40/10	14'-8"	13'-9"	12'-11"	17'-0"	16'-6"	15'-6"	18'-10"	18'-7"	17'-8"
		40/15	14'-0"	13'-4"	12'-4"	16'-2"	15'-10"	14'-9"	17'-11"	17'-7"	16'-10"
		50/10	13'-5"	12'-8"	12'-4"	15'-6"	15'-3"	14'-10"	17'-2"	17'-0"	16'-10"
		50/15	12'-10"	12'-7"	11'-10"	14'-11"	14'-8"	14'-2"	16'-6"	16'-3"	16'-2"

ROOF SPAN TABLE NOTES

- Maximum deflection is limited to L/180 at total load, L/240 at live load.
- Slope roof joists at least 1/4" over 12" to minimize ponding.
- Reductions in live loadings have been applied in accordance with 1994 UBC section 1605. Check to be sure local building codes allow these reductions.
- Maximum slope is limited to 12/12 for use of these tables.
- Tables are based on a simple span.
- For other loads, joist series or on-center spacings, contact your Boise Cascade supplier.

9-10 *Roof span charts for sizing BCI I-joists from Boise Cascade.* Courtesy Boise Cascade

TJI® Joist Residential Load Bearing Cantilever Tables

TJI® JOIST LOAD BEARING 24" CANTILEVER TABLE

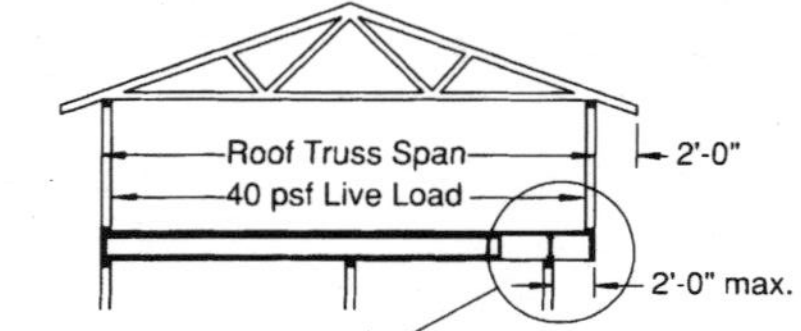

TJI® joists may be cantilevered up to a maximum of 2'-0" when supporting roof load, but may require reinforcement. Consult table and refer to footnotes to determine required reinforcement. See details E1, E2, E3 and E4 on page 8 for methods of reinforcement (detail E4 not used with TJI®/55 DF joists).

Numbers in charts refer to footnotes below.

0. No reinforcement required.
K. Web stiffener required each side of joist at bearing (detail E1 on page 8).
1. 3/4" x 48" reinforcement required on one side of joist (detail E2 on page 8) or, double the joists (detail E4 on page 8). Detail E4 not used with TJI®/55 DF joists.
2. 3/4" x 48" reinforcement required on both sides of joist (detail E3 on page 8) or, double the joists (detail E4 on page 8). Detail E4 not used with TJI®/55 DF joists.
X. Will not work. Reduce spacing of joists and recheck on table.

GENERAL NOTES:

- Assumes a 10 psf roof dead load and 60 plf wall load. Additional support may be required for other loadings.
- 3/4" reinforcement refers to 3/4" CDX plywood or other 3/4" exterior grade 48/24 span-rated sheathing that is cut to match the full depth of the joist. Install with face grain horizontal. Reinforcing member must bear fully on the wall plate. Minimum wall plate width is 3 1/2 inches.
- Calculations assume a bearing stress of 480 psi.
- Tables assume roof truss with 24" soffit.

	Roof Total Load	30 PSF			40 PSF			50 PSF		
	Joist Spacing (Roof Truss Span)	16" o.c.	19.2" o.c.	24" o.c.	16" o.c.	19.2" o.c.	24" o.c.	16" o.c.	19.2" o.c.	24" o.c.
9½" TJI®/Pro™	24'	0	0	0	0	0	X	0	X	X
	26'	0	0	0	0	0	X	X	X	X
	28'	0	0	X	0	X	X	X	X	X
	30'	0	0	X	0	X	X	X	X	X
	32'	0	0	X	0	X	X	X	X	X
	34'	0	0	X	X	X	X	X	X	X
	36'	0	0	X	X	X	X	X	X	X
9½" TJI®/15 DF	24'	0	0	0	0	0	1	0	1	X
	26'	0	0	0	0	1	1	1	1	X
	28'	0	0	1	0	1	1	1	1	X
	30'	0	0	1	0	1	1	1	1	X
	32'	0	0	1	0	1	X	1	X	X
	34'	0	0	1	1	1	X	1	X	X
	36'	0	1	1	1	1	X	1	X	X
9½" TJI®/25 DF	24'	0	0	0	0	0	1	0	1	1
	26'	0	0	0	0	0	1	0	1	1
	28'	0	0	0	0	0	1	0	1	1
	30'	0	0	0	0	0	1	1	1	X
	32'	0	0	1	0	1	1	1	1	X
	34'	0	0	1	0	1	1	1	1	X
	36'	0	0	1	0	1	1	1	1	X
11⅞" TJI®/Pro™	26'	0	0	0	0	1	1	1	1	X
	28'	0	0	1	0	1	1	1	1	X
	30'	0	0	1	0	1	X	1	X	X
	32'	0	0	1	0	1	X	1	X	X
	34'	0	0	1	1	1	X	1	X	X
	36'	0	1	1	1	1	X	1	X	X
	38'	0	1	1	1	1	X	X	X	X

	Roof Total Load	30 PSF			40 PSF			50 PSF		
	Joist Spacing (Roof Truss Span)	16" o.c.	19.2" o.c.	24" o.c.	16" o.c.	19.2" o.c.	24" o.c.	16" o.c.	19.2" o.c.	24" o.c.
14" TJI®/25 DF	26'	0	0	0	0	0	K	0	K	1
	28'	0	0	0	0	0	1	0	1	1
	30'	0	0	K	0	K	1	K	1	1
	32'	0	0	K	0	K	1	K	1	1
	34'	0	0	K	0	K	1	K	1	1
	36'	0	0	K	0	1	1	1	1	1
	38'	0	0	1	K	1	1	1	1	1
	40'	0	K	1	K	1	1	1	1	1
	42'	0	K	1	K	1	1	1	1	1
14" TJI®/35 DF	26'	0	0	0	0	0	0	0	0	1
	28'	0	0	0	0	0	K	0	K	1
	30'	0	0	0	0	0	K	0	K	1
	32'	0	0	0	0	0	1	0	K	1
	34'	0	0	0	0	0	1	0	1	1
	36'	0	0	0	0	K	1	K	1	1
	38'	0	0	K	0	K	1	K	1	1
	40'	0	0	K	0	K	1	K	1	1
	42'	0	0	K	0	K	1	1	1	1
14" TJI®/55 DF	26'	0	0	0	0	0	0	0	0	0
	28'	0	0	0	0	0	0	0	0	0
	30'	0	0	0	0	0	0	0	0	0
	32'	0	0	0	0	0	0	0	0	0
	34'	0	0	0	0	0	0	0	0	0
	36'	0	0	0	0	0	0	0	0	1
	38'	0	0	0	0	0	0	0	0	1
	40'	0	0	0	0	0	0	0	0	1
	42'	0	0	0	0	0	0	0	0	1

Joist	Roof Truss Span									
11⅞" TJI®/15 DF	26'	0	0	1	0	1	1	1	1	1
	28'	0	0	1	0	1	1	1	1	1
	30'	0	0	1	1	1	1	1	1	1
	32'	0	0	1	1	1	1	1	1	1
	34'	0	1	1	1	1	1	1	1	X
	36'	0	1	1	1	1	1	1	1	X
	38'	0	1	1	1	1	1	1	1	X
11⅞" TJI®/25 DF	26'	0	0	0	0	0	K	0	K	1
	28'	0	0	0	0	0	1	0	1	1
	30'	0	0	K	0	K	1	K	1	1
	32'	0	0	K	0	K	1	K	1	1
	34'	0	0	K	0	K	1	K	1	1
	36'	0	0	K	0	1	1	1	1	1
	38'	0	0	1	K	1	1	1	1	1
11⅞" TJI®/35 DF	26'	0	0	0	0	0	0	0	0	1
	28'	0	0	0	0	0	K	0	K	1
	30'	0	0	0	0	0	K	0	K	1
	32'	0	0	0	0	0	1	0	K	1
	34'	0	0	0	0	0	1	0	1	1
	36'	0	0	0	0	K	1	K	1	1
	38'	0	0	K	0	K	1	K	1	2
11⅞" TJI®/55 DF	26'	0	0	0	0	0	0	0	0	0
	28'	0	0	0	0	0	0	0	0	0
	30'	0	0	0	0	0	0	0	0	0
	32'	0	0	0	0	0	0	0	0	1
	34'	0	0	0	0	0	0	0	0	1
	36'	0	0	0	0	0	0	0	0	1
	38'	0	0	0	0	0	0	0	0	1
16" TJI®/25 DF	26'	0	0	0	0	0	K	0	K	1
	28'	0	0	0	0	0	1	0	1	1
	30'	0	0	K	0	K	1	K	1	1
	32'	0	0	K	0	K	1	K	1	1
	34'	0	0	K	0	K	1	K	1	1
	36'	0	0	K	0	1	1	1	1	1
	38'	0	0	1	K	1	1	1	1	1
	40'	0	K	1	K	1	1	1	1	1
	42'	0	K	1	K	1	1	1	1	1
16" TJI®/35 DF	26'	0	0	0	0	0	0	0	0	1
	28'	0	0	0	0	0	K	0	K	1
	30'	0	0	0	0	0	K	0	K	1
	32'	0	0	0	0	0	1	0	K	1
	34'	0	0	0	0	0	1	0	1	1
	36'	0	0	0	0	K	1	K	1	1
	38'	0	0	K	0	K	1	K	1	1
	40'	0	0	K	0	K	1	K	1	1
	42'	0	0	K	0	K	1	1	1	1
16" TJI®/55 DF	26'	0	0	0	0	0	0	0	0	0
	28'	0	0	0	0	0	0	0	0	0
	30'	0	0	0	0	0	0	0	0	0
	32'	0	0	0	0	0	0	0	0	0
	34'	0	0	0	0	0	0	0	0	0
	36'	0	0	0	0	0	0	0	0	0
	38'	0	0	0	0	0	0	0	0	0
	40'	0	0	0	0	0	0	0	0	1
	42'	0	0	0	0	0	0	0	0	1

9-11 *Cantilever sizing chart for I-joists.* Courtesy Trus Joist MacMillan

Microllam™ LVL and Parallam® PSL Headers and Beams

FLOOR BEAM SIZING TABLE

GENERAL NOTES:

1. Table assumes a residential floor loading of 40 psf live load and 12 psf dead load with beam deflection limited to L/360 at live load. For other loading conditions refer to allowable uniform load tables on page 22 or contact your Trus Joist MacMillan representative for assistance.
2. The beam weight is accounted for in the table and does not need to be included in the dead load calculations.
3. Table assumes a continuous floor joist span and a simple or continuous beam span.
4. Beam selections are based on the lower of 1.8E WS Microllam™ LVL or 2.0E WS Parallam® PSL design values.
5. Reduction in live load has been applied in accordance with UBC 2306, NBC 1115.1 and SBC 1203.2.
6. Minimum beam support to be double trimmers (3" bearing) at ends and 7½" bearing at intermediate supports of continuous spans. In shaded areas, support beams with triple trimmers (4½" bearing) at ends and 11¼" bearing at intermediate supports of continuous spans. See page 27 for information on Parallam® PSL columns and posts.
7. Beam widths of 3½" and 5¼" may be one piece or multiple pieces as shown in the following chart:

BEAM DEPTH	BEAM WIDTH 3½"	BEAM WIDTH 5¼"
5½" & 7¼"	Two 1¾"	Three 1¾"
9½"-18"	One 3½" or Two 1¾"	One 5¼" or Three 1¾" or One 3½" & One 1¾"

Multiple member beams must be properly connected together. See pages 19 and 20 for connection details.

Column Spacing	FLOOR JOIST SPAN (Use ½ the sum of the joist spans on both sides of the beam.) 11'	12'	13'	14'	15'	16'	17'	18'	20'
10'	3½" x 9½"	3½" x 9½"	5¼" x 9½" 3½" x 11⅞"	5¼" x 9½" 3½" x 11⅞"	5¼" x 9½" 3½" x 11⅞"	5¼" x 9½" 3½" x 11⅞"	5¼" x 9½" 3½" x 11⅞"	5¼" x 9½" 3½" x 11⅞"	3½" x 11⅞"
12'	3½" x 11⅞"	3½" x 11⅞"	3½" x 11⅞"	5¼" x 11⅞" 3½" x 14"	5¼" x 11⅞" 3½" x 14"	5¼" x 11⅞" 3½" x 14"	5¼" x 11⅞" 3½" x 14"	5¼" x 11⅞" 3½" x 14"	5¼" x 11⅞" 3½" x 14"
14'	5¼" x 11⅞" 3½" x 14"	5¼" x 11⅞" 3½" x 14"	5¼" x 11⅞" 3½" x 14"	3½" x 14"	5¼" x 14" 3½" x 16"	5¼" x 14" 3½" x 16"	5¼" x 14" 3½" x 16"	5¼" x 14" 3½" x 16"	5¼" x 14" 3½" x 16"
16'	5¼" x 14" 3½" x 16"	5¼" x 14" 3½" x 16"	5¼" x 14" 3½" x 16"	5¼" x 14" 3½" x 16"	5¼" x 14" 3½" x 16"	5¼" x 16" 3½" x 18"	5¼" x 16" 3½" x 18"	5¼" x 16" 3½" x 18"	5¼" x 16" 3½" x 18"
18'	5¼" x 16" 3½" x 18"	5¼" x 16" 3½" x 18"	5¼" x 16" 3½" x 18"	5¼" x 16" 3½" x 18"	5¼" x 16" 3½" x 18"	5¼" x 16"	5¼" x 16"	5¼" x 18"	5¼" x 18"
20'	5¼" x 16" 3½" x 18"	5¼" x 16"	5¼" x 18"	5¼" x 18"	5¼" x 18"	5¼" x 18"	5¼" x 18"	5¼" x 18"	

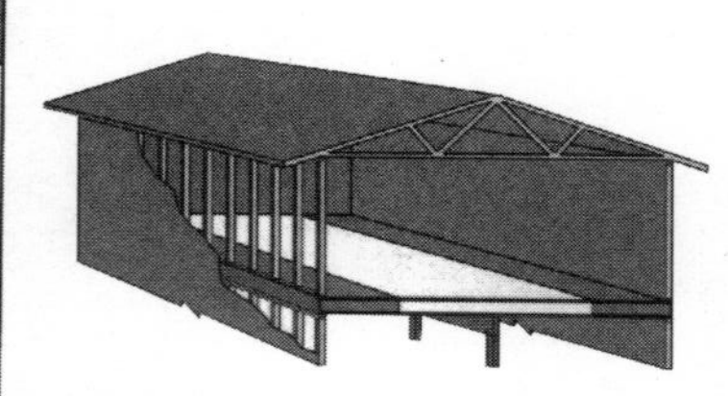

9-12 *Another sizing chart for floors, in this case for beams.* Courtesy Trus Joist MacMillan

RIDGE BEAM SIZING TABLE

Roof Load PSF	NON-SNOW (125%)		SNOW (115%)			
COLUMN SPACING	20LL + 12DL	20LL + 25DL	25LL + 12DL	30LL + 12DL	40LL + 12DL	50LL + 12DL
10'	5¼" x 7¼" 2 11/16" x 9½"	5¼" x 7¼" 3½" x 9½" 2 11/16" x 9½"	5¼" x 7¼" 3½" x 9½" 2 11/16" x 9½"	5¼" x 7¼" 3½" x 9½" 2 11/16" x 9½"	3½" x 9½" 2 11/16" x 11⅞"	5¼" x 9½" 3½" x 11⅞" 2 11/16" x 11⅞"
12'	3½" x 9½" 2 11/16" x 9½"	5¼" x 9½" 3½" x 11⅞" 2 11/16" x 11⅞"	3½" x 9½" 2 11/16" x 11⅞"	5¼" x 9½" 3½" x 11⅞" 2 11/16" x 11⅞"	5¼" x 9½" 3½" x 11⅞" 2 11/16" x 11⅞"	3½" x 11⅞" 2 11/16" x 14"
14'	5¼" x 9½" 3½" x 11⅞" 2 11/16" x 11⅞"	3½" x 11⅞" 2 11/16" x 14"	5¼" x 9½" 3½" x 11⅞" 2 11/16" x 11⅞"	3½" x 11⅞" 2 11/16" x 14"	5¼" x 11⅞" 3½" x 14" 2 11/16" x 14"	5¼" x 11⅞" 3½" x 14"
16'	3½" x 11⅞" 2 11/16" x 14"	5¼" x 11⅞" 3½" x 14" 2 11/16" x 14"	5¼" x 11⅞" 3½" x 14" 2 11/16" x 14"	5¼" x 11⅞" 3½" x 14" 2 11/16" x 16"	5¼" x 14" 3½" x 16" 2 11/16" x 16"	5¼" x 14" 3½" x 18"
18'	5¼" x 11⅞" 3½" x 14" 2 11/16" x 14"	5¼" x 14" 3½" x 16" 2 11/16" x 16"	3½" x 14" 2 11/16" x 16"	5¼" x 14" 3½" x 16"	5¼" x 14" 3½" x 18"	5¼" x 16"
20'	5¼" x 14" 3½" x 16" 2 11/16" x 16"	5¼" x 16" 3½" x 18"	5¼" x 14" 3½" x 16"	5¼" x 16" 3½" x 18"	5¼" x 16"	5¼" x 18"
22'	5¼" x 14" 3½" x 16"	5¼" x 16" 3½" x 18"	5¼" x 16" 3½" x 18"	5¼" x 16"	5¼" x 18"	
24'	5¼" x 16" 3½" x 18"	5¼" x 18"	5¼" x 18"	5¼" x 18"		

GENERAL NOTES:

1. Table assumes beam supports a tributary width of 18 feet.
2. The beam weight is accounted for in the table and does not need to be included in the dead load calculations.
3. Table is based on worst case of simple or continuous span.
4. Deflection limited to L/240 at live load or L/180 at total load.
5. Beam selections are based on the lower of 1.8E WS Microllam™ LVL or 2.0E WS Parallam® PSL design values.
6. Reduction in live load has been applied in accordance with UBC 2306, NBC 1110.2 and SBC 1203.6, for the beam sizes listed in the non-snow (125%) columns.
7. Minimum beam support to be double trimmers (3" bearing) at ends and 7½" bearing at intermediate supports of continuous spans. In shaded areas, support beams with triple trimmers (4½" bearing) at ends and 11¼" bearing at intermediate supports of continuous spans.
8. For loading conditions not shown refer to allowable uniform load tables on page 22 or contact your Trus Joist MacMillan representative for assistance.
9. Beam widths of 3½" and 5¼" may be one piece or multiple pieces as shown in the following chart:

	BEAM WIDTH	
BEAM DEPTH	3½"	5¼"
5½" & 7¼"	Two 1¾"	Three 1¾"
9½"-18"	One 3½" or Two 1¾"	One 5¼" or Three 1¾" or One 3½" & One 1¾"

When top loaded, fasten multiple pieces together with a minimum of two rows of 16d nails at 12" on-center. **Use three rows of 16d nails at 12" on-center for 14", 16" and 18" beams.** When side-loaded see page 20 for connection of multiple pieces.

Non-shaded portion indicates area of load on beam.

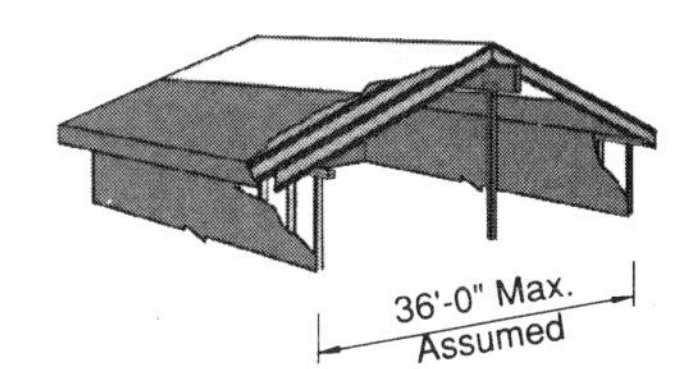

9-13 *A similar table for sizing roof beams.* Courtesy Trus Joist MacMillan

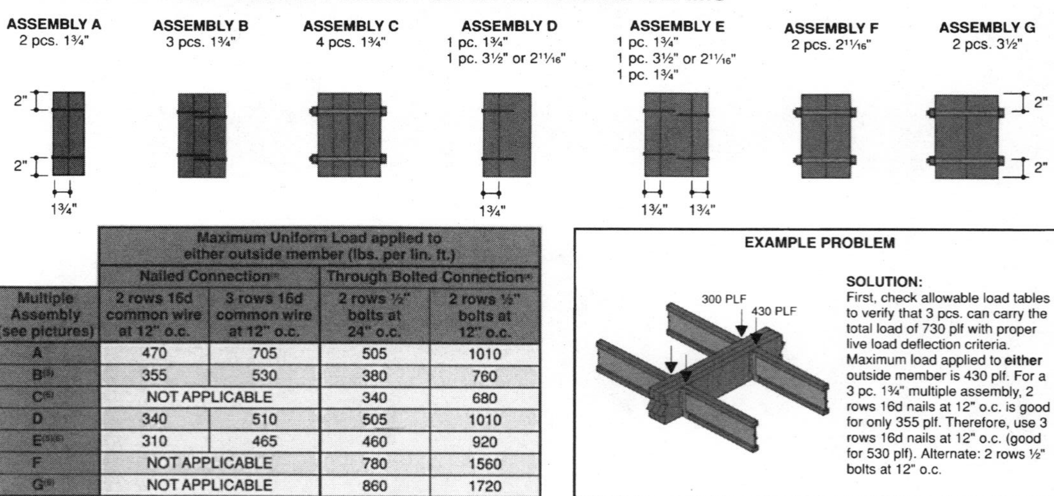

Microllam™ LVL and Parallam® PSL Headers and Beams

MULTIPLE MEMBER CONNECTIONS FOR SIDE-LOADED BEAMS

Multiple Assembly (see pictures)	Maximum Uniform Load applied to either outside member (lbs. per lin. ft.)			
	Nailed Connection[3]		Through Bolted Connection[4]	
	2 rows 16d common wire at 12" o.c.	3 rows 16d common wire at 12" o.c.	2 rows ½" bolts at 24" o.c.	2 rows ½" bolts at 12" o.c.
A	470	705	505	1010
B[5]	355	530	380	760
C[6]	NOT APPLICABLE		340	680
D	340	510	505	1010
E[3][6]	310	465	460	920
F	NOT APPLICABLE		780	1560
G[6]	NOT APPLICABLE		860	1720

EXAMPLE PROBLEM

SOLUTION:
First, check allowable load tables to verify that 3 pcs. can carry the total load of 730 plf with proper live load deflection criteria. Maximum load applied to **either** outside member is 430 plf. For a 3 pc. 1¾" multiple assembly, 2 rows 16d nails at 12" o.c. is good for only 355 plf. Therefore, use 3 rows 16d nails at 12" o.c. (good for 530 plf). Alternate: 2 rows ½" bolts at 12" o.c.

GENERAL NOTES:

1. Verify adequacy of beam in uniform load tables, pages 22 and 23. **Size assembly for the lowest grade member.** Example: Two member assembly with 1¾" 1.8E WS Microllam™ LVL and 3½" 2.0E WS Parallam® PSL would be sized from the *1¾" 1.8E WS MICROLLAM™ LVL* tables, multiply value in table x 3.0 to verify beam capacity.
2. Values listed are for 100% stress level. Increase 15% for snow loaded roof conditions or 25% for non-snow roof conditions, where code allows.
3. *Nailed Connection* values may be doubled for 6" o.c. or tripled for 4" o.c. nail spacing.
4. Bolts are to be material conforming to ASTM standard A307 (machine bolts). Bolt holes are to be the same diameter as the bolt, and located 2" from the top and bottom of the member. Washers should be used under head and nut.
5. For a three-piece member, the nailing specified is from each side.
6. 7" wide beams should only be side-loaded when loads are applied to both sides of the members (to minimize rotation).
7. Beams wider than 7" require special consideration by the design professional.

9-14 *This chart describes options for connecting multiple-beam members to increase load-carrying capacity.* Courtesy Trus Joist MacMillan.

FILLER AND BACKER BLOCK SIZES

	9½" or 11⅞" TJI®/Pro™	9½" or 11⅞" TJI®/15 DF	9½" or 11⅞" TJI®/25 DF	14" or 16" TJI®/25 DF
Filler Block (Detail H6)	1⅛" net	1⅛" net	2x6	2x8
Backer Block (Detail H6)	½" or ⅝"	½" or ⅝"	⅝" or ¾"	⅝" or ¾"

	11⅞" TJI®/35 DF	14" or 16" TJI®/35 DF	11⅞" TJI®/55 DF	14" or 16" TJI®/55 DF
Filler Block (Detail H6)	2x6+½" plywood	2x8+½" plywood	2-2x6	2-2x8
Backer Block (Detail H6)	1" net	1" net	2x6	2x8

Filler and backer block length should accommodate required nailing without splitting.

9-15 *Filler and backer block sizing information for use with I-joists.* Courtesy Trus Joist MacMillan

Your dealer can perform all these calculations and cost comparisons even more quickly, using the design and specification software provided by the various manufacturers. For a computerized design layout, more information, or specific information on calculating and ordering engineered lumber products, consult your lumber dealer or contact the manufacturer directly. (See Appendix E at the end of the book.)

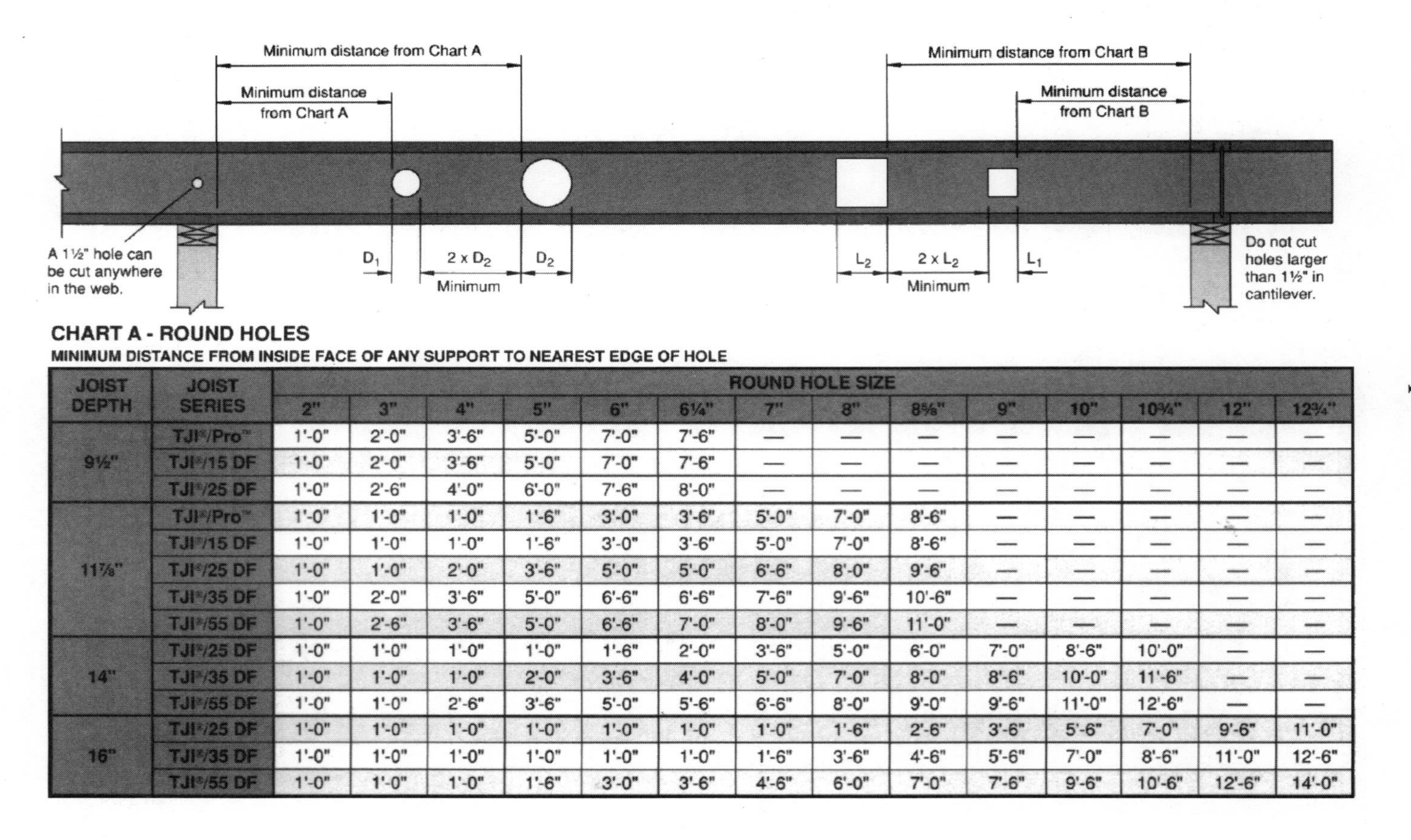

CHART A - ROUND HOLES

MINIMUM DISTANCE FROM INSIDE FACE OF ANY SUPPORT TO NEAREST EDGE OF HOLE

JOIST DEPTH	JOIST SERIES	ROUND HOLE SIZE													
		2"	3"	4"	5"	6"	6¼"	7"	8"	8⅝"	9"	10"	10¾"	12"	12¾"
9½"	TJI®/Pro™	1'-0"	2'-0"	3'-6"	5'-0"	7'-0"	7'-6"	—	—	—	—	—	—	—	—
	TJI®/15 DF	1'-0"	2'-0"	3'-6"	5'-0"	7'-0"	7'-6"	—	—	—	—	—	—	—	—
	TJI®/25 DF	1'-0"	2'-6"	4'-0"	6'-0"	7'-6"	8'-0"	—	—	—	—	—	—	—	—
11⅞"	TJI®/Pro™	1'-0"	1'-0"	1'-0"	1'-6"	3'-0"	3'-6"	5'-0"	7'-0"	8'-6"	—	—	—	—	—
	TJI®/15 DF	1'-0"	1'-0"	1'-0"	1'-6"	3'-0"	3'-6"	5'-0"	7'-0"	8'-6"	—	—	—	—	—
	TJI®/25 DF	1'-0"	1'-0"	2'-0"	3'-6"	5'-0"	5'-0"	6'-6"	8'-0"	9'-6"	—	—	—	—	—
	TJI®/35 DF	1'-0"	2'-0"	3'-6"	5'-0"	6'-6"	6'-6"	7'-6"	9'-6"	10'-6"	—	—	—	—	—
	TJI®/55 DF	1'-0"	2'-6"	3'-6"	5'-0"	6'-6"	7'-0"	8'-0"	9'-6"	11'-0"	—	—	—	—	—
14"	TJI®/25 DF	1'-0"	1'-0"	1'-0"	1'-0"	1'-6"	2'-0"	3'-6"	5'-0"	6'-0"	7'-0"	8'-6"	10'-0"	—	—
	TJI®/35 DF	1'-0"	1'-0"	1'-0"	2'-0"	3'-6"	4'-0"	5'-0"	7'-0"	8'-0"	8'-6"	10'-0"	11'-6"	—	—
	TJI®/55 DF	1'-0"	1'-0"	2'-6"	3'-6"	5'-0"	5'-6"	6'-6"	8'-0"	9'-0"	9'-6"	11'-0"	12'-6"	—	—
16"	TJI®/25 DF	1'-0"	1'-0"	1'-0"	1'-0"	1'-0"	1'-0"	1'-0"	1'-6"	2'-6"	3'-6"	5'-6"	7'-0"	9'-6"	11'-0"
	TJI®/35 DF	1'-0"	1'-0"	1'-0"	1'-0"	1'-0"	1'-0"	1'-6"	3'-6"	4'-6"	5'-6"	7'-0"	8'-6"	11'-0"	12'-6"
	TJI®/55 DF	1'-0"	1'-0"	1'-0"	1'-6"	3'-0"	3'-6"	4'-6"	6'-0"	7'-0"	7'-6"	9'-6"	10'-6"	12'-6"	14'-0"

CHART B - SQUARE OR RECTANGULAR HOLES
MINIMUM DISTANCE FROM INSIDE FACE OF ANY SUPPORT TO NEAREST EDGE OF HOLE

JOIST DEPTH	JOIST SERIES	SQUARE OR RECTANGULAR HOLE SIZE*													
		2"	3"	4"	5"	6"	6¼"	7"	8"	8⅝"	9"	10"	10¾"	12"	12¾"
9½"	TJI®/Pro™	1'-0"	2'-0"	4'-0"	6'-0"	—	—	—	—	—	—	—	—	—	—
	TJI®/15 DF	1'-0"	2'-0"	4'-0"	6'-0"	—	—	—	—	—	—	—	—	—	—
	TJI®/25 DF	1'-0"	2'-6"	4'-6"	6'-6"	—	—	—	—	—	—	—	—	—	—
11⅞"	TJI®/Pro™	1'-0"	1'-0"	2'-0"	4'-6"	7'-6"	7'-6"	8'-0"	9'-0"	—	—	—	—	—	—
	TJI®/15 DF	1'-0"	1'-0"	2'-0"	4'-6"	7'-6"	7'-6"	8'-0"	9'-0"	—	—	—	—	—	—
	TJI®/25 DF	1'-0"	1'-6"	3'-6"	5'-6"	8'-0"	8'-6"	8'-6"	9'-6"	—	—	—	—	—	—
	TJI®/35 DF	1'-0"	2'-6"	4'-6"	6'-6"	9'-0"	9'-0"	9'-6"	9'-6"	—	—	—	—	—	—
	TJI®/55 DF	3'-6"	4'-6"	6'-0"	7'-6"	9'-6"	9'-6"	10'-0"	10'-6"	10'-6"	—	—	—	—	—
14"	TJI®/25 DF	1'-0"	1'-0"	1'-6"	4'-0"	6'-0"	6'-6"	8'-6"	10'-0"	10'-6"	11'-0"	11'-6"	12'-0"	—	—
	TJI®/35 DF	1'-0"	1'-0"	2'-6"	4'-6"	7'-0"	7'-6"	9'-6"	11'-0"	11'-0"	11'-6"	12'-0"	12'-6"	—	—
	TJI®/55 DF	2'-6"	4'-0"	5'-6"	7'-6"	9'-0"	9'-6"	10'-6"	12'-0"	12'-0"	12'-0"	13'-0"	13'-0"	—	—
16"	TJI®/25 DF	1'-0"	1'-0"	1'-0"	1'-6"	4'-0"	4'-6"	6'-6"	9'-6"	11'-0"	11'-6"	12'-6"	13'-0"	14'-0"	14'-6"
	TJI®/35 DF	1'-0"	1'-0"	1'-0"	2'-6"	5'-0"	5'-6"	7'-6"	10'-6"	12'-0"	12'-6"	13'-0"	13'-6"	14'-6"	15'-0"
	TJI®/55 DF	1'-0"	2'-6"	4'-6"	6'-0"	8'-0"	8'-6"	10'-0"	12'-0"	13'-6"	13'-6"	14'-0"	14'-6"	15'-0"	16'-0"

* Rectangular holes based on measurement of longest side.

GENERAL NOTES:

1. If more than one hole is to be cut in the web, the length of the uncut web between holes must be twice the length of the longest dimension of the largest adjacent hole. Holes may be located vertically anywhere within the web.
2. TJI® joists are manufactured with 1½" perforated knockouts in the web at approximately 12" on-center along the length of the joist.
3. The distances in the hole charts are based on uniformly loaded joists using maximum loads shown for any of the tables listed within this guide. **For other load conditions or hole configurations not included in these charts, contact your Trus Joist MacMillan representative.**

DO

Full depth rectangular holes are also possible. Contact your Trus Joist MacMillan representative for assistance.

9-16 *The placement of holes in the web of the I-joist is crucial in order to maintain the joist's strength rating. Charts such as this describe the hole locations in detail.* Courtesy Trus Joist MacMillan

10

Storing, handling, and installing structural engineered lumber products

It's pretty unlikely that you have ever received an order of 2 × 4s and 2 × 10s from the local lumberyard, started unbanding and stacking the material on the site, and come across a complete set of instructions included with the lumber, showing in detail how to properly use the lumber to frame a house. It might seem a ridiculous thought at first, but that's pretty much what you'll find with an order of engineered lumber. And whether you're new on the job or an old hand at framing, a clear set of directions can certainly simplify matters when it comes to installation.

The manufacturer's installation instructions give you a pretty good overview of the "dos and don'ts" of working with engineered lumber material. You'll see instructions on how and where to drill holes, cut and nail various types of members in a variety of applications, install blocking, install web stiffeners and support blocks, and much more.

You can find the instructions just about everywhere, as well. The dealers hand them out at the lumberyard, they're included with just about every order, they're in the design manuals, and you'll find them on the back of your computerized plans. There are even videotapes available to further explain and illustrate the installation processes. And the nice thing is that the process is the same from one manufacturer to the next. If you've been installing TJI I-joists made by Trus Joist MacMillan and then arrive at a job site where Louisiana-Pacific's LPI I-joists are specified, the installation requirements and techniques are going to be the same.

As with everything having to do with contractors and instruction sheets, all these specific directions might at times seem more a hindrance than a help. Unlike the solid sawn lumber you're used to working with, engineered lumber really must be installed in a certain manner in order to be safe and to structurally perform as designed. Once you get used to how to work with the material—which is actually quite easy—the instructions will become second nature.

At first, though, it might all seem a little confusing. Don't get frustrated; just read through the instructions—it's all visual, since the manufacturers know that contractors hate to read instructions—and discuss them with your crew. Handle each framing situation as it arises. The instructions cover lots of situations, many of which you might not encounter on your particular job. Just work through the details until you understand them.

This chapter examines the instructions for storing, handling, and installing structural engineered lumber products in more detail. Even though you'll have the visual instructions in front of you, studying this chapter before working with these materials on the job site should help clarify a few things, and should also speed you up a little.

Job-site storage

Once you've selected and ordered the joists and other material you need for your project, they will be delivered to your job site by the dealer. Some products arrive wrapped in a protective water-resistant paper (Fig. 10-1); others will just be bare and banded like a normal delivery of solid lumber.

If possible, unload the entire banded unit of material using a forklift. The material should be placed on level ground, on top of stickers placed approximately 10 feet apart (Fig. 10-2). The stickers need to be high enough to permit easy withdrawal of the forklift forks, and to keep the material up above the normal level of mud and water on the site. If forklift unloading isn't available, carefully unband the material on the truck, then hand-unload it and place it on the stickers.

The I-joists need to be stored in a vertical (upright) manner (Fig. 10-3), in the same orientation they'll be installed. They can be set next

10-1 *Most types of engineered lumber beams arrive on-site wrapped in a protective paper. It's a good practice to leave them covered until the building is water-tight.*

to one another, or interlocked so the bottom flange of one rests on top of the bottom flange of the ones preceding and following it. Stacking in this manner, which is how they come from the manufacturing plant, saves space and helps keeps the units upright. Do not store the I-joists flat, either individually or as a unit, and be sure you have adequate supports under them.

HANDLING AND STORAGE

- Keep LPI Joists and Gang-Lam LVL beams dry.
- Where possible, keep them in wrapped bundles.
- Don't stack bundles more than 10' high.
- Use forklifts and cranes carefully to avoid damaging joists.
- Don't use joists and beams for other purposes - such as ramps, planks, etc.

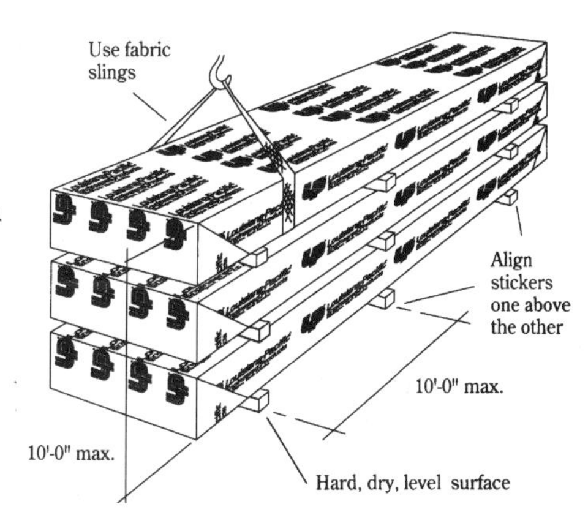

10-2 *Handling and lifting instructions for I-joists and engineered lumber beams.* Courtesy Louisiana-Pacific

10-3 *To prevent damage, store I-joists upright until you're ready to use them.*

Glulams, beams, headers, and other solid material also need to be set on stickers and kept upright. The taller dimension of the beam should be vertical, oriented how it will be installed. Provide a sufficient number of stickers or other temporary braces under the beams to prevent them from sagging.

Do not place any heavy loads on the beams or joists when they're on the ground. Don't use them as temporary scaffolding, or as a temporary work platform. Exercise care in handling any of the material to avoid damage, and take equal care with it once it's sitting on the ground. Bumping into a unit of I-joists with a forklift or a truck, for example, could damage the flanges or web material and greatly reduce the joist's ability to carry the load it was designed for. If you seriously damage any material while it's stored on-site, contact your dealer or a manufacturer's representative and have the material checked before installing it.

If you need to store the material for more than a couple of days before installation, it needs to be covered and protected both from sun and water (Fig. 10-4). You can use the wrapping material that covered the order when it was delivered to you, and you can usually get more from your dealer since they typically have some left over from units they've unloaded. If that material is damaged or unavailable, use 6-millimeter clear plastic sheeting. This weight is tough enough to pre-

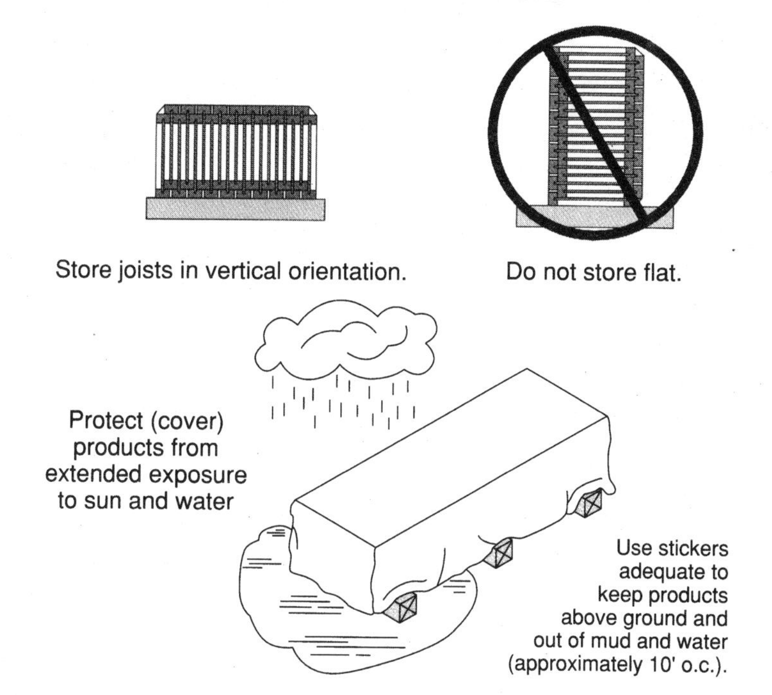

10-4 *Some additional manufacturer's instructions for protecting the material.* Courtesy Trus Joist MacMillan

vent rips and tears, and clear is better than black in preventing the buildup of excess moisture around the material, especially if it's stored in warm, wet conditions where humidity is trapped under the plastic.

If a situation arises where any engineered lumber product that's been delivered to your job site will have to be stored for a very long period of time, contact your dealer or the product representative. They might be able to take the material back to their yard for safer storage until you need it, or help you with specific instructions for long-term, on-site storage.

Handling during installation

There are a few precautions you need to be aware of and observe while installing any engineered lumber product, especially I-joists. Some installations can be quite unstable and dangerous until they're completed and braced, so you really need to be aware of what the dangers are and take all the necessary steps that the manufacturer calls for.

When setting a beam or an I-joist temporarily in place over its supports, be sure it's placed upright, with the deeper dimension oriented vertically. Beams are sized to handle a load over a specific distance between supports, and can sag dangerously if set in place on their side instead of up and down. I-joists, which get a great deal of strength from their design and construction of flanges and a vertical web, have very little strength when set on their sides, and can sag to the degree that they will slip off their supports. Always store everything upright, even if you're just setting it down momentarily prior to making a permanent installation.

Once the I-joists are in place, *never* walk on them or set anything on them until they're braced, as shown in Fig. 10-5 (see *Bracing I-joists*, later in the chapter, for specific instructions on how to install bracing). I-joists are very tall in relation to the width of the flanges and, even after they've been nailed to the plates, walking on them will cause them to roll over or lean outward in unpredictable directions, making it extremely easy to lose your balance and fall.

For the same reason, never place a load of any type on newly installed I-joists until they're properly and completely braced. Setting as little as a single sheet of plywood on top of unbraced I-joists can easily cause them to act like a line of dominoes and topple over. Once that happens, the joists will be lying on their sides—an orientation, as described previously, where they have virtually no strength—and they will be unable to support the weight of the plywood. The joists will either sag down and slip off their supports or simply snap, dropping the load.

Once the beams or joists are properly in place and either braced or sheathed, it's still important not to exceed their designed load limits by placing heavy weights in unsupported areas. If you have a set of I-joists, for example, that are designed to handle 10 pounds per square foot of dead load under normal installed conditions, setting a couple of pallets of brick or a couple of units of plywood on them, even temporarily, can overload the joists and cause them to fail.

All construction materials need to be stacked directly over walls, beams, posts, or other adequate supports. If conditions make this impossible, you need to install temporary walls or beams beneath the joists so they're directly beneath the load of materials, and these temporary supports need to stay in place until the load is removed.

Installing I-joists for floor framing

Wood I-joists are as easy to install as the solid sawn lumber you're probably used to—actually easier in many ways—using conventional

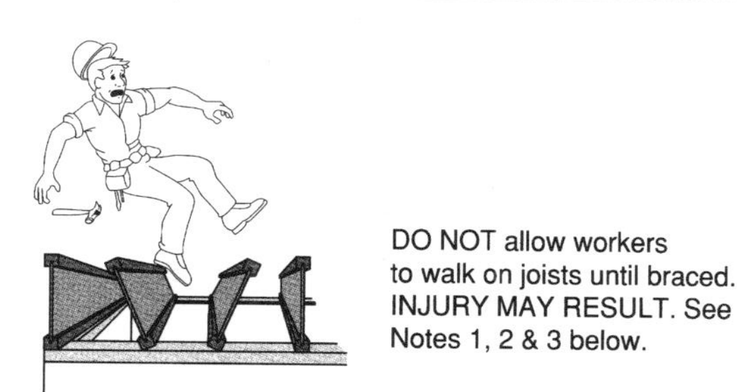

WARNING

JOISTS ARE UNSTABLE UNTIL BRACED LATERALLY

DO NOT allow workers to walk on joists until braced. INJURY MAY RESULT. See Notes 1, 2 & 3 below.

BRACING INCLUDES:

- BLOCKING
- HANGERS
- STRUT LINES
- SHEATHING

DO NOT stack building materials on unsheathed joists. Stack only over beams or walls. See Note 4 below.

WARNING NOTES:

Lack of concern for proper bracing during construction can result in serious accidents. Under normal conditions if the following guidelines are observed, accidents will be avoided.

1. All blocking, hangers, rim boards and rim joists at the end supports of the TJI® joists must be completely installed and properly nailed.
2. Lateral strength, like a braced end wall or an existing deck, must be established at the ends of the bay. This can also be accomplished by a temporary or permanent deck (sheathing) nailed to the first 4 feet of joists at the end of the bay.
3. Temporary strut lines of 1x4 (min.) must be nailed to a braced end wall or sheathed area as in Note 2 and to each joist. Without this bracing, buckling sideways or roll over is highly probable under light construction loads – like a worker and one layer of unnailed sheathing.
4. Sheathing must be totally attached to each TJI® joist before additional loads can be placed on the system.
5. Ends of cantilevers require strut lines on both the top and bottom flanges.
6. The flanges must remain straight within a tolerance of ½" from the true alignment.

10-5 *I-joists are very unstable until braced and sheathed. All the manufacturers offer similar precautions and warnings about how unstable the joists can be when first installed.* Courtesy Trus Joist MacMillan

hand, air, and power tools. This section describes some generic installation procedures, but remember to follow the manufacturer's specific installation instructions for bearing, blocking, nail size and spacing, and other installation details (Fig. 10-6). Failure to follow these instructions could result in an installation that is weak or unsafe, which will reduce the load-bearing capacity of the materials. Incorrect installation might also void your warranty; read the fine print for specific details.

Wood I-joists are lightweight, and in most lengths can easily be handled by one person. To cut the joist to length, lay it on its side on the ground or over supports. Measure and mark a square cutting line, then, using a circular saw, simply cut along the waste side of the line as you would with any conventional lumber (Fig. 10-7). Long pieces of waste need to be supported as they are cut off, to avoid pinching the saw blade and creating a potential kickback. Because flanges are higher than the intermediate web material, you might find it difficult to cut a clean, straight line. To solve this problem, you can purchase an inexpensive plastic cutting guide (Fig. 10-8) that fits snugly between the flanges of 9½- and 11⅞-inch joists, the two most common sizes. The saw guide fills in the lower web space, making it easier to run the saw over the joist. It also serves as a fence for guiding the saw, which greatly simplifies getting a 90° cut end.

If you have only a couple of joists to cut and you don't want to invest in the saw guide, you can make one up out of a piece of plywood. Select a plywood thickness that matches the distance between the face of the web and the face of the flange, so the saw will move smoothly over it. Cut the plywood to length so it fits snugly but not tightly between the flanges. The plywood cutting guide is reusable, but you'll need to have different ones to fit different joist sizes.

Measure the joists and cut them accurately to length. For an installation where you'll be placing the joists directly onto a sill plate or the wall top plate, remember to allow for the thickness of the rim joist at each end. A minimum bearing length of 1¾ inches is required at both ends of the joists for most residential or single-story applications.

For commercial applications or for installation on the first floor of a multifloor building, a 2¼-inch minimum bearing surface is required. A full 3½ inches of bearing surface is necessary where the joists cross over intermediate supports and, in some instances, 5¼ inches of bearing surface might be necessary in order to achieve the full rated strength of the joist.

FLOOR DETAILS

- Verify capacity and fastening of hangers and connections.
- Some wind or seismic loads may require different or additional details and connections.

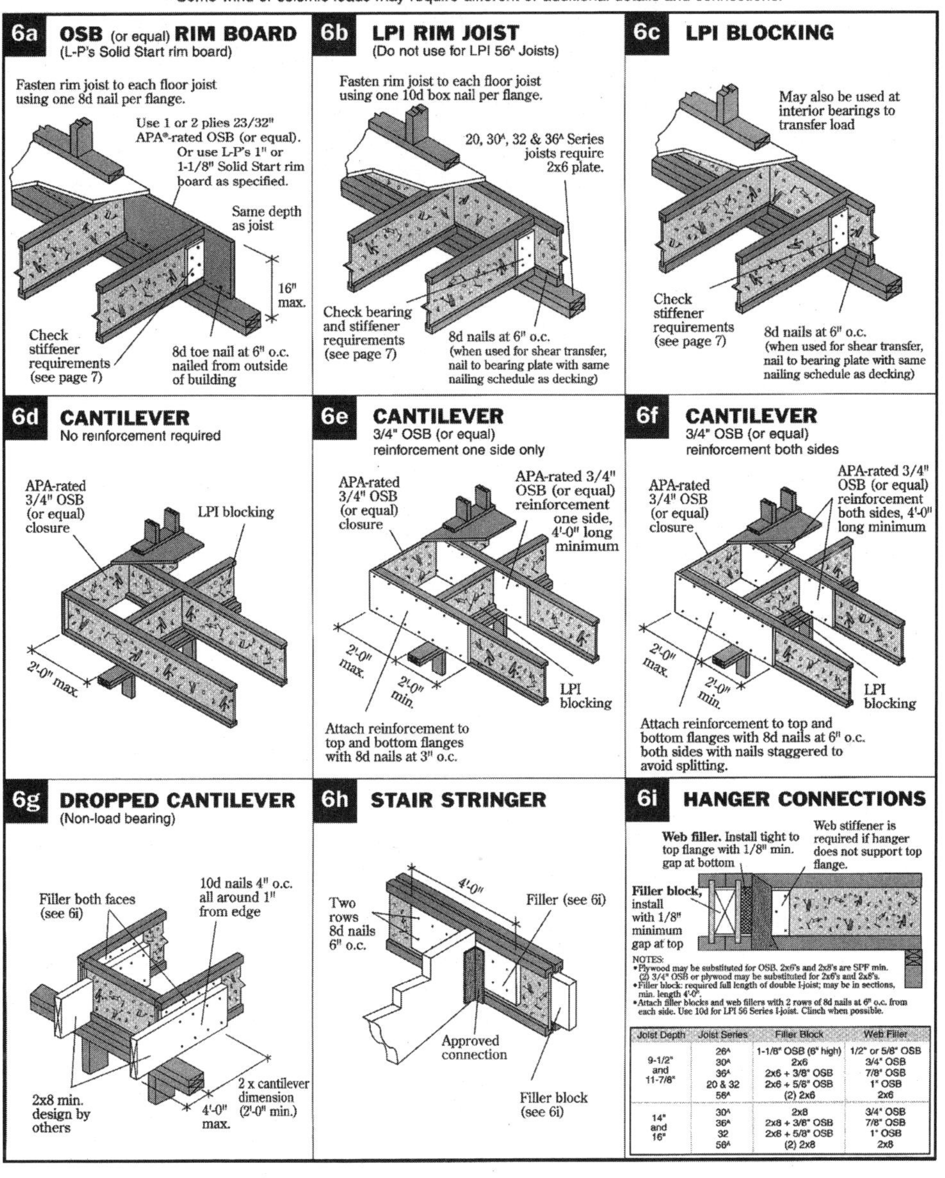

Joist Depth	Joist Series	Filler Block	Web Filler
9-1/2" and 11-7/8"	26^A	1-1/8" OSB (6" high)	1/2" or 5/8" OSB
	30^A	2x6	3/4" OSB
	36^A	2x6 + 3/8" OSB	7/8" OSB
	20 & 32	2x6 + 5/8" OSB	1" OSB
	56^A	(2) 2x6	2x6
14" and 16"	30^A	2x8	3/4" OSB
	36^A	2x8 + 3/8" OSB	7/8" OSB
	32	2x8 + 5/8" OSB	1" OSB
	56^A	(2) 2x8	2x8

10-6 *Standard manufacturer's instructions and details for framing with engineered lumber, in this instance floor framing.* Courtesy Louisiana-Pacific

10-7 *Cutting an I-joist with an electric circular saw.* Courtesy Trus Joist MacMillan

10-8 *Trus Joist MacMillan's unique sawing guide simplifies the cutting process. They're available at most lumberyards that carry TJI I-joists.* Courtesy Trus Joist MacMillan

After cutting the joist to length, lift it into position. Align it with your layout lines as you would any conventional joist layout (Fig. 10-9), and secure the joist in place by nailing through the bottom flange into the plate (Fig. 10-10). Use two hand- or air-driven 10d or 12d box nails at each end of the joist, and two at each intermediate support. Keep the nails at least 1½ inches back from the end of the joist in order to minimize the chances of splitting the bottom flange material. Do not use 16d sinker nails, as their larger diameter can split the flange.

10-9 *I-joists installed and aligned, awaiting the rim joist. Note the cantilevered joists.*

10-10 *Securing an I-joist with an air-driven nail gun.* Courtesy Trus Joist MacMillan

Install the rim joist next (Fig. 10-11). You can use solid sawn lumber if desired or, better yet, one of the laminated strand lumber (LSL) rim boards designed for this purpose (Fig. 10-12). LSL has the advantage of being dry and straight, and it's available in widths that exactly

10-11 *Installing the rim board.*

10-12 *A rim board of TimberStrand LSL, which is an exact match for the depth of the floor joists.* Courtesy Trus Joist MacMillan

match the depth of the joists, saving you the additional hassle of ripping or shimming solid lumber to size.

Secure the rim board by first nailing it to the ends of the joists, again using 10d or 12d box hand- or air-driven nails. Install one nail through the rim joist into both the top and bottom flanges of each joist, then toenail the rim joist to the plates (Fig. 10-13). No additional bridging or mid-span blocking is required to complete the installation, another nice feature of engineered lumber.

Hangers

Wood I-joists can also be installed into metal hangers at one or both ends (Fig. 10-14). Use a hanger style that's approved by the manufacturer for the particular installation and joist size and type (Fig. 10-15). For a complete listing of hanger types and their uses, refer to Chapter 7 or to the instructions and applications given in the booklets from either the I-joist or hanger manufacturers.

Attach the hanger to the supporting member, using an approved hanger nail and making certain that you use all the holes in the hanger. Failure to fill all the holes reduces the strength of the hanger below its rated load characteristics. Install the I-joist into the hanger, and make certain it's seated all the way in. Bend the hanger tabs over the bottom flange of the I-joist, and secure the tabs to the joist flange with one nail in each side.

10-13 *Securing the rim board by toenailing it to the plates.*
Courtesy Trus Joist MacMillan

10-14 *Hanger installation for engineered lumber columns, beams, and I-joists.* Courtesy Trus Joist MacMillan

If you're using a hanger that does not have bend-tabs, web stiffeners might be necessary (see *Web stiffeners and blocks*, later in the chapter). Attach the hanger to the supporting member, attach the web stiffeners to the end of the joist, then install the joist in the hanger. Seat the joist completely into the hanger, and secure the joist to the hanger by installing all required fasteners.

Cantilevers

I-joists can be cantilevered out past their supports in order to form bays or other projections. You need to carefully follow a number of specific design and installation guidelines, however, in order to ensure structural stability. Each of the manufacturers offer a number of detailed tables describing a variety of cantilever applications. You need to study the charts and decide which is the proper detail for your installation, or consult with your dealer.

Cantilevers need to be parallel with the run of the joists, so you might need to turn your joist layout at some point in order to accommodate the cantilever. In general, in an application where the cantilever will not support a concentrated load, you can cantilever the joists up to one-third the length of the joist's total back

TJI® JOIST FRAMING CONNECTORS

C1 – TOP MOUNT SINGLE JOIST HANGER

DEPTH	JOIST	HANGER	
		RESIDENTIAL	COMM./MULTI-FAMILY
9½"	TJI®/15 DF	ITT29.5	ITT29.5*
	TJI®/25 DF	ITT9.5	ITT9.5*
11⅞"	TJI®/15 DF	ITT211.88	ITT211.88*
	TJI®/25 DF	ITT11.88	ITT11.88*
	TJI®/35 DF	ITT3511.88	ITT3511.88*
	TJI®/55 DF	MIT11-2**	MIT11-2*
14"	TJI®/25 DF	ITT14	ITT14*
	TJI®/35 DF	ITT3514	MIT3514*
	TJI®/55 DF	MIT414**	MIT414*
16"	TJI®/25 DF	ITT16	ITT16*
	TJI®/35 DF	MIT3516	MIT3516*
	TJI®/55 DF	MIT416**	MIT416*

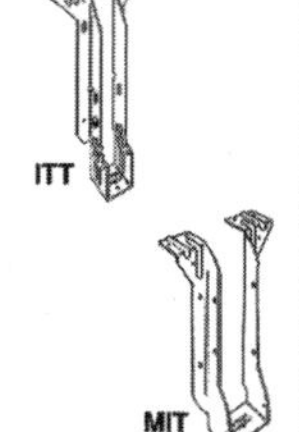

* Requires use of web stiffeners.
** Requires use of web stiffeners when the joist reaction exceeds 1200 pounds.

C2 – TOP MOUNT DOUBLE JOIST HANGER

DEPTH	JOIST	HANGER	
		RESIDENTIAL	COMM./MULTI-FAMILY
9½"	TJI®/15 DF	WP29.5-2	WP29.5-2*
	TJI®/25 DF	MIT9-2	MIT9-2*
11⅞"	TJI®/15 DF	WP211.88-2	WP211.88-2*
	TJI®/25 DF	MIT11-2	MIT11-2*
	TJI®/35 DF	WP3511.88-2**	WP3511.88-2*
	TJI®/55 DF	WPI411.88-2**	WPI411.88-2*
14"	TJI®/25 DF	MIT414	MIT414*
	TJI®/35 DF	WP3514-2**	WP3514-2*
	TJI®/55 DF	WPI414-2**	WPI414-2*
16"	TJI®/25 DF	MIT416	MIT416*
	TJI®/35 DF	WP3516-2**	WP3516-2*
	TJI®/55 DF	WPI416-2**	WPI416-2*

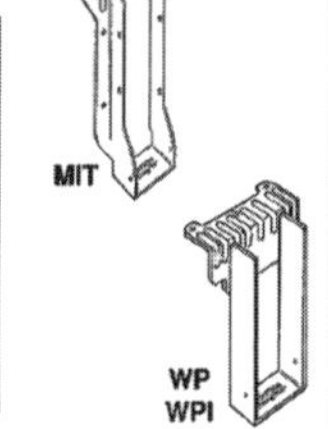

* Requires use of web stiffeners.
** Requires use of web stiffeners when the joist reaction exceeds 2000 pounds.

C3 – FACE MOUNT SINGLE JOIST HANGER

DEPTH	JOIST	HANGER	
		RESIDENTIAL	COMM./MULTI-FAMILY
9½"	TJI®/15 DF	IUT29	LSSU210*
	TJI®/25 DF	IUT9	LSSUI25*
11⅞"	TJI®/15 DF	IUT211	IUT211*
	TJI®/25 DF	IUT11	IUT11*
	TJI®/35 DF	IUT3512	IUT3512*
	TJI®/55 DF	IUT412**	U410*
14"	TJI®/25 DF	IUT14	IUT14*
	TJI®/35 DF	IUT3514	IUT3514*
	TJI®/55 DF	IUT414**	U414*
16"	TJI®/25 DF	IUT14*	IUT14*
	TJI®/35 DF	IUT3514*	IUT3514*
	TJI®/55 DF	IUT414*	U414*

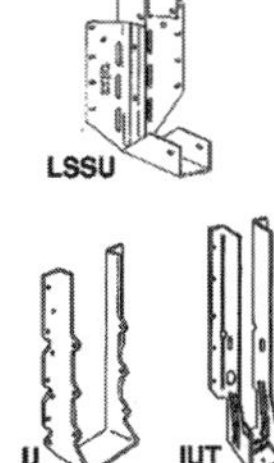

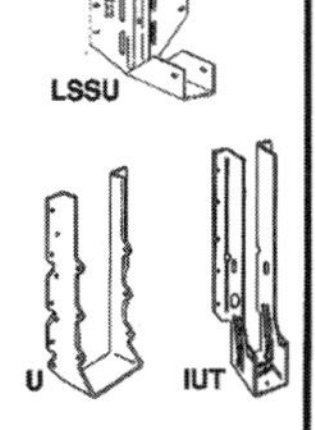

* Requires use of web stiffeners.
** Requires use of web stiffeners when the joist reaction exceeds 1200 pounds.

C4 – FACE MOUNT DOUBLE JOIST HANGER

DEPTH	JOIST	HANGER
9½"	TJI®/15 DF	HU210-2*
	TJI®/25 DF	U410*
11⅞"	TJI®/15 DF	HU212-2*
	TJI®/25 DF	U414*
	TJI®/35 DF	U3512-2*
	TJI®/55 DF	HU412-2*
14"	TJI®/25 DF	U414*
	TJI®/35 DF	U3512-2*
	TJI®/55 DF	HU414-2*
16"	TJI®/25 DF	U414*
	TJI®/35 DF	U3512-2*
	TJI®/55 DF	HU414-2*

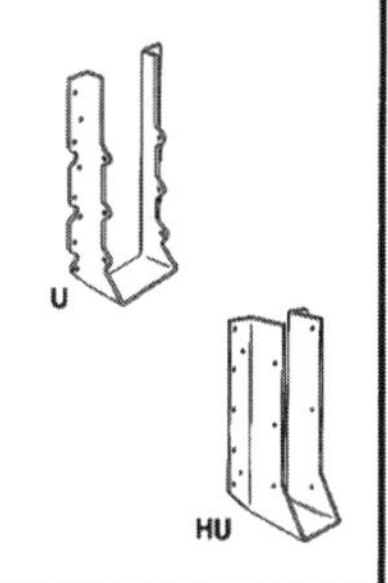

* Requires use of web stiffeners.

- Some hangers shown have less capacity than the capacity of the TJI® joists. For single joist applications beyond those shown in the span charts and all double joist applications, these hangers will need to be checked to assure adequate capacity.
- Hanger capacities will need to be checked if hanger is supported from other than Trus Joist MacMillan products or Douglas fir-larch or southern pine species support members. Contact your Trus Joist MacMillan representative for assistance.
- Hangers can only achieve maximum capacity if all nail holes are filled with the proper nails.
- In some cases, the hangers shown may have greater capacity when used in conjunction with certain supporting member categories or support member criteria.
- Leave 1/16" clearance between end of TJI® joist and support member.
- The hangers listed above are manufactured by Simpson Strong-Tie® Company, Inc. For additional hanger information, or possible KANT-SAG® hangers, please refer to the appropriate manufacturer's evaluation report(s).

10-15 *A selection chart to help you choose the proper hanger for specific applications.* Courtesy Trus Joist MacMillan

span. In other words, if the span of the joist is 15 feet, the cantilever could be as much as 5 feet.

For load-bearing applications, the joists can typically be cantilevered up to 2 feet past the last point of support. Anything beyond that will usually require additional support. You can easily and effectively close off the ends of the cantilever in either a bearing or nonbearing application by using LSL rim boards, solid lumber, or ¾-inch plywood.

Web stiffeners and blocks

Different types of framing applications might require the installation of web stiffeners, backer blocks, filler blocks, or squash blocks

(Fig. 10-16). All of these blocks have specific uses and installation criteria, which should be spelled out on your framing detail sheets. The blocks are all fabricated and installed at the job site—as opposed to being furnished or installed by the manufacturer—but they must comply with the manufacturer's requirements.

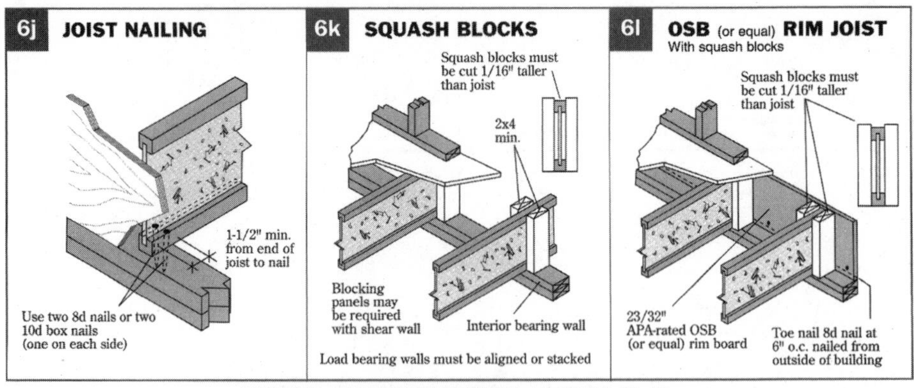

WEB STIFFENER DETAILS

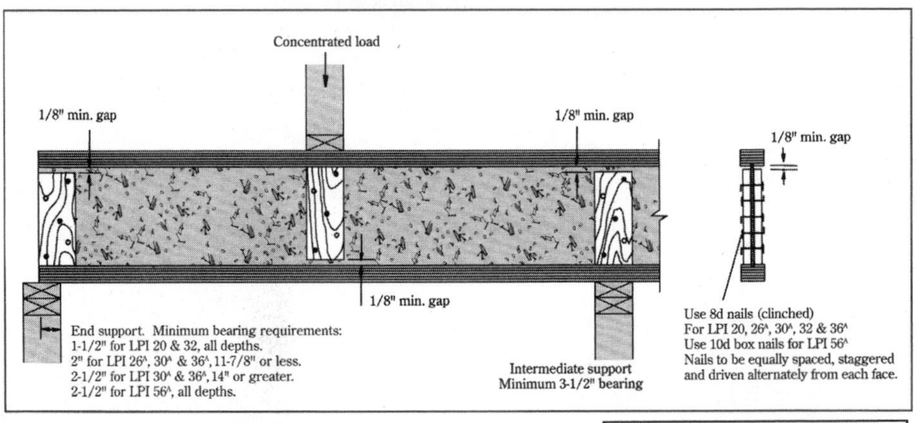

NOTES:

- Web stiffeners at bearings must have a minimum 1/8" gap at the top flange, but be tight and fully bearing on the bottom flange. Top flange concentrated load stiffeners must be tight to the top flange and have a minimum 1/8" gap at the bottom flange (see drawing above).
- APA-rated OSB (or equal) stiffener is to be a minimum of:
  15/32" thick for LPI 26[A].
  23/32" thick for LPI 20, 30[A], 32, & 36[A].
  Use 2x stiffeners (1-1/2" thick) for LPI 56[A].
- Minimum width to be same as bearing width 3-1/2" minimum.
- Web stiffeners are required on both faces of the joist.

	JOIST DEPTH			
	9-1/2"	11-7/8"	14"	16"
MAXIMUM STIFFENER HEIGHT	6-3/8"	8-3/4"	10-7/8"	12-7/8"

WEB STIFFENER NAIL REQUIREMENTS

	JOIST DEPTH			
	9-1/2"	11-7/8"	14"	16"
JOIST SERIES	NAILS PER FACE	NAILS PER FACE	NAILS PER FACE	NAILS PER FACE
26[A]	3	3	N/A	N/A
30[A]	3	3	4	4
20	3	3	N/A	N/A
32	3	3	3	3
36[A]	N/A	4	5	6
56[A]	N/A	4	5	6

10-16 *Details for building and installing web stiffeners.* Courtesy

The blocks are typically made of plywood or OSB, or solid sawn 2 × 4s. Plywood and OSB material must be sheathing grade (the same grade as what you're using for the subfloor and sheathing, so you can use job-site scraps in most instances), and it must be installed with the face grain of the block parallel to the long axis of the I-joist. The 2 × 4s must be construction grade or better—once again, the same material that's already on the job site—and must be installed perpendicular to the long axis of the joist.

In all the applications described in this section, you will find that it's considerably easier to install the blocks using an air-driven nail as opposed to hand-nailing. Attempting to hold the plywood blocks in place on each side of the joist while at the same time driving nails all the way through them by hand is virtually impossible, unless you first clamp the blocks together to hold them temporarily.

Web stiffeners are small blocks of ½-, ⅝-, or 1-inch plywood or OSB ripped to a minimum of 2 5⁄16 inches wide. They can also be, depending on the type of I-joist and the size of the flanges, 2 × 4 blocks. They are installed at the end of the I-joist, between the I-joist flanges and against the sides of the web on both sides of the joist (Fig. 10-17). Web stiffeners are necessary in certain installations in order to strengthen and stiffen the web, to increase the nailing surface, or both. They are common with certain types of joist hanger installations, and might also be necessary in certain long-span applications.

10-17 *Web stiffeners installed on an I-joist.*

Follow the specific installation instructions provided by the manufacturer to see exactly when and where web stiffeners are required.

Web stiffeners should be installed tight against the bottom flange of the I-joist, but with a gap of between ⅛ inch and about 2 inches between the top of the stiffener and the top flange. Place one web-stiffener block on each side of the joist, flush with the end. Drive three 8d box nails through both blocks and the I-joist web. The nails will penetrate all the way through, so it's necessary to then clinch them over, which also locks them and the blocks in place. Keep the nails at least 1 inch up from the top and bottom of the web-stiffener block to prevent splitting.

If the joist you're using requires 2 × 4 blocks instead of plywood, you must again cut the blocks ⅛ inch to 2 inches short of the top flange, and install them tight against the bottom flange. Secure the blocks with three 16d box nails, two driven from one side and one from the other side to equalize the shear strength and better prevent splitting. Keep the nails about 1½ inches from the ends of the blocks, again to prevent splitting. The 16d nails will not penetrate all the way through, so clinching is not necessary.

Backer blocks are used in certain applications where one I-joist meets another at a right angle, and they're installed in order to give an additional bearing surface for the nails that hold the hanger in place at that intersection. A backer block is usually plywood or OSB, installed with the face grain parallel with the long axis of the joist. For most applications, the backer block needs to extend 6 to 12 inches past the intersecting joist on each side. For stair stringer applications (where the stringer is attached to the I-joist), the backer block should be at least 4 feet long. Secure the block in place with 10 to 15 10d nails, clinched if possible.

You use *filler blocks* to fill the space between webs when using doubled I-joists. They are necessary to help tie the two joists together, and to stiffen and reinforce the paired joists to accommodate intersecting loads.

Depending on the size and type of the I-joist, filler blocks can be 2 × 6 or 2 × 8 solid lumber, 1⅛-inch plywood or OSB, or a combination of these materials. The length of the block varies with the installation, anywhere from a minimum of 4 feet long to the entire length of the doubled joists; refer to the manufacturer's specific instructions for your application. Secure the block to the joists using 10 to 15 10d box nails, clinched, or, for larger joists, approximately 15 16d nails. Again, refer to the manufacturer's specific instructions.

The fourth type of block is a *squash block* or *load support block*. Its purpose is to provide additional load-bearing support where upper-floor walls or concentrated loads such as posts and columns bear on the I-joists below. Squash blocks are made from 2 × material, of a width equal to the width of the load it's supporting. For example, if the squash block will help support a second-floor wall that's framed out of 2 × 6 lumber, then the block should be 2 × 6 as well. For posts or columns, you might need to build up several blocks in order to match the overall size of the supported post.

The squash block is different from the other types of blocks in that it directly supports the load above it, relieving the weight from bearing directly on top of the I-joist. The other three type of blocks are used primarily to strengthen and support the web. Because of this, the squash block is attached to both sides of the joist, cut to be 1⁄16 inch longer than the depth of the joist. This allows the block to take the weight of the load above it and transfer it down to the foundation or lower supports, without being so tall as to disrupt the level of the floor.

Bracing I-joists

As mentioned earlier, it is extremely important to properly brace I-joists to prevent them from rolling, prior to walking on them or placing any sort of load on them (Fig. 10-18). Here are the bracing instructions issued by Trus Joist MacMillan, which are typical for all manufacturers:

Warning notes:

Lack of concern for proper bracing during construction can result in serious accidents. Under normal conditions if the following guidelines are observed, accidents will be avoided:

1. All blocking, hangers, rim boards, and rim joists at the end supports of the TJI joists must be completely installed and properly nailed.

2. Lateral strength, like a braced end wall or an existing deck, must be established at the ends of the bay. This can also be accomplished by a temporary or permanent deck (sheathing) nailed to the first 4 feet of joists at the end of the bay.

3. Temporary strut lines of 1 × 4 (min.) must be nailed to a braced end wall or sheathed area as in note 2 and to each joist. Without this bracing, buckling sideways or roll over is highly probable under light construction loads—like a worker and one layer of unnailed sheathing.

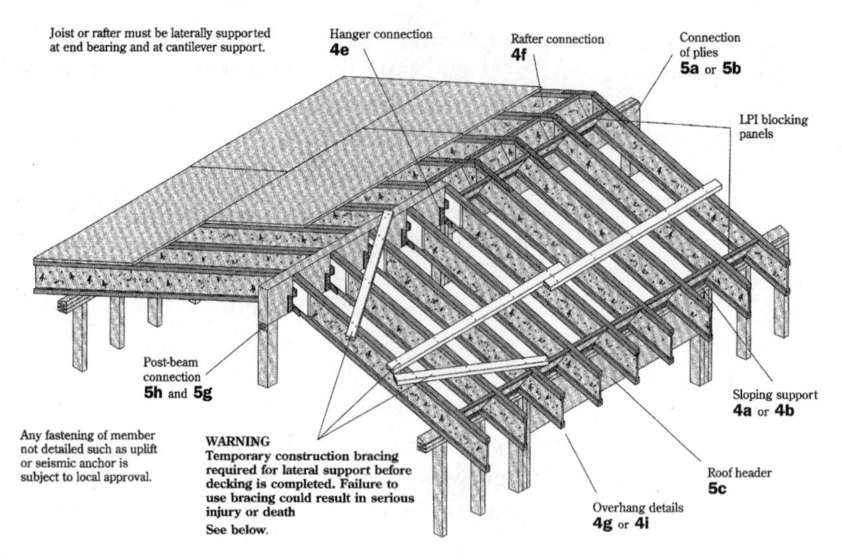

TEMPORARY BRACING

- Use at least 1x4 temporary bracing members nailed to each LPI Joist with two 8d nails.
- Keep them parallel and no more than 6'-0" apart for LPI 26^A^ series, 8'-0" apart for other LPI Joist series.
- Use long pieces, not short blocks, lap the ends to keep a continuous line of bracing.
- To prevent endwise movement of the continuous 1x4 lines of bracing, anchor them at the ends and at 25'-0" intervals into a stable end wall or an area braced by sheathing or diagonal bracing.
- Remember, the continuous 1x4 bracing is not effective unless attached to the braced area.
- Use particular care removing temporary bracing when applying sheathing. Remove the bracing as the sheathing is attached.

NOTE: Some wind or seismic loads may require different or additional details and connections.

10-18 *The proper way to brace a roof until the sheathing is applied.* Courtesy Trus Joist MacMillan

4. Sheathing must be totally attached to each TJI joist before additional loads can be placed on the system.

5. Ends of cantilevers require strut lines on both the top and bottom flanges.

6. The flanges must remain straight within a tolerance of ½" from the true alignment.

Simply put, you need to first establish a solid, secure point on the structure, then brace off of that. You can achieve that by installing solid sheathing or diagonal bracing at one end of the floor to hold that section rigid, then extending 1 × 4s off of that. The 1 × 4s should be placed on top of the joists about 8 feet apart, and should be secured to each I-joist with two 8d to 12d box nails. Be certain that the I-joists are properly placed on their intended center-to-center spacing before nailing the 1 × 4s in place. For cantilevered areas, place the 1 × 4s (referred to as the *strut lines*) on both the top and the bottom of the joists.

You need to install the temporary bracing immediately after completing installation of the I-joists and all necessary rim joists and blocking, but before installing the subfloor sheathing material. The bracing needs to remain in place as the subfloor is being installed, and should be removed a row at a time only as you reach it with the subflooring.

Installing subflooring

Virtually all contractors are using sheet subfloor these days, as opposed to the 1 × and 2 × lumber of days past (Fig. 10-19). Subflooring in sheets is fast, smooth, and easy to install, and, properly placed down, it provides a strong floor that is very resistant to racking or any type of lateral movement.

Subfloor installation is the same over engineered lumber I-joists as it is over joists of solid sawn lumber. Be certain your joists are square and level. Check the building using diagonal measurements, string lines, 3:4:5 measurements, or a combination of all three to ensure that everything is square before placing the sheathing.

The engineered lumber manufacturers strongly recommend that you use construction or subfloor adhesive, and all their strength, load, and deflection calculations are based on a subfloor that has been both glued and nailed. Initially, apply a bead of adhesive to the top flanges of the I-joists in one corner, covering only an area equivalent to one sheet of subfloor. Immediately install the first subfloor sheet, move it square to the corner of the building, press it solidly into the adhesive, and nail it in place. Use 8d hand- or air-driven nails

10-19 *Installing sheet subflooring.* Courtesy Trus Joist MacMillan

on a maximum spacing 18 inches apart, and nail off the entire sheet before proceeding to the next one. You can use 14-gauge air-driven staples in place of the nails, providing the staples are long enough to penetrate a minimum of 1 inch into the joist flange.

Working out from that first corner, apply only enough adhesive for one to two sheets of subfloor at a time. In hot weather, limit the adhesive to one sheet at a time to prevent premature drying or skinning of the adhesive bead. Immediately install the subfloor panel, interlocking the tongue-and-groove joints and allowing a gap of about ⅛ inch—about the thickness of a nail—to remain between the panels. For best results and to achieve the strongest bond between the subfloor and the joists, immediately nail off the entire subfloor panel before moving on.

Leave the temporary 1 × 4 braces in place on the I-joists while you install the subfloor. This keeps the floor framing system rigid for safety, and also keeps the joists on their proper spacing, which speeds up installing the flooring panels. Carefully remove the temporary bracing boards and nails only as you reach each row with the sheathing.

Engineered lumber wall framing

There are a number of engineered lumber products that will work very well in many framing applications all around the structure. Laminated veneer lumber (LVL) and parallel strand lumber (PSL) products work very well for beams and headers of virtually any span, and the newer laminated strand lumber (LSL) is great for beams and headers of short to medium spans. PSL beams stain out remarkably well, with a dramatic and unusual grain pattern that looks great left as an exposed beam.

In addition, Trus Joist MacMillan has recently introduced their TimberStrand LSL Premium Studs for use in wall framing. The studs are available in nominal 2 × 4 and 2 × 6 sizes to match conventional solid sawn lumber, and in standard precut lengths of 92⅝ and 104⅝ inches. LSL studs are cut and installed in the same manner as regular solid lumber.

According to the manufacturer, these studs have currently been rated as an acceptable replacement for stud-grade hem-fir, SPF, SPF (south), and other species having design values lower than the tested values for the LSL . The LSL material can also be used "as plate material with SPF (south) stud-grade applications...."

Studs manufactured from laminated strand lumber are a great improvement over conventional solid sawn lumber when it comes to moisture content, dimensional stability, and uniformity. They might well represent the next generation of framing products used regularly on the job site, but at present there are some limitations on their use, as previously described. Be sure to consult with your local building officials, your lumber dealer, or directly with the manufacturer to confirm their suitability for any specific construction application.

Garage-door headers

Anyone who has ever framed a house has wrestled with that big stick of wood over the garage door, only to have it sag later on and cause the door to bind. A couple of alternatives to that situation are the use of LVL or LSL garage-door headers (Fig. 10-20). These are 3½-inch-wide beams designed specifically for use over wide-span garage-door openings of up to 18 feet, 3 inches wide. They are solid, straight, and some of them, such as Trus Joist MacMillan's Big Red Garage Door Header, are manufactured with a slight camber to improve load and deflection performance.

10-20 *An example of an engineered lumber garage-door header. Note the doubled trimmers at each end.* Courtesy Trus Joist MacMillan

Engineered lumber headers require double trimmers at each end for most span applications (3 inches of bearing surface). Certain long-span applications under heavy snow-load conditions might require triple headers (4½ inches of bearing surface). Consult with the manufacturer's load and span charts for complete details.

Installing I-joists for roof framing

As you have probably noticed in working through some of the sizing exercises in the previous chapter, wood I-joists have become a very popular material for roof framing. I-joists are strong enough to handle long spans without bowing or needing additional bracing, but are considerably lighter than solid sawn lumber (Fig. 10-21), which makes handling them on a roof framing project a much easier and safer undertaking.

I-joists work especially well in vaulted ceiling applications (Fig. 10-22). Their size allows them to span from ridge to wall plate without other interior supports in most applications, and they are deep enough to achieve a high level of insulation. The prepunched knockout holes in the joist webs greatly simplify the installation of wiring and small-diameter water pipes, and, depending on the installation, the knockouts can also provide cross-ventilation for the ceiling cavities.

At the ridge, the I-joists can be either attached to the ridge beam with hangers, or placed on top of the ridge. Refer to your installation plans for full details on the method that was designed and specified for your building (Fig. 10-23).

For a hanger installation, the ends of the joist need to be plumb-cut to match the installed angle of the joist. Install a web stiffener on each side of the joist that has been bevel-cut on the top and bottom to match the plumb cut of the I-joist. Once again, the web stiffener should be installed tight against the bottom flange of the I-joist and have a small gap at the top. Install the web stiffeners with three clinched-over nails, as described previously in the section *Web stiffeners and blocks.*

Use a variable-slope seat hanger, such as the Simpson Strong-Tie LSSU series. Place the hanger on the end of the joist and press it on firmly to seat it. Bend the seat of the hanger to match the angle of the joist, and nail the hanger to the joist as per the manufacturer's instructions. Lift the joist into place, align it properly, and attach the hanger flanges to the ridge beam. Be sure to use all the holes in the hanger to achieve its full rated strength.

10-21 *The light weight of an engineered lumber I-joist makes it considerably easier and safer to use on a roof. Note the pieces of LSL blocking between joists, which are notched to provide ventilation.*
Courtesy Trus Joist MacMillan

10-22 *I-joists being used for a vaulted ceiling application.* Courtesy Trus Joist MacMillan

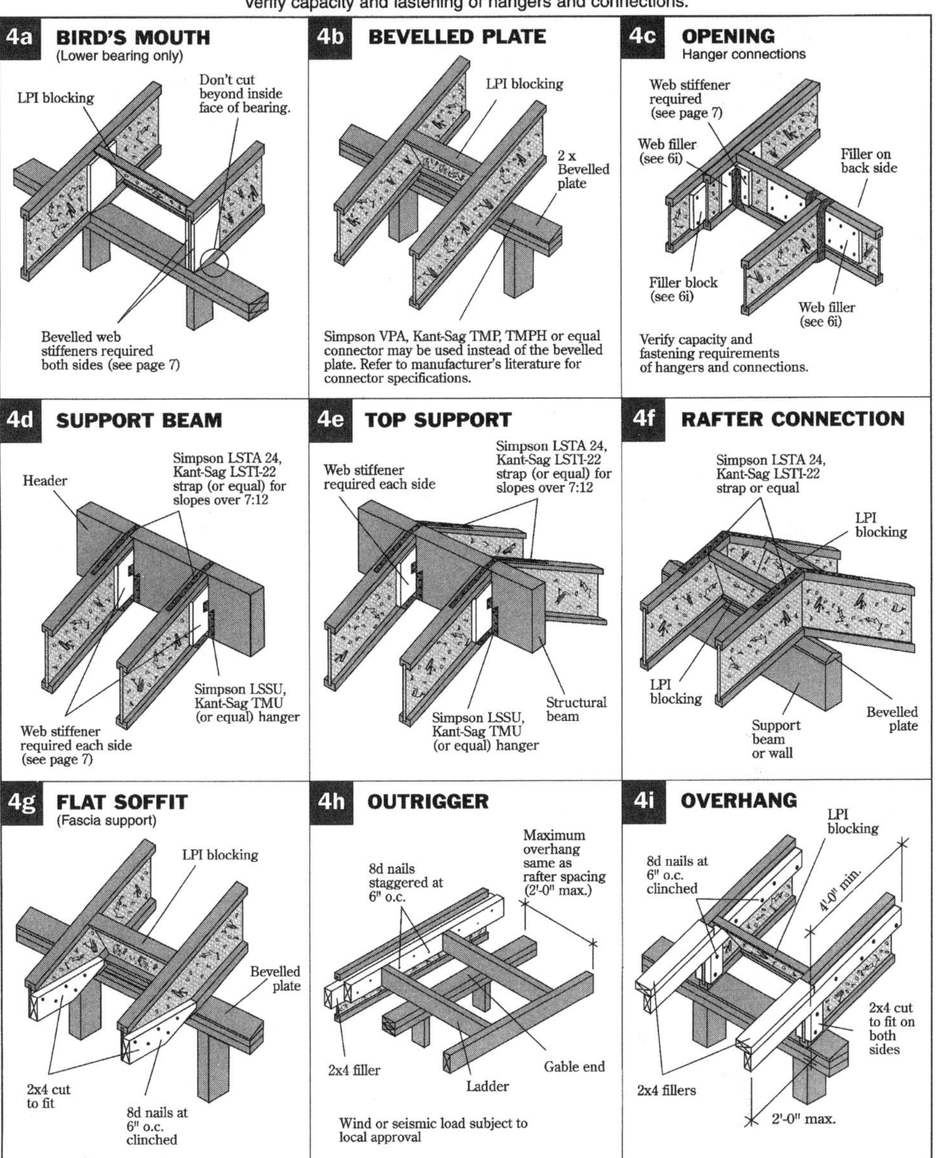

10-23 *Standard roof framing details.* Courtesy Louisiana-Pacific

For I-joists that overlap the ridge beam, you need to follow a different procedure. First, since I-joists cannot be cut at the top with the traditional birdmouth notch found on solid sawn lumber, you need to rip a board on a bevel and attach it to the top of the ridge beam. This is done to provide a full bearing support under the I-joists. Plumb-cut the ends of the I-joists to match the slope of the roof, and set them in place over the ridge; web stiffeners should not be required in most applications. Block between the pairs of rafters using solid lumber, I-joist material, or metal or wood X-bracing to prevent the joists from rolling.

If the pitch of the roof exceeds 7/12, you'll also need to use perforated metal strapping such as Simpson LSTA15 or MSTA36. Attach the strapping over the top flanges of the rafter pairs, nailing it to each I-joist with twelve 10d 1½-inch nails. Strapping is also a good idea on rafters that are installed in hangers, again if the pitch of the roof exceeds 7/12.

There are several different connection methods to choose from at the plate, and, when you order your I-joists, these details should be spelled out for you to accommodate your particular framing application. All the plate/I-joist options listed here also require that blocking be installed between the I-joists for lateral strength. This can be in the form of solid blocking, which also closes off the space between the joists to prevent anything from entering the attic space; or you could use wood or metal X-bracing and then close off the space with a flat soffit, siding, or other means. Some of the common plate connections are:

- Installing a beveled plate on top of the wall plate, with the bevel matching the angle of rafter. This gives full bearing support under the I-joist, and you can attach the joist directly to plate with two 10d nails.
- Birdmouth the I-joist at the plate, as in conventional solid lumber framing practices. There are specific limitations on how this birdmouth notch is to be cut, and it cannot extend past the plate into the building. The flange at the bottom of the I-joist needs to bear fully on the wall plate, and a web stiffener is typically called for at this point as well. Remember that you can make a birdmouth cut only at the low end of the joist.
- Connect the I-joist to the plate using a metal variable-slope seat connector, such as Simpson's VPA series. This is one of the easier connection methods, since it doesn't require notching the beam or shimming the plate, and the variable hanger adjusts to any of a variety of pitch angles.
- Make a plumb cut on the end of the I-joist, and stop the rafter in line with the plate. This requires a variable connector or a

beveled shim on the top plate to make the connection. Use this kind of connection when you don't want the I-joist to extend beyond the plate line for aesthetic reasons. Once the I-joist has been cut off, you can install a solid trim board over the cut ends if you don't want any overhang, or you can continue the overhang framing using conventional lumber and facia.

- For a flat, closed soffit, simply make a level cut on the bottom of the I-joist and a plumb cut on the end. Extend the cut far enough back along the joist so the joist's bottom has full bearing on the wall plate, but not so far back that it extends past the inside of the plate. Install beveled web stiffeners, then add a 2 × 4 block to the end of the joist that's cut at the same plumb and level cuts as the joist. Then attach your soffit panels and facia to the 2 × 4 block.
- If desired, use LSL lumber between the I-joists over the plate, since the LSL rim-board material will match the height of the joists. Nail the LSL to the top flange of the I-joist, then toenail it to the plates. If necessary to meet your ventilation requirements, you can cut a V-shaped notch into the top of the LSL block to provide good ventilation flow into the roof cavities.

Holes and cutouts in engineered lumber

Due to the nature of the materials and construction, engineered lumber products have very specific restrictions on placing holes and notches in the wood (Fig. 10-24). Each manufacturer publishes a chart (Fig. 10-25) that might look a little confusing at first, but is actually quite easy to follow. The charts specify exactly where in the material you can place your holes, and what size they can be. This might require a little more preplanning for plumbers and HVAC installers, but you'll probably find that it's a minor inconvenience. Here are some general rules to be aware of for I-joists:

- A hole of 1½ inches or less in diameter can be placed anywhere in the web. This is large enough for virtually all electrical wiring applications, as well as most water-line plumbing applications.
- 1½-inch knockouts are provided in the I-joist web approximately every 12 inches to eliminate a lot of the

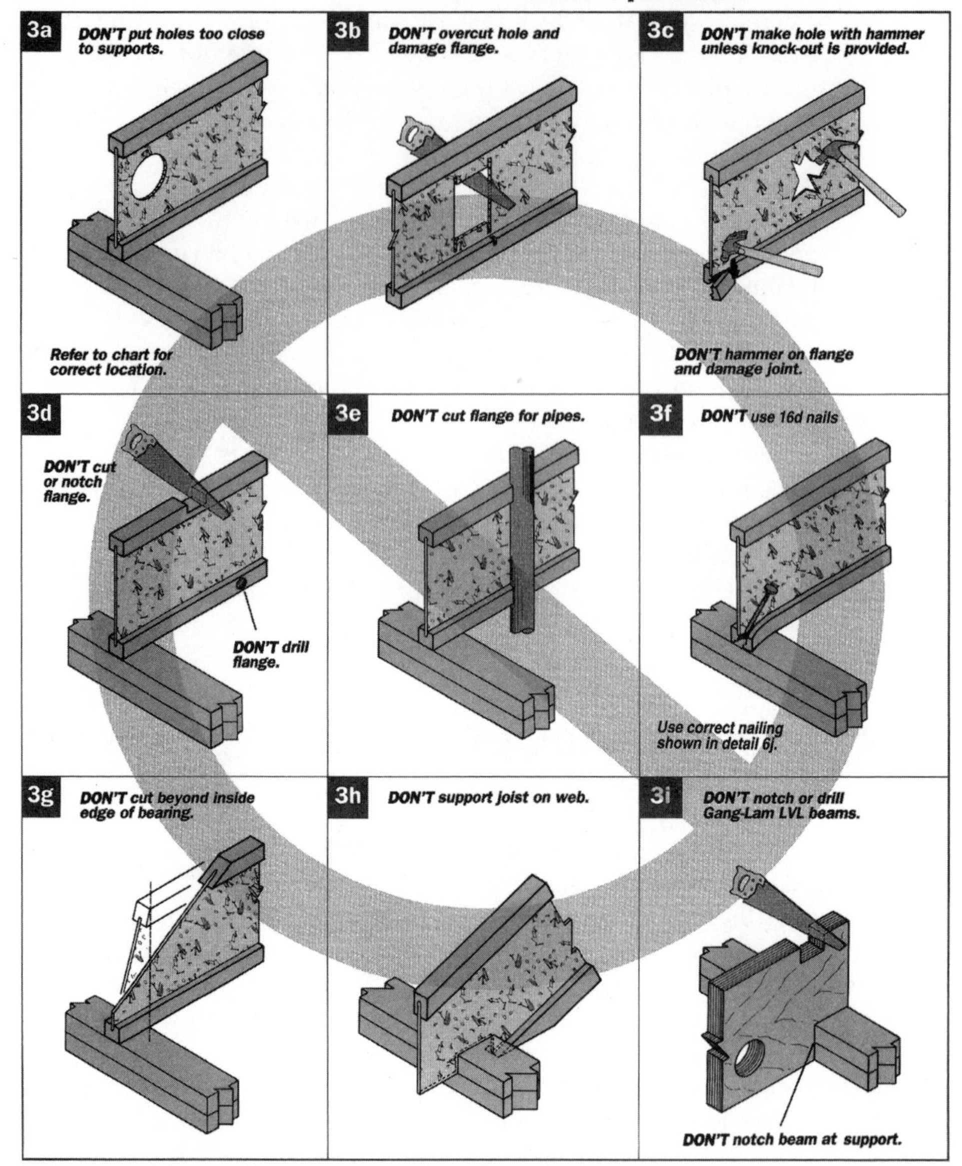

10-24 *In addition to specific ways of working with I-joists and LVL beams, there are also specific things you can't do, as this manufacturer's chart*

normal drilling required in solid sawn lumber (Figs. 10-26 and 10-27).

- When cutting or drilling multiple holes in the web, the length of the uncut web between the holes must be equal to at least twice the length of the largest adjacent hole. For example, if you want to drill a 2-inch hole and a 3-inch hole, you need a minimum of 6 inches of uncut web (2 times the size of the largest hole) remaining between the two holes.
- Holes can be located vertically anywhere in the web.
- Holes or notches of any size cannot be cut or drilled in the top or bottom flange at any location.

General rules for LVL and PSL members are:

- There is an area in any LVL or PSL member known as the "hole zone," where you can drill holes. This zone is located within the middle third of the distance between any two beam supports, and vertically within the center third of the beam. For example, if you had a beam that spanned 18 feet between supports, you could drill holes in the center 6 feet. If the beam was 12 inches deep, you could drill anywhere from 4 inches down from the top to 4 inches up from the bottom.
- Hole sizes are limited to a maximum of 1¾ inches in diameter for beam depths of 5½ inches, or up to a maximum of 2 inches in diameter for beam depths of 7¼ up to 18 inches.
- The distance between any two holes must be equal to or greater than twice the diameter of the larger hole. For example, if you wanted to drill two 2-inch holes in the beam, there would need to be at least 4 inches of solid beam between the holes.
- Square or rectangular holes are not permitted.
- The beam cannot be notched at any point along the top or bottom.

The general rules for LSL studs are:

- Notches of up to ⅞ inch for 2 × 4 studs or up to 1⅜ inch for 2 × 6 studs are allowable anywhere except in the middle third of the stud's length.
- Holes of up to 1⅜ inch in diameter for 2 × 4 studs or up to 2³⁄₁₆ inches in diameter for 2 × 6 studs can be drilled anywhere along the length of the stud, but it must be at least ⅝ inch back from either edge.
- Notches and holes cannot occur in the same cross-section of the stud.

WEB HOLE DETAILS

Warning: Do NOT cut or notch flanges.

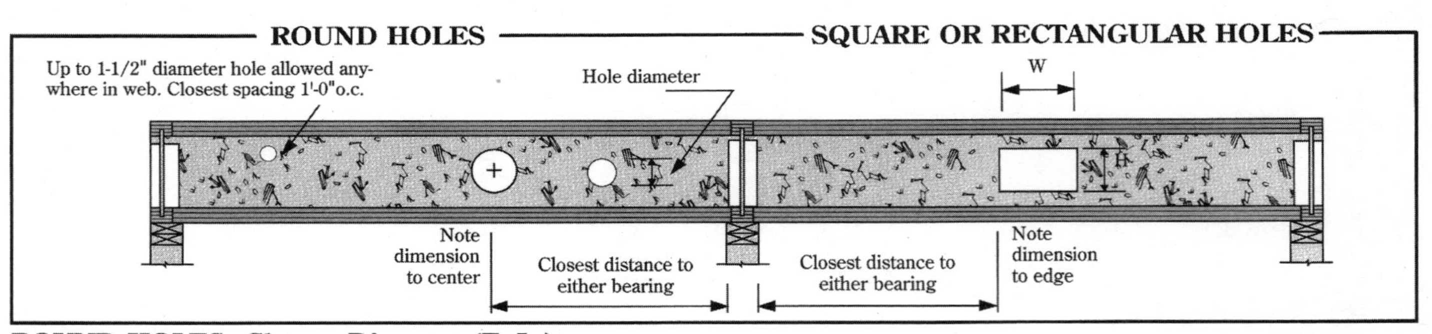

ROUND HOLES Closest Distance (Ft-In)

JOIST DEPTH	JOIST SERIES	HOLE DIAMETER										
		2"	3"	4"	5"	6"	7"	8"	9"	10"	11"	12"
9-1/2"	LPI 20 & 32	1'-0"	2'-0"	3'-0"	4'-0"	–	–	–	–	–	–	–
	LPI 26[A]	1'-0"	1'-0"	1'-11"	3'-3"	4'-6"	–	–	–	–	–	–
	LPI 30[A]	1'-0"	1'-0"	2'-5"	3'-10"	5'-4"	–	–	–	–	–	–
11-7/8"	LPI 20 & 32	1'-0"	2'-0"	3'-0"	4'-0"	5'-0"	5'-0"	6'-0"	–	–	–	–
	LPI 26[A]	1'-5"	2'-3"	3'-1"	3'-11"	4'-9"	5'-7"	6'-8"	–	–	–	–
	LPI 30[A]	1'-1"	1'-1"	1'-11"	2'-8"	3'-6"	4'-3"	5'-0"	–	–	–	–
	LPI 36[A] & 56[A]	1'-0"	1'-11"	2'-11"	3'-10"	4'-10"	5'-9"	7'-3"	–	–	–	–
14"	LPI 32	1'-0"	2'-0"	3'-0"	4'-0"	5'-0"	5'-6"	6'-6"	7'-0"	8'-0"	–	–
	LPI 30[A]	2'-2"	2'-10"	3'-5"	4'-0"	4'-8"	5'-3"	5'-10"	6'-6"	7'-1"	–	–
	LPI 36[A] & 56[A]	3'-10"	4'-4"	4'-9"	5'-2"	5'-8"	6'-1"	6'-6"	6'-11"	7'-5"	–	–
16"	LPI 32	2'-0"	3'-0"	4'-0"	5'-0"	6'-0"	6'-0"	7'-0"	8'-0"	8'-0"	8'-0"	9'-0"
	LPI 30[A]	3'-8"	4'-3"	4'-9"	5'-4"	5'-11"	6'-5"	7'-0"	7'-6"	8'-1"	8'-7"	9'-2"
	LPI 36[A] & 56[A]	3'-7"	4'-4"	5'-1"	5'-10"	6'-7"	7'-4"	8'-1"	8'-10"	9'-7"	10'-7"	12'-3"

SQUARE OR RECTANGULAR HOLES Closest Distance (Ft - In)

JOIST DEPTH	JOIST SERIES	LONGEST HOLE DIMENSIONS H or W										
		2"	3"	4"	5"	6"	7"	8"	9"	10"	11"	12"
9-1/2"	LPI 20	3'-1"	3'-7"	3'-10"	4'-3"	–	–	–	–	–	–	–
	LPI 32	1'-0"	2'-0"	2'-10"	3'-10"	–	–	–	–	–	–	–
	LPI 26^A	2'-4"	3'-2"	4'-1"	4'-11"	6'-2"	–	–	–	–	–	–
	LPI 30^A	2'-9"	3'-8"	4'-7"	5'-7"	6'-11"	–	–	–	–	–	–
11-7/8"	LPI 20	2'-4"	2'-11"	3'-7"	4'-3"	5'-0"	5'-9"	6'-3"	–	–	–	–
	LPI 32	1'-0"	2'-0"	2'-10"	3'-10"	6'-9"	7'-9"	8'-8"	–	–	–	–
	LPI 26^A	4'-1"	4'-8"	5'-3"	5'-10"	6'-5"	8'-2"	9'-8"	–	–	–	–
	LPI 30^A	4'-8"	5'-3"	5'-11"	6'-9"	8'-0"	9'-3"	10'-6"	–	–	–	–
	LPI 36^A & 56^A	6'-2"	7'-0"	7'-11"	8'-9"	9'-8"	10'-6"	12'-1"	–	–	–	–
14"	LPI 32	1'-0"	2'-0"	2'-10"	3'-10"	6'-9"	7'-9"	8'-8"	9'-8"	10'-7"	–	–
	LPI 30^A	2'-1"	3'-0"	3'-8"	4'-10"	5'-8"	6'-7"	7'-6"	9'-0"	11'-2"	–	–
	LPI 36^A & 56^A	3'-11"	4'-8"	5'-2"	6'-2"	6'-11"	7'-8"	9'-3"	11'-0"	12'-9"	–	–
16"	LPI 32	2'-0"	3'-0"	3'-10"	4'-10"	6'-9"	7'-9"	8'-8"	9'-8"	10'-7"	10'-7"	11'-6"
	LPI 30^A	3'-7"	4'-5"	5'-3"	6'-1"	6'-11"	7'-10"	8'-8"	9'-6"	10'-7"	12'-4"	14'-0"
	LPI 36^A & 56^A	5'-5"	6'-2"	6'-11"	7'-8"	8'-5"	9'-2"	9'-11"	11'-5"	13'-1"	14'-10"	14'-11"

DISTRIBUTOR:

LP, Louisiana-Pacific, Wood-E, Solid Start, LPI, and Gang-Lam are registered trademarks of Louisiana-Pacific Corporation. APA is a registered trademark of The Engineered Wood Association. PFS Corporation is a registered trademark of PFS Corporation.

NOTE: Louisiana-Pacific Corporation periodically updates and revises its product information. To verify that this version is current, contact the sales offices.

NOTES FOR HOLE CHARTS:

- Square and rectangular holes must be centered at mid height of web.
- Round holes do not need to be at mid height, but must not be closer than 1/2" from joist flange.
- Cut holes carefully. **DO NOT** overcut. **DO NOT** cut flanges.
- Maximum of three holes per span.
- The length of uncut web between holes must be at least twice the length of the longest adjacent hole dimension.
- Hole locations are for maximum uniformly distributed loads and simple or multiple spans as published in current L-P literature. For other conditions contact your L-P engineered products distributor.
- Larger hole sizes and closer proximity to supports for components carrying less than maximum load, may be determined by special design. Use L-P's Wood-E® Design Software, or contact your L-P engineered products distributor.

For more information of LPI Joists or Gang-Lam LVL, please contact one of our sales offices:

10-25 *Another chart for sizing and placing holes in the web of an I-joist.* Courtesy Louisiana-Pacific

10-26 *Prepunched knockouts in the web of an I-joist are convenient for installing plumbing pipes.*

10-27 *Prepunched knockouts in the web of an I-joist are also used for running electrical wires.*

Appendixes

A

Products by manufacturer's product name

Product name	Type of product	Manufacturer
1.3E Premium Studs	Laminated strand lumber wall studs	Trus Joist MacMillan
1.5E Header	Laminated strand lumber header	Trus Joist MacMillan
BC Calc	Software for sizing engineered lumber	Boise Cascade
BC Framer	Software for engineered lumber systems	Boise Cascade
BCI Joists	Wood I-joists	Boise Cascade
Bin Board	Plywood for the container industry	Georgia-Pacific
Com-Ply	Wood veneer/wood fiber composite panel	American Plywood Association*
Cutrite	Saw guide for wood I-joists	Trus Joist MacMillan
Dreamspace	Vinyl lumber	Thermal Industries
Ener-Grid	Polystyrene/cement building blocks	Ener-Grid

* Name of trade organization overseeing product standards for a variety of manufacturers.

Product name	Type of product	Manufacturer
Epoxy-Tie	Epoxy-based adhesive for anchors	Simpson Strong-Tie
FiberBond	Fiber-reinforced gypsum underlayment, wallboard, sheathing	Louisiana-Pacific
Flex-Trim	Flexible polymer composite trim	Flex Trim Industries
Formbeams	Laminated veneer lumber for concrete forms	Trus Joist MacMillan
FrameWorks	Engineered lumber framing system	Trus Joist MacMillan
FrameWorks Header	Laminated strand lumber header	Trus Joist MacMillan
G-P Lam	Laminated veneer lumber beams, headers	Georgia-Pacific
Gang-Lam S	Laminated veneer lumber beams	Louisiana-Pacific
Glulam	Glue-laminated structural wood beam	American Plywood Association*
HardiShake	Fiber-cement roofing shingles	James Hardie Products
HardiSlate	Fiber-cement roofing shingles	James Hardie Products
Inner-Seal	Oriented strand board sheathing, siding	Louisiana-Pacific
LPI Joists	Wood I-joists	Louisiana-Pacific
Masonite	Hardboard	Masonite Corporation
MaxiPlank	Fiber-cement siding	Maxi Tile
Micro=lam (original spelling)	Laminated veneer lumber headers, beams	Trus Joist MacMillan
Microllam	Laminated veneer lumber headers, beams	Trus Joist MacMillan

Product name	Type of product	Manufacturer
MSR Lumber	Machine Stress Rated lumber	Western Wood Products Assoc.*
Nailon	Nylon masonry anchor screw system	Simpson Strong-Tie
Parallam	Parallel strand lumber headers, beams, columns, posts	Trus Joist MacMillan
Performance Plus	Oriented strand board web for wood I-joists	Trus Joist MacMillan
Performance Rated	APA-rated plywood and OSB panels	American Plywood Association*
Ply Frame	Multicore plywood, industrial, furniture uses	Georgia-Pacific
Ply-Bead	Plywood with beaded grooves	Georgia-Pacific
PrimeTrim	Wood fiber composite trim, facia, molding	Georgia-Pacific
R-Control	Insulated building panels	AFM Corporation
Rustic Shingle	Aluminum roofing shingles	Classic Products
SeaFrame	Marine plywood	Georgia-Pacific
Silent Floor	Engineered lumber floor framing system	Trus Joist MacMillan
Simple Framing	Engineered lumber framing system	Boise CascadeSystem
Sleeve-All	Masonry anchor screw	Simpson Strong-Tie
SolId Start	Engineered wood framing system	Louisiana-Pacific
SpaceMaker Truss	Wood site-assembled roof truss system	Trus Joist MacMillan
SpecLam	Insulated building panels	AFM Corporation
Structurwood	Oriented strand board panels	Weyerhaeuser
Sturd-I-Floor	APA-rated sheet subfloor materials	American Plywood Association*

Product name	Type of product	Manufacturer
Sturdi-Wood	Oriented strand board panels	Weyerhaeuser
T-1-11	Grooved plywood siding pattern	American Plywood Association*
TecLam	Laminated veneer lumber	Tecton Laminates
TimberStrand	Laminated strand lumber rim boards, beams	Trus Joist MacMillan
TJ-Beam	Software for engineered lumber beam sizing	Trus Joist MacMillan
TJ-Dealer	Software for lumber dealers	Trus Joist MacMillan
TJ-Spec	Software for sizing engineered lumber	Trus Joist MacMillan
TJ-Xpert	Software for engineered lumber systems	Trus Joist MacMillan
TJI Joists	Wood I-joists	Trus Joist MacMillan
TJI/Pro	Wood I-joists	Trus Joist MacMillan
TLI Joists	Wood I-joists	Tecton Laminates
Top Notch Inner-Seal	Oriented strand board, tongue-and-groove subfloor panels	Louisiana-Pacific
Trex	Wood-polymer lumber products	Mobil Chemical
Versa-Lam	Laminated veneer lumber beams, headers	Boise Cascade
Versa-Lam Plus	Laminated veneer lumber beams, headers	Boise Cascade
Versa-Rim 98	Laminated veneer lumber rim boards	Boise Cascade
Wedge-All	Masonry anchor bolt	Simpson Strong-Tie
Wolmanized	Preservative-treated lumber	Hickson Corp.
Wood I Beam	Wood I-joist	Georgia-Pacific
Wood-E	Software for sizing engineered lumber	Louisiana-Pacific
ZzzzFlex	Flexible polymer composite trim	Flex Trim Industries

B

Products by type of product

Type of product	Product name	Manufacturer
Aluminum roofing shingles	Rustic Shingle	Classic Products
Computer software	BC Calc	Boise Cascade
	BC Framer	Boise Cascade
	TJ-Beam	Trus Joist MacMillan
	TJ-Dealer	Trus Joist MacMillan
	TJ-Spec	Trus Joist MacMillan
	TJ-Xpert	Trus Joist MacMillan
	Wood-E	Louisiana-Pacific
Engineered lumber framing systems	FrameWorks Simple Framing System	Trus Joist MacMillan Boise Cascade
	SolId Start	Louisiana-Pacific
Engineered lumber floor framing system	Silent Floor	Trus Joist MacMillan
Epoxy adhesive	Epoxy-Tie	Simpson Strong-Tie
Fiber-cement roofing	HardiShake	James Hardie Prod.
	HardiSlate	James Hardie Prod.
Fiber-cement siding	MaxiPlank	Maxi Tile
Fiber-reinforced gypsum	FiberBond	Louisiana-Pacific

*Name of trade organization overseeing product standards for a variety of manufacturers.

Type of product	Product name	Manufacturer
Flexible trim	Flex-Trim	Flex Trim Industries
	ZzzzFlex	Flex Trim Industries
Glue-laminated beams	Glulam	American Plywood Assoc.*
Hardboard	Masonite	Masonite Corporation
I-joists, wood	BCI Joists	Boise Cascade
	LPI Joists	Louisiana-Pacific
	TJI Joists	Trus Joist MacMillan
	TJI/Pro	Trus Joist MacMillan
	TLI Joists	Tecton Laminates
	Wood I Beams	Georgia-Pacific
I-joist saw guide	Cutrite	Trus Joist MacMillan
Insulated blocks/panels	Ener-Grid	Ener-Grid
	R-Control	AFM Corporation
	SpecLam	AFM Corporation
Laminated strand lumber	1.3E Premium Studs	Trus Joist MacMillan
	1.5E Header	Trus Joist MacMillan
	FrameWorks Header	Trus Joist MacMillan
	TimberStrand	Trus Joist MacMillan
Laminated veneer lumber	Formbeams	Trus Joist MacMillan
	G-P Lam	Georgia-Pacific
	Gang-Lam S	Louisiana-Pacific
	Microllam	Trus Joist MacMillan
	TecLam	Tecton Laminates
	Versa-Lam	Boise Cascade
	Versa-Lam Plus	Boise Cascade
	Versa-Rim 98	Boise Cascade
MSR lumber	Machine Stress Rated lumber	Western Wood Products Association*

Type of product	Product name	Manufacturer
Oriented strand board products	Blue Ribbon OSB	Georgia-Pacific
	Inner-Seal	Louisiana-Pacific
	Structurwood	Weyerhaeuser
	Sturdi-Wood	Weyerhaeuser
Parallel strand lumber	Parallam	Trus Joist MacMillan
Performance-rated plywood	Performance-Rated plywood	American Plywood Assoc.*
Site-assembled trusses	SpaceMaker Truss	Trus Joist MacMillan
Specialty plywood	Bin Board	Georgia-Pacific
	Com-Ply	American Plywood Assoc.*
	Ply Frame	Georgia-Pacific
	Ply-Bead	Georgia-Pacific
	SeaFrame	Georgia-Pacific
Vinyl lumber	Dreamspace	Thermal Industries
Wood fiber trim	PrimeTrim	Georgia-Pacific
Wood-polymer lumber	Trex	Mobil Chemical

C

Engineered lumber products listed by manufacturer

Manufacturer	Product name	Type of product
Boise Cascade	BC Calc	Computer software
	BC Framer	Computer software
	Simple Framing System	Engineered lumber framing system
	BCI Joists	Wood I-joists
	Versa-Lam	Laminated veneer lumber
	Versa-Lam Plus	Laminated veneer lumber
	Versa-Rim 98	Laminated veneer lumber
Georgia-Pacific	Wood I Beam	Wood I-joists
	G-P Lam	Laminated veneer lumber
	Bin Board	Specialty plywood
	Ply Frame	Specialty plywood
	Ply-Bead	Specialty plywood
	SeaFrame	Specialty plywood

Manufacturer	Product name	Type of product
	PrimeTrim	Wood fiber products
Kant-Sag	Kant-Sag Connectors	Metal hangers and connectors
Louisiana-Pacific	Wood-E	Computer software
	FiberBond	Fiber-reinforced gypsum products
	Gang-Lam S	Laminated veneer lumber
	Inner-Seal	Oriented strand board products
	SolId Start	Engineered lumber framing system
Masonite	Masonite	Hardboard products
Simpson Strong-Tie	Strong-Tie Connectors	Metal hangers and connectors
	Epoxy-Tie	Epoxy anchoring systems
Tecton Laminates	TecLam	Laminated veneer lumber
	TLI Joists	Wood I-joists
Trus Joist MacMillan	TJ-Beam	Computer software
	TJ-Dealer	Computer software
	TJ-Spec	Computer software
	TJ-Xpert	Computer software
	FrameWorks	Engineered lumber framing system
	Silent Floor	Engineered lumber floor-framing system
	TJI Joists	Wood I-joists
	TJI/Pro	Wood I-joists
	Cutrite	Wood I-joist saw guide
	1.3E Premium Studs	Laminated strand lumber
	1.5E Header	Laminated strand lumber
	FrameWorks Header	Laminated strand lumber

Manufacturer	Product name	Type of product
	TimberStrand	Laminated strand lumber
	Formbeams	Laminated veneer lumber
	Microllam	Laminated veneer lumber
	Parallam	Parallel strand lumber
	SpaceMaker Truss	Site-assembled roof trusses
Weyerhaeuser	Sturdi-Wood	Oriented strand board products
	Structurwood	Oriented strand board products

Note: This appendix is only a partial listing of the manufacturers' full product lines.

D

Sample guarantees and warranties

LPI® JOIST, GANG-LAM® LVL, SOLID START® RIM BOARD, AND TOP NOTCH® T&G FLOORING

LIMITED LIFETIME WARRANTY

Louisiana-Pacific Corporation (L-P) warrants that its LPI Joists, Gang-Lam LVL and Solid Start rim board are free from defects in materials, workmanship and design and, when installed and finished according to L-P's published installation instructions will perform in accordance with L-P's current published specifications for the lifetime of the home or building. L-P further warrants that its Top Notch tongue-and-groove flooring, when used in a flooring system application with LPI Joists and/or Gang-Lam LVL and L-P's Solid Start rim board, is free from defects in materials, workmanship and design and, when installed and finished according to L-P's published installation instructions will perform in accordance with L-P's current published specifications for the lifetime of the home or building.

LIMITATIONS

L-P MUST BE GIVEN A REASONABLE OPPORTUNITY TO INSPECT THE PRODUCT BEFORE IT WILL HONOR ANY CLAIMS UNDER THE ABOVE WARRANTY. IF AFTER INSPECTION AND VERIFICATION OF THE PROBLEM, L-P DETERMINES THAT THERE IS A FAILURE COVERED BY THE ABOVE WARRANTY, L-P WILL PAY TO THE OWNER OF THE STRUCTURE AN AMOUNT OF MONEY EQUAL TO THE REASONABLE COST OF LABOR AND MATERIALS REQUIRED TO REMOVE AND REPLACE OR REPAIR THE DEFECTIVE PRODUCT. THE PRODUCT MUST BE PROTECTED FROM EXPOSURE TO MOISTURE FROM WHATEVER SOURCE BY PROPER BUILDING STANDARDS FOR SIDING SYSTEMS, WALL SYSTEMS AND FLOOR SYSTEMS. EXPOSURE TO MOISTURE BEYOND INCIDENTAL EXPOSURE DURING NORMAL CONSTRUCTION PERIODS MAY CAUSE PRODUCT FAILURE AND WILL VOID THIS LIMITED LIFETIME WARRANTY. EXPOSURE TO STANDING WATER AND ACCUMULATIONS OF SNOW AND ICE WITHOUT REASONABLY PROMPT REMOVAL WILL VOID THIS LIMITED LIFETIME WARRANTY.

THE ABOVE WARRANTY SHALL APPLY ONLY IF THE LPI JOISTS, GANG-LAM LVL, L-P'S SOLID START RIM BOARD, AND TOP NOTCH T&G FLOORING ARE SUBJECTED TO NORMAL USE AND EXPOSURE. THE PRODUCT MUST BE STORED, HANDLED, AND INSTALLED IN A MANNER GENERALLY ACCEPTED IN THE INDUSTRY, AND IN ACCORDANCE WITH L-P'S CURRENT PUBLISHED INSTALLATION INSTRUCTIONS, AND IN COMPLIANCE WITH L-P'S PRODUCT DESIGN SPECIFICATIONS RELATING TO SPANS AND LOADING. FAILURE TO FOLLOW SUCH INSTRUCTIONS WILL VOID THIS WARRANTY.

DISCLAIMER

EXCEPT FOR THE EXPRESS WARRANTY AND REMEDY SET FORTH ABOVE, L-P DISCLAIMS ALL OTHER WARRANTIES AND GUARANTEES, EXPRESSED OR IMPLIED, INCLUDING IMPLIED WARRANTIES OF MERCHANTABILITY OR FITNESS FOR A PARTICULAR PURPOSE. NO OTHER WARRANTY OR GUARANTEE WILL BE MADE BY OR ON BEHALF OF THE MANUFACTURER OR THE SELLER OR BY OPERATION OF LAW WITH RESPECT TO THE PRODUCT OR ITS INSTALLATION, STORAGE, HANDLING, MAINTENANCE, USE, REPLACEMENT, OR REPAIR. NEITHER L-P NOR THE SELLER SHALL BE LIABLE BY VIRTUE OF ANY WARRANTY OR GUARANTEE, OR OTHERWISE, FOR ANY SPECIAL, INCIDENTAL OR CONSEQUENTIAL LOSS OR DAMAGE RESULTING FROM THE USE OF THE PRODUCT. L-P MAKES NO WARRANTY OR GUARANTEE WITH RESPECT TO INSTALLATION OF THE PRODUCT BY THE BUILDER OR THE BUILDER'S CONTRACTOR OR BY ANY OTHER INSTALLER.

Some states do not allow the exclusion or limitation of incidental or consequential damages, and in such states the above limitation or exclusion may not apply to you.

This warranty gives you specific legal rights, and you may also have other rights which vary from State to State.

For information on the above warranty contact:

Louisiana-Pacific Corporation
Customer Service
230 W. Superior Suite 550
Duluth, MN 55802

(800) 648-6893 (8am - 5pm Central Standard Time)

For more information please contact Louisiana-Pacific sales offices:

LPI Joists, Gang-Lam LVL
and Solid Start Rim Board
Sales & Engineering

Fernley, NV 89408
(800) 223-5647

Wilmington, NC 28401
(800) 999-9105

Top Notch T&G Flooring Sales

Chicago, IL 60173
(847) 517-8833

Conroe, TX 77305
(409) 756-0541

Hayden Lake, ID 83835
(208) 772-6011

Louisiana-Pacific®

D-1 *Manufacturer warranty for engineered lumber products.* Courtesy of Louisiana-Pacific.

COLORLOK®

15 YEAR LIMITED WARRANTY — FACTORY FINISH

Colorlok® Prefinished Hardboard Siding (the "Siding"), when installed according to Masonite's published application instructions and when properly maintained, is warranted for a period of 15 years not to require refinishing due to peeling, blistering, cracking or erosion of the factory-applied finish except for reasonable fade from normal weathering (10 N.B.S. units in eight years).

If Masonite, after inspection and verification, determines that the factory-applied finish failed under the terms of this warranty, the **sole** and **exclusive remedy** provided by Masonite will be to compensate the Owner for correcting the finish problem on the affected Siding limited to a maximum of twice the original purchase price of the affected Siding.

25 YEAR LIMITED WARRANTY — SUBSTRATE

Masonite® hardboard siding substrate, when properly installed and maintained according to Masonite's published application instructions, is warranted for a period of 25 years from the date of installation: (A) to meet, as of the date of purchase, the hardboard industry standard ANSI/AHA A135.6 as published by the American National Standards Institute; and (B) against hail damage, delamination, splitting or cracking of the substrate face under normal conditions of use and exposure when caused by substrate defects.

If Masonite, after inspection and verification, determines that the Siding failed under the terms of this limited warranty, the **sole** and **exclusive remedy** provided by Masonite will be to compensate the Owner for correcting the substrate problem on the affected siding limited to twice the original purchase price of the affected Siding.

GENERAL PROVISIONS AND LIMITATIONS

THE LIMITED WARRANTIES ARE SUBJECT TO THE FOLLOWING GENERAL PROVISIONS AND LIMITATIONS.

The limited warranties are effective only if there is proper storage, handling, installation and maintenance of the Siding in strict accordance with the instructions packaged with the particular Siding product and maintenance instructions.

Claims must be made in writing to Masonite within 60 days of the discovery of a problem and authorization obtained prior to beginning any repair or replacement work. Claims can be made by writing, certified or registered mail to Masonite, P.O. Box 808, St. Charles, IL 60174-0808; Attention: Product Performance Department. After receiving such notice of the alleged defect, Masonite will investigate the claim. Masonite must be given a reasonable opportunity to inspect and verify the claim. Under no circumstance will the total compensation under these limited warranties exceed twice the purchase price in total.

Masonite shall have no liability for defects or damage resulting from (A) misuse or abuse, (B) improper installation, including, but not limited to, inadequate protection against all sources of moisture accumulation within the wall cavity, (C) lack of proper maintenance, such as prolonged contact with accumulated water due to failure to maintain caulking, finish coatings, or other normal weather protection, (D) performance of coatings other than those covered by the limited warranties, (E) contact with harmful chemicals, fumes, vapors or air pollutants, (F) mildew, (G) settlement, shrinkage, or distortion of the structure, or (H) other causes beyond the control of Masonite, such as acts of God, fire and casualty. Masonite shall have no liability for the cost of removing affected Siding. The warranty period for any replacement Siding will expire upon the expiration of the warranty period for the original Siding.

DISCLAIMER OF IMPLIED WARRANTIES & LIMITATION OF REMEDIES

THE LIMITED WARRANTIES STATE THE ENTIRE LIABILITY OF MASONITE WITH RESPECT TO THE PRODUCTS COVERED BY THEM. MASONITE SHALL HAVE NO LIABILITY FOR ANY INCIDENTAL OR CONSEQUENTIAL DAMAGES. NO PERSON IS AUTHORIZED TO MAKE ANY REPRESENTATION OR WARRANTY ON BEHALF OF MASONITE EXCEPT AS EXPRESSLY SET FORTH ABOVE, AND ANY SUCH STATEMENT SHALL NOT BE BINDING ON MASONITE. Some states do not allow the exclusion or limitation of incidental or consequential damages, so the above limitation or exclusion may not apply to you.

EXCEPT AS EXPRESSLY SET FORTH ABOVE, MASONITE MAKES NO WARRANTY OF ANY KIND, EXPRESS OR IMPLIED, INCLUDING, WITHOUT LIMITATION, ANY IMPLIED WARRANTY OF MERCHANTABILITY OR FITNESS FOR A PARTICULAR PURPOSE. THE FOREGOING DISCLAIMER OF IMPLIED WARRANTIES SHALL NOT BE APPLICABLE TO SALES SUBJECT TO THE MAGNUSON-MOSS WARRANTY ACT, IN WHICH CASE THE DURATION OF ANY IMPLIED WARRANTY SHALL BE THE DURATION OF THE LIMITED WARRANTY OR SUCH SHORTER DURATION AS PROVIDED UNDER APPLICABLE STATE LAW. Some states do not allow limitations on how long an implied warranty lasts, so the above limitation may not apply to you. These limited warranties give you specific legal rights, and you may also have other rights which vary from state to state.

D-2 *Manufacturer warranty for hardboard siding.* Courtesy of Masonite Corporation.

LIMITED WARRANTY FOR L-P SIDING

Louisiana-Pacific

LIMITED 25-YEAR SIDING WARRANTY

Louisiana-Pacific Corporation ("L-P") warrants the Overlaid OSB lap and panel sidings, when installed and finished according to the published installation and finishing instructions and when properly maintained, for a period of 25 years from the date of installation against manufacturing defects under normal conditions of use and exposure.

LIMITATIONS

L-P MUST BE GIVEN A 60-DAY OPPORTUNITY TO INSPECT THE SIDING BEFORE IT WILL HONOR ANY CLAIMS UNDER THE ABOVE WARRANTY. IF AFTER INSPECTION AND VERIFICATION OF THE PROBLEM, L-P DETERMINES THAT THERE IS A FAILURE COVERED BY THE ABOVE WARRANTY, L-P WILL REFUND TO THE OWNER AN AMOUNT OF MONEY EQUAL TO TWICE THE RETAIL COST OF THE ORIGINAL SIDING MATERIAL. THE COST OF LABOR AND MATERIALS OTHER THAN SIDING ARE NOT INCLUDED. WARRANTY PAYMENTS WILL BE BASED UPON THE AMOUNT OF AFFECTED SIDING MATERIAL.

DURING THE FIRST FIVE (5) YEARS, L-P'S OBLIGATION UNDER THE ABOVE WARRANTY SHALL BE LIMITED TO TWICE THE RETAIL COST OF THE SIDING MATERIAL, WHEN ORIGINALLY INSTALLED ON THE STRUCTURE.

IF THE ORIGINAL SIDING COST CANNOT BE ESTABLISHED BY THE OWNER, THE COST SHALL BE DETERMINED BY L-P IN ITS SOLE AND REASONABLE DISCRETION.

DURING THE 6TH THROUGH 25TH YEAR, AS DETERMINED IN THE ABOVE MANNER, WARRANTY PAYMENTS SHALL BE REDUCED EQUALLY EACH YEAR SUCH THAT AFTER 25 YEARS FROM THE DATE OF INSTALLATION NO WARRANTY SHALL BE APPLICABLE.

THE ABOVE WARRANTY SHALL APPLY ONLY IF THE SIDING IS SUBJECTED TO NORMAL SIDING USE AND EXPOSURE. THE SIDING MUST BE STORED, HANDLED, INSTALLED, FINISHED AND MAINTAINED IN ACCORDANCE WITH L-P'S PUBLISHED INSTRUCTIONS. FAILURE TO FOLLOW SUCH INSTRUCTIONS WILL VOID THIS WARRANTY.

IMPORTANT NOTICE:

FAILURE TO INSTALL, FINISH AND MAINTAIN IN ACCORDANCE WITH L-P'S PUBLISHED INSTRUCTIONS MAY CAUSE DAMAGE TO THE SIDING AND WILL VOID THIS WARRANTY.

CONDITIONS COVERED BY THIS WARRANTY

- DELAMINATION OF THE OVERLAY FROM THE SUBSTRATE.
- RESIN SPOTS.
- SPOTS ON OVERLAY RESULTING FROM A MANUFACTURING PROCESS WHICH CANNOT BE COVERED WITH PAINT.
- FOLDED OR "POPPED" WAFERS/STRANDS THAT BREAK THE OVERLAY SURFACE.
- CRACKING, PEELING, CHIPPING OR FLAKING OF THE OVERLAY SURFACE.
- EXCESSIVE OR MISSING SEALANT ON EDGES AND/OR GROOVES.
- DIMENSIONAL VARIANCE FROM SPECIFICATIONS, AT THE TIME OF SALE.
- PATTERN VARIANCES FROM SPECIFICATIONS.

CONDITIONS NOT COVERED BY THIS WARRANTY

- INADEQUATE PROTECTION OF SIDING DURING STORAGE.
- FAILURES DUE TO MOISTURE IN THE WALL CAVITY OR CHIMNEY CHASES.
- FAILURES DUE TO INSUFFICIENT PAINT COVERAGE ON FACE AND EXPOSED EDGES.
- FAILURES DUE TO FAILURE OF THE PAINT SYSTEM.
- FAILURES RELATED TO MOLD, MILDEW AND/OR ALGAE ON PAINTED SIDING SURFACE.
- FAILURES DUE TO FACE NAILING ON LAP SIDING.
- FAILURES DUE TO INADEQUATE SPACING AND/OR SEALANT.
- FAILURES DUE TO UNCONTROLLED WATER RUNOFF OR INADEQUATE FLASHING.
- FAILURES DUE TO SIDING BEING IN DIRECT CONTACT WITH MASONRY AND/OR LESS THAN 6" FROM THE GROUND.
- FAILURES DUE TO SPRINKLERS SPRAYING ON THE SIDING.

DISCLAIMER: LOUISIANA-PACIFIC DISCLAIMS ALL WARRANTIES, EXPRESS OR IMPLIED, REGARDING UTILITY GRADE SIDING, INCLUDING IMPLIED WARRANTIES OF MERCHANTABILITY OR FITNESS FOR A PARTICULAR PURPOSE. THE FOREGOING EXPRESS WARRANTIES ARE APPLICABLE ONLY TO OUR A-GRADE PRODUCT AND NOT OUR UTILITY GRADE WHICH IS SOLD "AS IS AND WITH ALL FAULTS". EXCEPT FOR THE EXPRESS WARRANTY AND REMEDY SET FORTH ABOVE, L-P DISCLAIMS ALL OTHER WARRANTIES, EXPRESS OR IMPLIED, INCLUDING IMPLIED WARRANTIES OF MERCHANTABILITY OR FITNESS FOR A PARTICULAR PURPOSE. NO OTHER WARRANTY WILL BE MADE BY OR ON BEHALF OF THE MANUFACTURER OR THE SELLER OR BY OPERATION OF LAW WITH RESPECT TO THE PRODUCT OR ITS INSTALLATION, STORAGE, HANDLING, MAINTENANCE, USE, REPLACEMENT OR REPAIR. NEITHER L-P NOR THE SELLER SHALL BE LIABLE BY VIRTUE OF ANY WARRANTY, OR OTHERWISE, FOR ANY SPECIAL, INCIDENTAL OR CONSEQUENTIAL LOSS OR DAMAGE RESULTING FROM THE USE OF THE PRODUCT. L-P MAKES NO WARRANTY WITH RESPECT TO INSTALLATION OF THE PRODUCT BY THE BUILDER OR THE BUILDER'S CONTRACTOR, OR ANY OTHER INSTALLER. SOME STATES DO NOT ALLOW THE EXCLUSION OR LIMITATION OF INCIDENTAL OR CONSEQUENTIAL DAMAGES, AND IN SUCH STATES THE ABOVE LIMITATION OR EXCLUSION MAY NOT APPLY TO YOU. THIS WARRANTY GIVES YOU SPECIFIC LEGAL RIGHTS AND YOU MAY ALSO HAVE OTHER RIGHTS WHICH VARY FROM STATE TO STATE.

SEE REVERSE SIDE FOR WARRANTY CHECKLIST TO ENSURE SIDING PERFORMANCE AND VALIDITY OF WARRANTY.

111 SW Fifth Avenue
Portland, Oregon 97204

LP and Louisiana-Pacific are registered trademarks of Louisiana-Pacific Corporation.

NOTE: Louisiana-Pacific Corporation periodically updates and revises its product information. To verify that this version is current, contact one of the sales offices.

D-3 *Manufacturer warranty for OSB siding.* Courtesy of Louisiana-Pacific.

QUALITY GUARANTEE

We guarantee that the Trus Joist MacMillan products used in your home or project were manufactured to precise tolerances and are free from defects. In the unlikely event that your floor or roof system develops squeaks or any other problem due to a defect in our products, we will promptly remedy that problem at no cost to you.

In addition, if you call us with a problem that you believe may be caused by our products, our representative will contact you within one business day to evaluate the problem and help solve it. Guaranteed.

This guarantee is effective for the life of your home.

1-800-628-3997

A Part of the* FrameWorks® *Building System

D-4 *Manufacturer warranty for engineered lumber framing system.* Courtesy of Louisiana-Pacific.

CONGRATULATIONS!

Your builder has chosen the high quality Simple Framing System™ from Boise Cascade for your new residence.

The Simple Framing System™ incorporates the high quality, clean looking, BCI Joists,® Versa-Lam,® Versa-Lam Plus® and Versa-Rim 98® products manufactured by Boise Cascade for your structural floor and/or roof system.

Even though these structural members may cost a little more, your builder has provided you with the best components available in the industry today and we guarantee them.

LIFETIME GUARANTEED QUALITY AND PERFORMANCE.

BOISE CASCADE fully warrants its BCI Joists,® Versa-Lam,® Versa-Lam Plus® and Versa-Rim 98® Products to be free from defects in material and workmanship in accordance with our specifications.

Further, we guarantee that these products, when correctly installed and used, will meet or exceed our performance specifications for the life of the structure.

D-5 *Manufacturer warranty for engineered lumber framing system.* Courtesy of Boise Cascade.

E

For more information

ABT Building Products Corporation
One Neenah Center, Suite 600
Neenah, WI 54956-3070
Phone 414-751-4982, Fax 414-751-0370
http:/www.abtco.com
Siding and other building products

AFM Corporation
Box 246
Excelsior, MN 55331
Phone 612-474-0809, 800-255-0176
Insulated building panels

American Plywood Association
P.O. Box 11700
Tacoma, WA 98411-0700
Phone 206-565-6600, Fax 206-565-7265
Plywood, panel products, and engineered lumber (trade association)

Boise Cascade
BCI Joist & Versa-Lam Products
P.O. Box 2400
White City, OR 97503-0400
Phone 503-826-0200 and 800-232-0788, Fax 503-826-0218
Engineered lumber, other building products

CertainTeed Corporation
Vinyl Building Products Group
P.O. Box 860
Valley Forge, PA 19482
Siding, roofing, and other building products

Classic Products, Inc.
8510 Industry Park Dr.
P.O. Box 701
Piqua, OH 45356
Phone 513-773-9840 and 800-543-8938, Fax 513-773-9261
Aluminum shingles

Ener-Grid
Phone 602-386-2232
Polystyrene/cement building blocks

Flex Trim Industries
11479 Sixth Street
Rancho Cucamonga, CA 91730
Phone 909-944-6665, Fax 909-945-5258
Flexible trim molding

Georgia-Pacific
133 Peachtree Street, NE
Atlanta, GA 30303
Phone 404-652-4000
Engineered lumber, panel products, and other building products

James Hardie Building Products, Inc.
26300 La Alameda, Suite 250
Mission Viejo, CA 92691
Phone 800-9-HARDIE
Fiber cement products

Jeld-Wen
3303 Lakeport Blvd.
Klamath Falls, OR 97601
Hardboard doors, vinyl windows, and other building products

Louisiana-Pacific Corporation
111 SW Fifth Avenue
Portland, OR 97204
Phone 503-221-0800, Fax 503-796-0204
Engineered lumber, OSB siding, and panel products

Masonite Corporation
Building Products Group
One South Wacker Dr.
Chicago, IL 60606
Phone 312-750-0900, Fax 312-263-5808
Hardboard panel products, doors, and siding

MaxiTile, Inc.
17141 S. Kingsview Avenue
Carson, CA 90746
Phone 310-217-0316 and 800-338-8453, Fax 310-515-6851
Fiber cement products

Mobil Chemical Company
Composite Products Division
20 South Cameron Street
Winchester, VA 22601
Phone 800-BUY TREX
Wood-polymer composites

Outwater Plastics Industries, Inc.
P.O. Box 347, 52 Passaic St.
Wood Ridge, NJ 07075
Phone 201-365-2002 and 800-835-4400, Fax 201-778-8231 and 800-835-4403
Vinyl and other building products

Simpson Strong-Tie Company, Inc.
2600 International Street
Columbus, OH 43228
Phone 614-876-8060
Hangers and connectors

Thermal Industries, Inc.
301 Brushton Ave.
Pittsburgh, PA 15221-2168
Phone 412-244-6400 and 800-245-1540, Fax 412-244-6496
Vinyl lumber products

Therma-Tru
2806 North Reynolds Road
Toledo, OH 43615
Phone 419-537-1931
Fiberglass doors

Trus Joist MacMillan
200 East Mallard Road, P.O. Box 60
Boise, ID 83706
Phone 208-364-1256, Fax 208-364-1300
Engineered lumber and software

Western Wood Products Association
522 SW Fifth Ave.
Portland, OR 97204-2122
Phone 503-224-3930, Fax 503-224-3934
Lumber products (trade association)

Wolmanized Wood
Hickson Corporation
1955 Lake Park Dr., Suite 250
Smyrna, GA 30080
Preservative-treated lumber

Glossary

air-driven fasteners Nails, staples, or other fasteners that are driven into the wood with a special gun powered by compressed air.

backer blocks Blocks of wood or other material used to fill in and reinforce the web of a wood I-joist. Backer blocks are typically used where a hanger will be attached to the I-joist, allowing for the attachment and support of another structural member.

beam A horizontal or sloping member that is designed to carry vertical loads. A *simple span* is a member supported at both ends, a *continuous* is a single member supported at more than two bearing locations, and a *cantilever* is a member that has one or both supports away from the ends, one of which overhangs its support.*

bending strength The ability of a member, such as a beam, to resist the tendency to break when exposed to external forces such as roof or floor loads. The strength is achieved by the resisting couple action of the tension and compression stresses at the top and bottom of the beam.*

bearing stress The compressive stress exerted on an external surface of a member. This stress commonly occurs at a point of support, such as a beam hanger.

billet A block of engineered wood from which a variety of smaller pieces of structural composite lumber can be cut.†

box nail A type of nail with a smaller-diameter shank than a common nail of the same length. Because of the small shank diameter, box nails have less shear strength than common nails, but they are less likely to cause splits in the wood as they're driven in.

bridging A system of lateral braces placed between joists to distribute the load and keep the joists in position.†

camber A slight upward curve in a piece of engineered lumber, such as a glulam. A camber of uniform radius is intentionally built into the wood member to help reduce the chances of deflection (sagging) when a load is placed on the member.

† Definitions courtesy of Trus Joist MacMillan.

* Definitions courtesy of American Plywood Association.

cantilever An overhanging portion of a floor, typically unsupported.

center *See* on center.

column A type of supporting pillar that is long and relatively slender, usually placed in the center of a beam or load rather than at the edges or ends of the beam above.†

Com-Ply A structural sheet material comprised of a front and back face, a center ply of wood veneer oriented parallel to the long dimension of the sheet, and two inner plies of OSB.

commodity lumber *See* lumber.

combination number The identification used to describe the type of lamination layup in the glulam member, the associated allowable design stresses, its intended application, and if the lumber was visually or mechanically graded.*

compression Stress imposed on an object by forces that attempt to compact the object. When a post is supporting a beam, for example, the post is in compression.

compression parallel to grain A measurement of the internal compressive stress induced in a wood member when a load is applied to the end of the piece. This is normally the stress that occurs on a column.*

compression perpendicular to grain A measurement of the internal compressive stress induced in a wood member when a load is applied to the edge or face of the piece. This is also referred to as *bearing stress* for a joist, beam, or similar piece of wood as it bears or supports a load. The load tends to compress the fibers, and thus the bearing area must be sufficient to prevent crushing.*

concrete form A temporary structure or mold that supports concrete while it sets, gaining sufficient strength to be self-supporting. It is typically constructed of plywood and held in place by stakes or rods. Form I-joists are used to support concrete poured in a suspended floor application, typically in commercial buildings.†

connector *See* hanger.

core material The interior structure of a window, door, or other millwork product, typically clad (or wrapped) in vinyl, aluminum, or another composite material. Millwork is classified as building products such as trim moldings, door and window frames, doors, windows, blinds, and stair parts, but it does not include flooring, ceiling, or siding products.†

cross bracing Braces that cross from one column to the next to increase support and the load-bearing capacity.†

crown The width of the head of a staple or a bow along the length of a piece of solid sawn lumber.

dead load *See* load.

defect In wood, knots, checks, splits, crowns, twists, etc. that mar the wood and reduce its strength. These are minimized and randomized in engineered lumber.†

deflection The result of a structural member bent by its own weight or an applied weight or load. Also, the amount of displacement resulting from this bending.†

deflection limit The maximum amount a beam is permitted to deflect under load. Different deflection limits are normally established for live load and total load.*

design values Allowable stress values established for each glulam beam, described in terms of bending (F_b), horizontal shear (F_v), modulus of elasticity (E), and other stresses.*

dimensional lumber *See* lumber.

dimensionally stable Any building material that does not alter shape appreciably (as in warping or twisting) due to changes in temperature, humidity, and loading conditions.†

DL Dead load. *See* load.

engineered wood A broad term categorizing wood building products created by remanufacturing wood fiber from logs. Also called *engineered lumber* or *structural composite lumber*. The latter term differentiates engineered wood products manufactured for structural frame applications, as opposed to panel products such as plywood and oriented strand board, or nondimension wood building products.†

equilibrium moisture content Any piece of wood gives off or takes on moisture from the surrounding atmosphere until the moisture in the wood is equal to that in the atmosphere. The moisture content of wood at this point of balance is called the equilibrium moisture content, which is expressed as a percentage of the weight of the oven-dried wood.*

extreme fiber in bending A measure of the stress applied to the parallel fibers of a piece of wood under loaded conditions. When a load is applied to a piece of wood, it causes the wood to bend, producing tension in the fibers on one side and compression in the fibers on the opposite side.*

falldown By-products of a lesser grade or quality that result during the manufacture of a higher-quality stock. Also referred to as waste from piece cuts at the lumberyard or job site.†

filler blocks Blocks of wood or other material used to fill in the space between the webs of two wood I-joists, used when the two I-joists are combined parallel to one another, for additional structural strength.

fishtail or fishtail veneer The inch or two outermost section of a log, which is often wasted in traditional milling processes.

flanges The top and bottom supports on an engineered lumber I-joist that connect the central web and give the joist its stability.†

glulam A structural wood product in which individual pieces of solid sawn lumber are bonded together with adhesives to make a single piece in which the grains of all the constituent pieces are parallel. Typically used as a large-dimension beam.†

hanger Any of a class of hardware used in supporting or connecting members of similar or different materials, for example a stirrup strap or beam hanger for supporting the end of a beam or joist at a concrete foundation wall. Also referred to as a *connector*.†

hardboard A panel product manufactured from layered wood fibers that are pressed together and baked into a hard, smooth sheet. Standard hardboard has no other treatments applied; tempered hardboard has oil added to the fiber mat prior to baking, producing a harder, smoother, more abrasion-resistant surface.

HDO *See* high-density overlaid.

header A structural member extending horizontally between two posts or wall studs to support a window or door opening, including a garage door.†

high-density overlaid (HDO) Consists of a core sheet of Exterior-grade plywood, covered on both sides with a resin-impregnated fiber surface that is bonded to the plywood with a combination of heat and pressure. It has a number of industrial applications. *See also* medium-density overlaid (MDO).

horizontal shear The measure of the resistance of shearing stress along the longitudinal axis of a piece of wood. When a load is applied to a piece of wood supported at each end, there is a stress over each support that tends to slide the fibers across each other horizontally. The internal force that resists this action is the horizontal shear strength of the wood. The shearing action is maximum at the mid-depth of a simple span beam at the supports.*

I-joist A structural member whose cross section resembles the capital letter I. A joist is a long piece of lumber, 2 or 4 inches thick and 6 or more inches wide (deep), used horizontally as a support for a ceiling (as a rafter) or floor-frame system. An engineered wood I-joist includes top and bottom flanges or chords, the part of the joist that resists tension and compression. The flanges are grooved to accommodate a center panel or web, which is used mainly in resisting shear stress.†

knock-out holes Perforated holes in the web of an I-joist that can be "knocked out" to allow electrical wires and plumbing pipes to run through the joists rather than below them in a soffit.†

laminations Individual pieces of lumber that are glued together end to end for use in the manufacture of glue-laminated timber. These end-jointed laminations are then face-bonded together to create the desired member shape and size.*

laminated strand lumber (LSL) An engineered lumber product made by reducing a log to thin strands up to 12 inches long that are then bonded with adhesive to create a billet from which smaller-dimension pieces can be cut. Currently, the only brand-name LSL product available is TimberStrand LSL from Trus Joist MacMillan for use as rim boards, window and door headers, and millwork core material. In Europe, the same product brand name is Intrallam LSL.†

laminated veneer lumber (LVL) The generic name for wood that's manufactured by peeling logs into thin veneers, which are then placed parallel to each other, applied with a surface adhesive, and bonded under heat and pressure. LVL is a high-strength alternative to large-dimension commodity lumber.†

lateral movement Movement from side to side.

live load (LL) *See* load.

LL Live load. *See* load.

load The force or combination of forces that acts on a structural system or individual member. Dead load (DL) is the combined weight of the structural components, fixtures, and permanently attached equipment used to design and construct the building and its foundation. Live load (LL) is the superimposed weight on the structural components by the use and occupancy of the building, not including the wind load, earthquake load, and dead load.†

LSL *See* laminated strand lumber.

lumber The product of cutting or sawing logs into wood structural members to be used in framing a house or building. *Solid sawn lumber* refers to lumber milled directly from logs rather than reprocessed or engineered into structural members. *Commodity lumber* refers to wood that's dependent on commodity or traditional sources of logs (such as Douglas fir or southern yellow pine) and thus the economic market for such resources. *Dimension lumber* typically refers to lumber that measures 2 to 12 inches thick or 2 or more inches wide, such as a 2 × 4 or 2 × 12.†

LVL *See* laminated veneer lumber.

MDF *See* medium-density fiberboard.

MDO *See* medium-density overlaid.

medium-density fiberboard (MDF) A dense, solid material manufactured from clean sawdust or other wood fibers, compressed under high heat and pressure. MDF is commonly used to manufacture moldings, trim, furniture, counters, and a number of other items.

medium-density overlaid (MDO) A sheet of exterior-grade plywood of a specific quality that has a surface coating of resin-impregnated wood fibers, creating a hard, abrasion-resistant surface that is weatherproof and excellent for painting. *See also* high-density overlaid (HDO).

Melamine Brand name for a type of panel product that consists of a core of particleboard or MDF to which a colored foil coating is applied on one or both sides. Melamine is a very popular product for cabinets, furniture, shelving, and a number of other applications.

modulus of elasticity (MOE) A measurement of a member's stiffness, determined by the ratio of the amount a material will deflect in proportion to the applied load. A member with a large MOE will deflect less under an applied load than a member with a smaller MOE when other conditions are equal. MOE is often referred to as the E value.*

moisture content The amount of water contained in wood, usually expressed as a percentage of the weight of oven-dried wood.*

o.c. *See* on center.

old growth Also called *virgin timber*, the term refers to trees in a mature, naturally established forest that have not previously been harvested. Old-growth trees are usually large, straight, and free of knots.†

on center The measured distance between the centers of two parallel members in a structure, used to define the spacing between wall studs, joists, and rafters (typically referred to as *on-center spacing*).†

open-web truss A heavy-duty joist in which the web is a zigzag or crisscross pattern of wood or steel tubing, instead of a solid plate or web connecting the chords or flanges.†

oriented strand board (OSB) Composite wood panels made of narrow strands of fiber, oriented lengthwise and crosswise in layers bonded with a resin adhesive under heat and pressure. Depending on the resin used, OSB can be suitable for interior or exterior applications.†

OSB *See* oriented strand board.

parallel strand lumber (PSL) An engineered lumber product manufactured by peeling logs into veneer sheets that are then cut into thin strands. The strands are laminated together with their grains parallel to one another. Parallel-laminated veneer is used in furniture and cabinetry to provide flexibility over curved surfaces and in the production of laminated-veneer products such as Parallam PSL from Trus Joist MacMillan, currently the only manufacturer of brand-name, structural-grade, parallel strand lumber. Parallam PSL is an alternative to large-dimension commodity lumber (4 inches and larger) for use as structural posts, columns, beams, and window and door headers.†

perimeter beam A beam or band of wood attached to the exposed ends or edges of floor joists.†

plated truss A structural wood truss assembly in which the joints of the truss are held together and reinforced with pressed-on steel plates. *See also* truss.

ply A single layer of a material, as the plies in a sheet of plywood.

plywood A flat panel made up of several thin sheets of wood (veneers or plies). The grain of each ply, or layer, is at right angles to the ones adjacent to it. The veneer sheets are bonded under heat and pressure. Plywood is used as roof and floor decking, sidewall sheathing, siding, and in many other capacities, and can be cut (or ripped) into rim board or filler material for window and door headers.†

point load A load or weight that is concentrated on a relatively limited area, such as a column or a piece of equipment that's bearing on a floor.

post A structural member used in a vertical position to support a beam or other structural member in a building. In lumber, most grading rules define a post as having dimensions of 5 inches in thickness by 5 or more inches in width, with the width not more than 2 inches greater than the thickness. Posts are also used to brace a structure temporarily, often referred to as a *shore*, *prop*, or *jack*.†

post and beam A structural framing technique in which horizontal members, or beams, are supported by large posts and columns rather than a system of wall studs. This type of construction creates a more open floor space, not interrupted or dictated by intersecting walls. Also called *timber frame construction*.†

pressure-treated lumber Lumber impregnated with various chemicals, such as preservatives and fire retardants. The process of pressure-treating forces the chemicals into the structure of the wood using high pressure. A common type of treatment is CCA (chromated copper arsenate), a chemical approved by the EPA and USDA. Once applied to the wood, CCA is inert and very unlikely to leach or off-gas.†

PSF Pounds per square foot.

purlin In roofs, a horizontal member supporting the common rafters.†

racking strength A measure of the ability of a structure, such as a wall, to withstand horizontal forces acting parallel to the structure.*

rafters In roofs, a series of sloping parallel members used to support a roof covering.†

ridge beam In roofing structures, a horizontal timber to which the peak ends of the rafters are fastened.†

rim board A piece of plywood, engineered lumber, or solid sawn lumber surrounding the perimeter of the floor framing, to which the ends of the floor joists are attached. Also called a *rim joist.*

second growth Timber or trees that have grown in areas following the harvest or destruction of the original timber resource.†

sheathing A structural or nonstructural covering over framing members such as rafters or wall studs. Sheathing can provide a solid membrane to resist racking and lateral movement in the structure, and a base for the application of other building materials, including siding and roofing.

solid sawn lumber *See* lumber.

span The distance between two supporting members. A multiple span occurs when a beam or joist spans across more than one supporting member (or column).†

squash blocks Blocks of 2 × 4, 2 × 6, or other sizes of solid sawn lumber, attached to the sides of a wood I-joist to help support a concentrated load.

stickers Pieces of wood or other material that are set either between stacks of lumber or a stack of lumber and the ground, to provide a space for air circulation and to allow access for forklifts or other lifting equipment.

subflooring The structural material covering floor joists, locking the floor framing system together and providing a solid, flat base for the application of finished flooring. Common subfloor materials include plywood, Com-Ply, OSB, and tongue-and-groove lumber.

sustainable yield In forest harvesting, sustainable yield is a practice ensuring that harvest does not exceed growth.†

tension Stress imposed on an object by stretching forces, which attempt to elongate the object. When a load is lifted by a rope, for example, the rope is in tension.

tension parallel to grain A measurement of the strength of wood when tension is applied in the same direction as that of the wood grain.*

tension perpendicular to grain A measurement of the strength of wood when tension is applied across the direction of the wood grain. Glulam members should be designed to avoid inducing tension perpendicular to grain stresses.*

timber Wood in standing trees that has the potential to be made into construction lumber.†

truss A structural component composed of a combination of members, usually in a triangular arrangement, to form a rigid framework often used to support a roof.†

unsupported length The distance between the end supports, or intermediate bracing of a column or beam.*

uplift The tendency of a material or structure to lift up off its supports or fasteners, either through external forces such as wind, or as the result of weight being placed on an opposing area.

veneer Wood that has been peeled, sawn, or sliced into sheets of a given thickness, usually less than ¼ inch.†

web stiffeners A block of wood, plywood, or other solid material added to a wood I-joist to stiffen the I-joist's web and provide additional support for a hanger or other connector.

weed species A term used to describe abundant groves of trees that have grown naturally in areas where larger trees were once harvested. They are typically too crooked and small—both in diameter and length—to be used as timber for construction lumber. However, such weed species as poplar and aspen have recently become marketable due to advances in engineered lumber technology and the production of structural-grade lumber.†

wood fiber A term generally referring to the individual elements of wood. Also used to denote a wood resource in many forms or sizes.†

Index

Aluminum shingles, 100
American Plywood Association (APA), 57, 63, 82
APA (*see* American Plywood Association)

Backer blocks, 281–285
Best management practices, 19
Bracing for I-joist installation, 285–287
Bridges, engineered lumber, 50–52
Bridging, metal, 176–177

Cantilevers, floor framing, 258–259, 280–281
Com-ply, 87–88
Cost comparisons, lumber, 21–23

Dead load: in floor framing, 230–231
  in roof framing, 245–246
Deflection, 231–233
  and sizing floor framing, 234–245
  and sizing roof framing, 245–256

Engineered lumber, 8
  computerized design, 211–219
  finishing, 80
  vs. lumber for callbacks, 23
  product manufacturers, 309–311, 319–321
  products by manufacturer's name, 301–304
  products by type, 305–307
  in remodeling, 20
  sizing for floors, 234–245
  sizing for roofs, 245–256
  storage, 79, 268–271
  systems, 45–49, 214–219
  in wall framing, 288–289
  warranties and guarantees, 314–318
Epoxy-Tie adhesive, 207–210

Fiber cement: shingles, 102–103
  siding, 101–102
  underlayment, 103
Fiber-reinforced gypsum: underlayment, 104–106
  wallboard and sheathing, 106
Fiberglass doors, 104
Filler blocks, 281–285
Finger-jointed lumber, 42–43
Floor framing: beams for, 260–261
  deflection in, 234–245
  design considerations, 226–228
  engineered lumber in, 228–229
  live and dead loads on, 230–231

Garage door headers, 289–290
Glulams, 9, 31–36
  appearance grades, 35–36
  camber in, 35
  grade stamps, 36
  laminations in, 34–35
Grade stamps: glulams, 36
  machine rated lumber, 40
  plywood, 62
  plywood siding, 83
  structural glued lumber, 43

Hangers: with Bend-Tab, 159
  codes and abbreviations, 157–58
  column bases, 181–182
  column caps, 182–185
  concealed flange, 198, 200
  fasteners for, 205–207
  for glulams, 173
  for I-joists, 162–165, 167–173
  installation, 205–207
  with Positive Angle Nailing, 158
  saddle, 201

Hangers: *continued*
- skewed, 169, 174, 198–199
- sloped, 198–199
- terminology, 156–157
- for trusses, 185–193
- variable-pitch adjustable, 175
- welded, 171–172

Hardboard, 67
- as board siding, 93–5
- as doors, 106–111
- in plyron, 66–67
- as sheet siding, 88–90
- as trim, 106–111

HDO (*see* Plywood)
High-density overlaid plywood (*see* Plywood)

I-joists, 44
- bracing, 285–287
- cantilevering in floor framing, 258–259, 280–281
- computerized design for, 211–219
- cutting guides for, 274, 276
- filler and backer blocks for, 263, 281–285
- handling, 271–273
- hangers for, 162–165, 167–173, 175–176, 279–280
- holes in, 264–265, 295–300
- installing in floor framing, 272–279
- installing in roof framing, 290–295
- manufacturing, 146–151
- sizing for floor framing, 234–245
- sizing for roof framing, 245–256
- squash blocks for, 284–285
- storage of, 268–271
- web stiffeners for, 281–285

Inner-Seal siding, 77
- *See also* Oriented strand board as board siding

Insulated structural building panels, 111–112

Laminated strand lumber (LSL), 15, 37
- garage door headers, 289–290
- manufacturing, 140–144

Laminated veneer lumber (LVL), 14, 37
- garage door headers, 289–290
- manufacturing, 132–139

Live load: deflection, 232–233
- floor framing, 230
- roof framing, 245–246

Load support blocks (*see* Squash blocks)
LSL (*see* Laminated strand lumber)
Lumber: costs, 2, 6, 21
- engineered (*see* Engineered lumber)
- vs. engineered lumber for call-backs, 23
- grading, 3
- machine stress rated, 39
- structural glued, 42

LVL (*see* Laminated veneer lumber)

Machine stress rated lumber, 39
- grade stamps for, 40

Masonite (*see* Hardboard)
Material Safety Data Sheet (MSDS) 30, 32–33
MDF (*see* Medium density fiberboard)
MDO (*see* Plywood)
Medium density fiberboard (MDF), 70
- siding, 87
- trim, 112–114

Medium density overlaid plywood (MDO) [*see* Plywood]
Melamine, 72
Microllam (*see* Laminated veneer lumber)
Moldings and trim: hardboard, 106–111
- medium density fiberboard, 112–114
- polymer composite (flexible), 115–116
- vinyl, 119–123

MSDS (*see* Material Safety Data Sheet)

Old-growth trees, 2–8
Open-web trusses, 49–50
Oriented strand board (OSB), 73
- board siding, 91, 95–97
- I-joists, 146–151
- manufacturing, 144–145
- panel siding, 90

Oriented strand lumber, 16
OSB (*see* Oriented strand board)

Parallam (*see* Parallel strand lumber)
Parallam PSL wood bridges, 50–52
Parallel strand lumber, 15, 39
- bridges, 50–52
- manufacturing, 139–140

Particleboard, 74
- electrically conductive, 76
- outgassing in, 75
- underlayment, 75

Plywood, 57
- exposure grades, 59

Plywood *continued*
fiberglass-reinforced-plastic, 67
grade stamps, 62
grading, 58–60
high-density overlaid, 69
medium-density overlaid, 70
performance-rated panels, 63
sanded, 64
siding, 82
finishes for, 84
grade stamps for, 83
span ratings, 62
specialty panels, 66
species, 61
Sturd-I-floor, 64
Polymer composite (flexible) trim, 115–116
Polystyrene building elements, 116–117
PSL (*see* Parallel strand lumber)

Remodeling, engineered lumber considerations, 20–21
Repetitive usage, 9
Rim boards, 278–279
Roof framing, sizing engineered lumber for, 245–256
Roofing: aluminum shingles, 100
fiber cement shingles, 102

Second-growth trees, 10, 23
and laminated veneer lumber, 140–144
and parallel strand lumber, 139
and oriented strand board, 145
Siding: board-style, 90
com-ply, 87
fiber cement boards and panels, 101–102
hardboard, 88
oriented strand board, 90
plywood, 82
grade stamps for, 83
vinyl, 123–124
wood-based sheets, 82
Silent Floor System, 49
Software: BC Calc (Boise Cascade), 221–222
Simpson Strong-Tie, 222–223
TJ-Beam (Trus Joist MacMillan), 221
TJ-Dealer (Trus Joist MacMillan), 221
TJ-Spec (Trus Joist MacMillan), 221
TJ-Xpert (Trus Joist MacMillan), 212–220

Software *continued*
Wood-E (Louisiana-Pacific), 222
SpaceMaker trusses, 52–53
Squash blocks, 284–285
Structural glued lumber, 42
Studs, Timberstrand LSL, 53–55
Sturd-I-floor (*see* Plywood)
Subfloors: com-ply, 87–88
installing, 287–288
oriented strand board for, 74
plywood for, 63–64
weather protection, 79

T-1-11, 83
(*See also* Plywood, siding)
Timber: best management practices, 19
sources of, 18
TimberStrand (*see* Laminated strand lumber; Studs)
TJI (*see* I-joists)
Trees: in engineered lumber, 9–15
forest environment, 16–18
harvesting as crop, 24
old growth, 2–8
second growth, 10, 23, 139–145
sources for timber, 18–20
Trex (wood-polymer lumber), 127–130
Trim (*see* Moldings and trim)
Trusses: hangers for, 185–193
open web, 49–51
SpaceMaker, 52–53

Underlayment: fiber cement sheets, 103
fiber-reinforced gypsum sheets, 104–106
particleboard, 75

Vinyl: architectural products, 119–123
fencing, 117–118
lumber, 121–122
moldings, 119–123
siding, 123–124
windows, 124–127

Web stiffeners, 281–285
Western Wood Products Association (WWPA), 39
Wood-polymer lumber, 127–130
WWPA (*see* Western Wood Products Association)

About the Author

Paul Bianchina is a general contractor and small business owner, with extensive experience in building, remodeling, and insurance restoration. He has contributed numerous articles to publications in the how-to construction field, and currently writes *Home Improvement*, a weekly home repair and remodeling advice column for *The Oregonian*, in Portland, OR. He is the author of nine other books, including *The Illustrated Dictionary of Building Materials and Techniques*, Second Edition, *Add-A-Room: A Practical Guide to Expanding Your Home*, *Forms and Documents-for the Builder*, and *Attic, Basement, and Garage Conversion*, all published by McGraw-Hill.